Lecture Notes in Physics

For information about Vols. 1–151, please contact your bookseller or Springer-Verlag.

Vol. 152: Physics of Narrow Gap Semiconductors. Proceedings, 1981. Edited by E. Gornik, H. Heinrich and L. Palmetshofer. XIII, 485 pages. 1982.

Vol. 153: Mathematical Problems in Theoretical Physics. Proceedings, 1981. Edited by R. Schrader, R. Seiler, and D.A. Uhlenbrock. XII, 429 pages. 1982.

Vol. 154: Macroscopic Properties of Disordered Media. Proceedings, 1981. Edited by R. Burridge, S. Childress, and G. Papanicolaou. VII, 307 pages. 1982.

Vol. 155: Quantum Optics. Proceedings, 1981. Edited by C.A. Engelbrecht. VIII, 329 pages. 1982.

Vol. 156: Resonances in Heavy Ion Reactions. Proceedings, 1981. Edited by K.A. Eberhard. XII, 448 pages. 1982.

Vol. 157: P. Niyogi, Integral Equation Method in Transonic Flow. XI, 189 pages. 1982.

Vol. 158: Dynamics of Nuclear Fission and Related Collective Phenomena. Proceedings, 1981. Edited by P. David, T. Mayer-Kuckuk, and A. van der Woude. X, 462 pages. 1982.

Vol. 159: E. Seiler, Gauge Theories as a Problem of Constructive Quantum Field Theory and Statistical Mechanics. V, 192 pages. 1982.

Vol. 160: Unified Theories of Elementary Particles. Critical Assessment and Prospects. Proceedings, 1981. Edited by P. Breitenlohner and H.P. Dürr. VI, 217 pages. 1982.

Vol. 161: Interacting Bosons in Nuclei. Proceedings, 1981. Edited by J.S. Dehesa, J.M.G. Gomez, and J. Ros. V, 209 pages. 1982.

Vol. 162: Relativistic Action at a Distance: Classical and Quantum Aspects. Proceedings, 1981. Edited by J. Llosa. X, 263 pages. 1982.

Vol. 163: J. S. Darrozes, C. Francois, Mécanique des Fluides Incompressibles. XIX, 459 pages. 1982.

Vol. 164: Stability of Thermodynamic Systems. Proceedings, 1981. Edited by J. Casas-Vázquez and G. Lebon. VII, 321 pages. 1982.

Vol. 165: N. Mukunda, H. van Dam, L.C. Biedenharn, Relativistic Models of Extended Hadrons Obeying a Mass-Spin Trajectory Constraint. Edited by A. Böhm and J.D. Dollard. VI, 163 pages. 1982.

Vol. 166: Computer Simulation of Solids. Edited by C.R.A. Catlow and W.C. Mackrodt. XII, 320 pages. 1982.

Vol. 167: G. Fieck, Symmetry of Polycentric Systems. VI, 137 pages, 1982.

Vol. 168: Heavy-Ion Collisions. Proceedings, 1982. Edited by G. Madurga and M. Lozano. VI, 429 pages. 1982.

Vol. 169: K. Sundermeyer, Constrained Dynamics. IV, 318 pages. 1982.

Vol. 170: Eighth International Conference on Numerical Methods in Fluid Dynamics. Proceedings, 1982. Edited by E. Krause. X, 569 pages. 1982.

Vol. 171: Time-Dependent Hartree-Fock and Beyond. Proceedings, 1982. Edited by K. Goeke and P.-G. Reinhard. VIII, 426 pages. 1982.

Vol. 172: Ionic Liquids, Molten Salts and Polyelectrolytes. Proceedings, 1982. Edited by K.-H. Bennemann, F. Brouers, and D. Quitmann. VII, 253 pages. 1982.

Vol. 173: Stochastic Processes in Quantum Theory and Statistical Physics. Proceedings, 1981. Edited by S. Albeverio, Ph. Combe, and M. Sirugue-Collin. VIII, 337 pages. 1982.

Vol. 174: A. Kadić, D.G.B. Edelen, A Gauge Theory of Dislocations and Disclinations. VII, 290 pages. 1983.

Vol. 175: Defect Complexes in Semiconductor Structures. Proceedings, 1982. Edited by J. Giber, F. Beleznay, J. C. Szép, and J. László. VI, 308 pages. 1983.

Vol. 176: Gauge Theory and Gravitation. Proceedings, 1982. Edited by K. Kikkawa, N. Nakanishi, and H. Nariai. X, 316 pages. 1983.

Vol. 177: Application of High Magnetic Fields in Semiconductor Physics. Proceedings, 1982. Edited by G. Landwehr. XII, 552 pages. 1983.

Vol. 178: Detectors in Heavy-Ion Reactions. Proceedings, 1982. Edited by W. von Oertzen. VIII, 258 pages. 1983.

Vol. 179: Dynamical Systems and Chaos. Proceedings, 1982. Edited by L. Garrido. XIV, 298 pages. 1983.

Vol. 180: Group Theoretical Methods in Physics. Proceedings, 1982. Edited by M. Serdaroğlu and E. İnönü. XI, 569 pages. 1983.

Vol. 181: Gauge Theories of the Eighties. Proceedings, 1982. Edited by R. Raitio and J. Lindfors. V, 644 pages. 1983.

Vol. 182: Laser Physics. Proceedings, 1983. Edited by J. D. Harvey and D. F. Walls. V, 263 pages. 1983.

Vol. 183: J. D. Gunton, M. Droz, Introduction to the Theory of Metastable and Unstable States. VI, 140 pages. 1983.

Vol. 184: Stochastic Processes – Formalism and Applications. Proceedings, 1982. Edited by G. S. Agarwal and S. Dattagupta. VI, 324 pages. 1983.

Vol. 185: H. N. Shirer, R. Wells, Mathematical Structure of the Singularities at the Transitions between Steady States in Hydrodynamic Systems. XI, 276 pages. 1983.

Vol. 186: Critical Phenomena. Proceedings, 1982. Edited by F.J.W. Hahne. VII, 353 pages. 1983.

Vol. 187: Density Functional Theory. Edited by J. Keller and J.L. Gázquez. V, 301 pages. 1983.

Vol. 188: A. P. Balachandran, G. Marmo, B.-S. Skagerstam, A. Stern, Gauge Symmetries and Fibre Bundles. IV, 140 pages. 1983.

Vol. 189: Nonlinear Phenomena. Proceedings, 1982. Edited by K. B. Wolf. XII, 453 pages. 1983.

Vol. 190: K. Kraus, States, Effects, and Operations. Edited by A. Böhm, J. W. Dollard and W. H. Wootters. IX, 151 pages. 1983.

Vol. 191: Photon Photon Collisions. Proceedings, 1983. Edited by Ch. Berger. V, 417 pages. 1983.

Vol. 192: Heidelberg Colloquium on Spin Glasses. Proceedings, 1983. Edited by J. L. van Hemmen and I. Morgenstern. VII, 356 pages. 1983.

Vol. 193: Cool Stars, Stellar Systems, and the Sun. Proceedings, 1983. Edited by S. L. Balliunas and L. Hartmann. VII, 364 pages. 1984.

Vol. 194: P. Pascual, R. Tarrach, QCD: Renormalization for the Practitioner. V, 277 pages. 1984.

Lecture Notes in Physics

Edited by H. Araki, Kyoto, J. Ehlers, München, K. Hepp, Zürich
R. Kippenhahn, München, H. A. Weidenmüller, Heidelberg
and J. Zittartz, Köln

Managing Editor: W. Beiglböck

230

Macroscopic Modelling of Turbulent Flows

Proceedings of a Workshop Held at INRIA,
Sophia-Antipolis, France, December 10–14, 1984

Edited by U. Frisch, J. B. Keller,
G. Papanicolaou and O. Pironneau

Springer-Verlag
Berlin Heidelberg GmbH

Editors

Uriel Frisch
Observatoire de Nice
B.P. 139, F-06003 Nice

Joseph B. Keller
Stanford University, Mathematics Department
Stanford, CA 94305, USA

George C. Papanicolaou
New York University, Courant Institute of Mathematical Sciences
251 Mercer Street, New York, NY 10012, USA

Olivier Pironneau
INRIA, Domaine de Voluceau – Rocquencourt
B.P. 105, F-78153 Le Chesnay Cedex

ISBN 978-3-540-15644-4 ISBN 978-3-540-39520-1 (eBook)
DOI 10.1007/978-3-540-39520-1

Originally published by Springer-Verlag Berlin Heidelberg New York Tokyo in 1985

2153/3140-543210

P R E F A C E

During the week of December 10, 1984 a workshop on **Macroscopic Modelling of Turbulent Flows** was held in Nice, France. This workshop was organized under the auspices of the USA - France bilateral scientific program administered by the **National Science Foundation** and the **Commissariat National de la Recherche Scientifique**. The **Institut National de Recherche en Informatique et Automatique (INRIA)** provided additional support and assistance for the organization of this meeting. The scientific organizers were U. FRISCH and O. PIRONNEAU from France and J.B. KELLER and G. PAPANICOLAOU from the USA.

The purpose of this workshop was to assess the present state of our understanding of model equations for turbulent flows based on analytical, numerical and experimental methods.

The lectures were organized around the following topics:

Turbulence modelling in engineering
Turbulence modelling by spectral methods
Asymptotic methods for turbulent flows
Phase turbulence
Turbulence modelling in meteorology
Recent experiments with turbulent flows.

Naturally, the presentations at the workshop and the contributed papers cover only some aspects of current interest in these topics. There was, however, substantial exchange and interaction between specialists in numerical, experimental and more mathematical areas.

U. FRISCH, J.B. KELLER
G. PAPANICOLAOU, O. PIRONNEAU

Monday Dec. 10

9h00 -10h00 :	Registration
10h00-10h15 :	Opening session
Chairman :	G. PAPANICOLAOU
10h15-11h00 :	U. FRISCH, Z. SHE, O. THUAL "Transport and homogenization (is turbulence elastic ?)"
11h30-12h15 :	R. CAFLISH "Flow of bubbly liquids"
12h15-13h00 :	L. TARTAR "Some aspects of homogenization theory"
Chairman :	C.W. VAN ATTA
14h30-15h15 :	P. PERRIER "Large and small scales related to the computation of transient to fully developped turbulent flows"
15h15-15h45 :	B. AUPOIX "Eddy viscosity model for homogeneous turbulence"
16h15-17h00 :	E. SIGGIA "Blow-up of solutions to Navier-Stokes and Euler equations"
17h00-17h45 :	J.P. BENQUE, M.D. LAURENCE "Macrosimulations of turbulence in the physical space ; analysis of spectral transfert rate of energy"

Tuesday Dec. 11

Chairman :	C.E. LEITH
9h30-10h15 :	S. CHILDRESS "Vortex stability and inertial range cascades"
10h15-10h45 :	J.P. BERTOGLIO "Subgrid modelling for sheared turbulence"
11h15-12h00 :	C.W. VAN ATTA "Comparison of experimental data with modelling results for some homogeneous and inhomogeneous flows"
12h00-12h30 :	Ph. ROY, K. DANG "Direct and large eddy simulation of homogeneous turbulence submitted to constant mean velocity gradient"
Chairman :	J. FERZIGER
14h00-14h45 :	P.L. SULEM, C. SULEM, O. THUAL, M. MENEGUZZI "Numerical simulation of 3-dimensional convection"
14h45-15h15 :	J.P. CHOLLET "Spectral closures and eddy viscosity"
15h45-16h30 :	D.D. KNIGHT "Modelling of three-dimensional shock wave. Turbulent boundary layer interactions"
16h30-17h00 :	B. ROUX "Numerical and theoretical study of different regimes occuring in horizontal fluid layers differentially heated"
20h45-22h15 :	C.E. LEITH Round table on subgrid modelling

Wednesday Dec. 12

Chairman	: Y. POMEAU
9h30-10h15	: E. HOPFINGER "Freely evolving turbulence in rotating fluids"
10h15-11h00	J.R. HERRING "Vorticity structure in 3-D rotating turbulence"
11h30-12h15	: J.C. ANDRE "Small scaled atmospheric turbulence and its interaction with larger scale flows"
Chairman	: J.C. ANDRE
14h30-15h15	: P. CLAVIN "Self turbulizing flame fronts"
15h15-16h00	: J.H. FERZIGER "Macroscopic modelling of turbulent flows and fluid mixtures"
16h30-17h15	: G. PAPANICOLAOU "Nonlinear equations with rapidly varying coefficients"
17h00-18h30	: J.B. KELLER Round table on homogenization

Thursday Dec. 13

Chairman	: P.L. SULEM
9h30-10h15	: D.W. McLAUGHLIN "Coherence and chaos in nonlinear solitary waves"
10h15-10h45	: P. COULLET "Phase instabilities near bifurcations"
11H15-12h00	: Y. POMEAU "The Kuramoto-Shivashinski equation as a system model with many degrees of freedom"
12h00-12h30	: S. ZALESKI "Anomalous diffusion of conserved quantities and phase chaos"
Chairman	: J.R. HERRING
14h00-14h45	: E.A. SPIEGEL "Why and how to derive amplitude-evolution equations ?"
14h45-15h15	: S. FAUVE "Competing instabilities in a rotating layer of mercury : a codimension two bifurcation"
15h45-16h15	: C.E. LEITH "Modelling of turbulent mixing induced by Rayleigh-Taylor instability"
16h15-17h00	: P. MANNEVILLE "Liapounov exponant for the Kuramoto-Sivanshinski model"
20h45-22h15	: E.A. SPIEGEL Round table on phase turbulence and dynamical systems.

Friday Dec. 14

Chairman : E. HOPFINGER

10h00-10h30 : C. BASDEVANT, R. SADOURNY
"Vortices and vortex couples in 2-D turbulence"
10h30-11h15 : J.C. McWILLIAMS
"Vorticity structures in 3D rotating turbulence"
11h45-12h15 : M.E. BRACHET, P.L. SULEM, M. MENEGUZZI
"Numerical simulation of 2 dimensional turbulence"
12h15-13h00 : J.B. KELLER
Conclusion

14h30-16h00 : Round table (general discussion)

END OF THE WORKSHOP

TABLE OF CONTENTS

HOMOGENIZATION AND VISCO-ELASTICITY OF TURBULENCE

Z.S. SHE
Observatoire de Meudon, 92195 Meudon Principal Cedex, France
U. FRISCH
C.N.R.S. Observatoire de Nice, B.P. 139, 06003 Nice Cedex, France
and
O. THUAL
C.N.R.M. 42, avenue Coriolis, 31057 Toulouse Cedex, France

ABSTRACT

A multiple-scale analysis (homogenization) is applied to study the stability of steady cellular solutions of the one-dimensional Kuramoto-Sivashinsky equation with 2π-periodic boundary conditions. It is found that these solutions exhibit visco-elastic behaviour under very large wavelength perturbations. This elasticity property is then extended to Navier-Stokes turbulence. It is suggested that two-dimensional flame fronts and various turbulent flows (e.g. solar granulation and cloud streets) may display elasticity. Inclusion of elasticity into engineering turbulence modelling is also discussed.

1. INTRODUCTION

On very large scales turbulence may be considered as a material having diffusive properties. This idea when introduced by G.I. TAYLOR, L. PRANDTL and others, was based on an interesting analogy with transport in non-turbulent matter that results from molecular collisions. However, ordinary matter also exhibits elastic behaviour when a vector perturbation, e.g. a deformation, is applied. Elastic reponse of homogeneous turbulence to a uniform strain has been noticed by TOWNSEND (1956,1976) in connection with "Rapid Distorsion theory", and also discussed by MOFFATT (1967), RIVLIN (1957), CROW (1968) and others. But this Rapid Distorsion Theory appears quite unrealistic since one must assume that the strain-induced distorsion is so quick that non-linear interactions are negligible. In fact a "Slow Distorsion Theory" in which the turbulence adiabatically adjusts to the strain, would be more appropriate in most circumstances.

A systematic developement of a "Slow Distorsion Theory" has been made by FRISCH, SHE and THUAL (1984), using the method of homogenization (LARSEN, 1980 ; BENSOUSSAN, LIONS and PAPANICOLAOU and PIRONNEAU,1983 ; CHACON and PIRONNEAU, 1984 ; FRISCH, 1983). Only the essentials of this work are here outlined. It is simple to begin with the one-dimensional model of KURAMOTO (1978) and SIVASHINSKY (1977) (KS model), and then to extend it to 3-D Navier-Stokes turbulence.

Section 2 is devoted to the KS model, for which the governing equation is

$$\partial_t u + u\,\partial_x u + \partial_x^2 u + \nu\,\partial_x^4 u = 0 \qquad (1.1)$$

The coefficient $\nu > 0$ is called superviscosity. We assume 2π-periodicity in x. For small ν, the KS equation has steady cellular solutions of period $2\pi/N$ with $N=O(\nu^{-1/2})$. A two-scales analysis (homogenization) allows us to study their stability with respect to large-scale perturbations. This leads to wave-like elastic behaviour on short time scales and to diffusive-like behaviour on larger ones. In Section 3 we give the physical interpretation of the elasticity of cellular motion, and we obtain the transport coefficients by numerical techniques. We show that a "visco-elastic" window appears, in which the cellular motions are stable and perturbations undergo visco-elastic damping. We briefly discuss elastic effects for 3-D Navier-Stokes turbulence in Section 4. Section 5 is the conclusion, in which we indicate possible inclusion of elastic effects in practical modelling of turbulence.

2. THE KS MODEL AND THE HOMOGENIZATION TECHNIQUE

We rewrite the KS equation (1.1) in Fourier space

$$\partial_t \hat{u}(t,K) + iK \sum_{p+q=K} \hat{u}(t,p)\,\hat{u}(t,q) = (K^2 - \nu K^4)\,\hat{u}(t,K) \qquad (2.1)$$

$$K = 0, \pm 1, \pm 2, \cdots$$

When $\nu > 1$ there is only the solution U=0 which is stable ; when ν crosses 1, a bifurcation occurs to a nonvanishing solution of the steady KS equation

$$\partial_x (U^2/2) + \partial_x^2 u + \nu\,\partial_x^4 u = 0 \qquad (2.2)$$

or

$$\partial_x (u^2/2) + \partial_x^2 u + (1-\eta)\,\partial_x^4 u = 0 \qquad (2.3)$$

In (2.3) we introduce $\eta = 1-\nu$, the bifurcation parameter. The solutions are determined up to a translation. Henceforth u(x) denotes the particular 2π-periodic solution of (2.2) with u(0)=0 and $\partial_x u(0)>0$ which is old in x. An approximate expression of u(x) may be found by bifurcation techniques. When η is small, we obtain

$$u(x) = 2\sqrt{12\eta}\,\sin x + 2\eta \sin 2x + O(\eta^{3/2}) \qquad (2.4)$$

These steady solutions exist up to the appeařence of second bifurcation, a Hopf bifurcation, at $\eta = \eta_c \doteq 0.75$

Now consider the case of very small superviscosity ν. For any integer number N, we make the following change of variables in (2.2)

$$\begin{aligned} u &\leftarrow N\tilde{u}(Nx) \\ \nu &\leftarrow \tilde{\nu}/N^2 \\ x &\leftarrow \tilde{x}/N \end{aligned} \qquad (2.5)$$

Then we obtain the rescaled steady KS equation :

$$\partial_{\tilde{x}}(\tilde{u}^2/2) + \partial_{\tilde{x}}^2\tilde{u} + \tilde{\nu}\partial_{\tilde{x}}^4\tilde{u} = 0 \qquad (2.6)$$

or (2.7)

$$\partial_{\tilde{x}}(\tilde{u}^2/2) + \partial_{\tilde{x}}^2\tilde{u} + (1-\eta)\partial_{\tilde{x}}^4\tilde{u} = 0$$

where the rescaled quantities $\tilde{u}, \tilde{x}, \tilde{\nu}$ are order of one and $\eta = 1-\tilde{\nu}$. From (2.6) or (2.7) it follows that : for any N, such that

$$1/\sqrt{2\nu} < N < 1/\sqrt{\nu} \quad (\text{i.e. } 0<\eta<0.75) \qquad (2.8)$$

the steady KS equation (2.2) has a $2\pi/N$ periodic solution in which only Fourier modes $\pm$ N, $\pm$ 2N, ... are excited. When ν is very small, there may be a number of integers N_1, N_2,... satisfying (2.8), i.e. many competing solutions U_{N1}, U_{N2},...

Are these solutions stable ? For this, we restrict ourselves to the most dangerous perturbations, those of scale much larger than $2\pi/N$ (for more general perturbations see FRISCH et al, 1984). An asymptotic formalism is appropriate, in which the expansion parameter is the inverse of the basic wave number

$$\varepsilon = 1/N \qquad (2.9)$$

The cellular solution, corresponding to the superviscosity $\nu = \varepsilon^2\tilde{\nu} = \varepsilon^2(1-\eta)$ is then (see 2.5)

$$U_\varepsilon(x) = \varepsilon^{-1}u(x/\varepsilon) \qquad (2.10)$$

Stability is governed by the linearized KS equation :

$$\partial_t W + \partial_x (u_\varepsilon W) + \partial_x^2 W + \varepsilon^2 (1-\eta)\, \partial_x^4 W = 0 \qquad (2.11)$$

A multiple-Scale homogenization formalism involves the following substitutions in (2.11)

$$\partial_x \leftarrow \partial_x + \varepsilon^{-1} \partial_y \qquad (2.12)$$

$$\partial_t \leftarrow \partial_t + \varepsilon^{-1} \partial_\tau \qquad (2.13)$$

Where y, τ are fast variables and x,t are slow ones. Note that a correct choice of scale is crucial in such an expansion formalism. (2.12) and (2.13) are based on a physical consideration : we expect that the cellular motion on a scale 0 (ε) with a velocity amplitude $0(\varepsilon^{-1})$ (see (2.10))gives rise to a diffusivity 0(1) which, on space scales 0(1) has dynamical times also 0(1) corresponding to choice of slow variables (x,t). However we must also consider the shorter $0(\varepsilon)$ time-scale to describe the (new) elastic effects.

We now expand W in powers of ε

$$W = W_0 + \varepsilon W_1 + \varepsilon^2 W_2 + \ldots \qquad (2.14)$$

Substitution of (2.12-2.14) to (2.11) and identification of various powers of yields the following equations (only the first four are written)

$$A W_0 = 0 \qquad (2.15)$$

$$A W_1 + \partial_\tau W_0 + B W_0 = 0 \qquad (2.16)$$

$$A W_2 + \partial_t W_0 + \partial_\tau W_1 + B W_1 + C W_0 = 0 \qquad (2.17)$$

$$A W_3 + \partial_t W_1 + \partial_\tau W_2 + B W_2 + C W_1 + D W_0 = 0 \qquad (2.18)$$

Here we use the following definitions

$$A = \partial_y (u_\varepsilon \cdot \;) + \partial_y^2 + (1-\eta)\, \partial_y^4 \qquad (2.19)$$

$$B = (u_\varepsilon \cdot \; + 2 \partial_y + 4 (1-\eta)\, \partial_y^3 \;)\, \partial_x = \tilde{B}\, \partial_x \qquad (2.20)$$

$$C = (1 + 6 (1-\eta)\, \partial_y^2 \;)\, \partial_x^2 = \tilde{C}\, \partial_x^2 \qquad (2.21)$$

$$D = 4 (1-\eta)\, \partial_y \partial_x^3 = \tilde{D}\, \partial_x^3 \qquad (2.22)$$

where $u_\varepsilon \cdot$ means "multiplication by $u_\varepsilon(y)$"

The following properties of the operator A are very important

(i) A has a one-dimensional null-space containing

$$\zeta = \partial_y u_\varepsilon \tag{2.23}$$

Indeed $A\zeta=0$ follows from (2.2) and (2.19).

(ii) The null-space of the adjoint A^+ of A reduces to the constants (in y) as follows from its expression:

$$A^+ = -u\,\partial_y + \partial_y^2 + (1-\eta)\,\partial_y^4 \tag{2.24}$$

Note that the null-space of A^+ is orthogonal to that of A. since $\langle\zeta\rangle = \langle\partial_y u\rangle = 0$.

(iii) The constant 1, representative vector of null-space of A^+, is mapped by A into the null-space of A

$$A\,1 = \zeta \tag{2.25}$$

This implies that the generalized eigenspace associated to eigenvalue zero has dimension 2, and that there is a Jordan bloc $\begin{pmatrix} 0 & 1 \\ 0 & 0 \end{pmatrix}$.
Now eq. (2.15-2.18) are studied by successive applications of solvability conditions to equations of the form $Af=g$. From (2.15), we get

$$W_0 = \lambda(t,\tau,x)\,\zeta \tag{2.26}$$

Here λ is an arbitrary function of t,τ,x that will be determined by subsequent conditions (and so are μ, ρ appearing later). The solvability condition of (2.16) is that $(\partial_\tau W_0+BW_0)$ be orthogonal to the null-space of A^+, 1 ; this is expressed by

$$\partial_\tau \langle W_0\rangle + \langle BW_0\rangle = 0 \tag{2.27}$$

where $\langle g\rangle$ denotes the inner product of g with 1, defined as follows

$$\langle g\rangle = (2\pi)^{-1}\int_0^{2\pi} g(y)\,dy \tag{2.28}$$

We see that (2.27) is identically satisfied. So the solution of (2.16) reads

$$W_1 = - A_{ps}^{-1} \left[(\zeta \, \partial_\tau \lambda) + (\partial_x \lambda) \tilde{B} \zeta \right] + \mu \zeta \tag{2.29}$$

With μ, arbitrary function of t, τ, x and A_{ps}^{-1} , a pseudo-inverse of A. The solvability condition for (2.17) gives

$$\partial_t \langle W_0 \rangle + \partial_\tau \langle W_1 \rangle + \langle B W_1 \rangle + \langle C W_0 \rangle = 0 \tag{2.30}$$

which reduces to

$$\partial_\tau \langle W_1 \rangle + \langle B W_1 \rangle = 0 \tag{2.31}$$

Combining equation (2.19) with (2.31), we obtain an equation for (t, τ, x)

$$-\langle Aps^{-1} \zeta \rangle \partial_\tau^2 \lambda - \langle \tilde{B} \, Aps^{-1} \tilde{B} \zeta \rangle \partial_x^2 \lambda = 0 \tag{2.32}$$

or

$$\partial_\tau^2 \lambda = c^2 \partial_x^2 \lambda \tag{2.33}$$

To get (2.33), we have used the condition $A1 = \zeta$ and the notation

$$c^2 = - \langle \tilde{B} \, Aps^{-1} \tilde{B} \zeta \rangle \tag{2.34}$$

Therefore we obtain a wave-like equation (2.33) describing the fast-time dynamics of large scale perturbations. If $c^2 > 0$, then $\lambda(t, \tau, x)$ can be written as :

$$\lambda(t, \tau, x) = \varphi_+(t, x - c\tau) + \varphi_-(t, x + c\tau) \tag{2.35}$$

Where φ_+, φ_- are respectively right and left propagating waves. The physical interpretation of this wave like behaviour will be given in the next section.

To obtain the slow time dependence of λ, we solve (2.17) for W_2

$$W_2 = - Aps^{-1} \left(\partial_t W_0 + \partial_\tau W_1 + B W_1 + C W_0 \right) + \rho \zeta \tag{2.36}$$

The solvability condition for (2.18), after tedious algebraïc manipulation, is

$$\partial_\tau^2 \mu - \partial_x^2 \mu = S \tag{2.37}$$

in which S contains terms linear in λ and $\partial_t \lambda$ and their τ and x -

derivative up to third order. The non secularity condition for μ then gives two identical equations, namely

$$\partial_t \varphi_\pm (t, x) = d\, \partial_x^2 \varphi_\pm (t, x) \tag{2.38}$$

Here d, eddy diffusivity, is defined by

$$d = 1/2 \left\{ \langle A_{ps}^{-1} (\tilde{B} A_{ps}^{-1} \tilde{B} \zeta - \langle \tilde{B} A_{ps}^{-1} \tilde{B} \zeta \rangle) \rangle + \langle \tilde{B} A_{ps}^{-1} A_{ps}^{-1} \tilde{B} \zeta \rangle \right.$$
$$\left. + \langle \tilde{B} A_{ps}^{-1} \tilde{B} A_{ps}^{-1} \zeta \rangle - \langle A_{ps}^{-1} \tilde{C} \zeta \rangle - \langle \tilde{C} A_{ps}^{-1} \zeta \rangle \right\} \tag{2.39}$$

Equation (2.38) is of diffusive type. Stability requires simultaneous positivity of c^2 and d. We then have visco-elastic behaviour.

3. ELASTICITY OF CELLULAR SOLUTION, INTERPRETATION AND NUMERICAL EXPERIMENTS

We have seen that when a cellular solution $\varepsilon^{-1} u(y)$ (where $y = x/\varepsilon$) is subject to a perturbation

$$W(\tau, x) = \varepsilon \lambda (\tau, x) \partial_y u \tag{3.1}$$

The response on time scales 0 (ε) is governed by the wave equation (2.33) and on time scales 0 (1) by the diffusion equation (2.38) . For the interpretation we observe first that

$$\varepsilon^{-1} u\,(y + \varepsilon^2 \lambda(\tau, x)\,) = \varepsilon^{-1} u\,(y) + \varepsilon \lambda (\tau, x) \partial_y u + O(\varepsilon^2) \tag{3.2}$$

that is, a perturbation $\varepsilon \lambda \partial_y u$ is equivalent to an x and t- dependent <u>displacement</u> $\varepsilon^2 \lambda (t, x)$ of the cellular variable y. This is a suggestive of weakly damped elastic behaviour as we now demonstrate. A non uniform displacement may be viewed as a <u>straining</u> of the cellular structure ; we therefore expect a <u>stress</u> proportional to the <u>strain</u> $\partial_x \lambda$. It here suffices to consider time-independent (slow) straining. In the one-dimensional KS model the "Reynolds stress" has only one component, namely one half of the square of the "velocity" u. When a strain is applied, the velocity changes by an amount.

$$\delta u = \varepsilon W_0 + \varepsilon^2 W_1 + \cdots \tag{3.3}$$

The average change in the Reynolds stress is therefore

$$\delta_{stress} = \langle uW_0 \rangle + \varepsilon \langle uW_1 \rangle + O(\varepsilon^2) \qquad (3.4)$$

The first term on the r.h.s. vanishes. The second may be calculated from (2.29) ; thus

$$\delta_{stress} = -\langle u\, A_{ps}^{-1}\, \tilde{B}\, \zeta \rangle\, \partial_x \Lambda = C^2\, \partial_x \Lambda \qquad (3.5)$$

which is the constitutive equation for an elastic medium of shear modulus C^2.

Here we pause for a moment to answer the question : why did we get dynamics that are second order in time rather than first order ? The technical reason is that the null-space of A^+ and of A are orthogonal. This reflects the Jordan bloc $\begin{pmatrix} 0 & 1 \\ 0 & 0 \end{pmatrix}$ associated to the degenerated eigenvalue zero. The physical origin is just the invariance under both translation and Galiliean transformation. We mention that a more general treatment of such systems, including nonlinear effects, can also be made using ideas of phase dynamics and normal forms (see COULLET 1984, FAURE 1984 and their contributions to these Procedings).

Now it remains to calculate the transport coefficients C^2 and d. Analytic determination can be done by perturbation expansions in near the threshold of the first bifurcation (FRISCH et als, 1984). This however leads to antielastic behaviour ($C^2<0$) for small η. For finite η the coefficients C^2 and d must be calculated numerically. This requires numerical evaluation of quantities such as $A_{ps}^{-1}\, g$, the pseudo-inverse of A applied to a function g. These are obtained by finite time integration of P.D.E.'s of the form $\partial_t f + Af = g$ with restriction to the subspace of functions orthogonal to the two-dimensional generalized null-space of A spanned by ζ and 1. These integrations were done by an alias-free pseudo-spectral method (GOTTLIEB and ORSZAG, 1977) with a modified Leap-Frog temporal schema, called "Slaved Frog" (see Appendix A of FRISCH et als, 1984). In this manner the shear modulus C^2 and the eddy-diffusivity are obtained as a function of η. Simultaneous positivity of C^2 and d is finally obtained in the "visco-elastic window" ($\eta_1 < \eta < \eta_2$, $\eta_1 \simeq 0.300$, $\eta_2 \simeq 0.40$). Hence for any integer N such that

$$K_C\,(1-\eta_2)^{1/2} < N < K_C\,(1-\eta_1)^{1/2}, \quad K_C = \nu^{-1/2} \qquad (3.6)$$

$$\eta_1 \simeq 0.300 \qquad \eta_2 \simeq 0.40$$

there is a stable cellular solution of period $2\pi/N$.

We have also performed full simulation s of the KS equation in order to demonstrate the visco-elastic effects. An exemple is now shown in Fig.1 We have taken $\nu = 0.123$, and the initial condition $u_0(x) = 0.8 \cos x + 30 * \sin 7x$ in which a perturbation of wave number one is added to the cell of basic wave number seven. The time step is $\delta t = 0.0002$. The evolution is followed over 15000 time steps. Eventually relaxation to the unperturbed cellular solution is obtained. In between, there are weakly damped elastic oscillations, as predicted by the theory. In Fig.1 u(t,x) is represented at output times 100 times steps apart (each new output is slightly shifted).

4. ELASTICITY OF 3 - D NAVIER - STOKES TURBULENCE

It is clearly of interest to extend the formalisms we have developped in preceding sections to 3-D Navier - Stokes turbulence. The complexities of real turbulence are such that in this preliminary discussion we limit ourselves to stating the conditions, that are required for visco-elastic behaviour.

We assume that the turbulence is driven by a <u>deterministic</u> mecanism, represented by a linear operator $\mathcal{L}$ in the N-S eq.

$$\partial_t \underline{u} + \underline{u} \cdot \nabla \underline{u} = -\nabla p + \mathcal{L} \underline{u} + \nu \nabla^2 \underline{u} \qquad (4.1)$$

$$\nabla \cdot \underline{u} = 0 \qquad (4.2)$$

One possibility is the driving by a mean parallel shear flow $\underline{\bar{u}}$:

$$\mathcal{L} \underline{u} = -\underline{\bar{u}} \cdot \nabla \underline{u} - \underline{u} \cdot \nabla \underline{\bar{u}} \qquad (4.3)$$

where $\underline{\bar{u}}$ depends on a single coordiante (e.g. x_3).

In addition, the following assumptions have been found necessary for visco-elastic turbulence :

<u>A1</u> (Translation invariance). There is no explicit dependence on the x_1 coordinate and the turbulence is statistically homogeneous in the x_1 direction.

A2 . (Free turbulence). Boundaries are sufficiently removed to be ignored, but there may be conditions such as the vanishing of the velocity at $x_3 = \pm\infty$.

A3 . (Parity). The turbulence is statistically invariant under reflection in the x_2-x_3 plane.

We now make a few comments on why such assumption are needed. The linearized equation are

$$\partial_t \underline{W} + \underline{u}.\nabla \underline{W} + \underline{W}.\nabla \underline{u} = -\nabla p' + \mathcal{L}\underline{W} + \nu \nabla^2 \underline{W} \tag{4.4}$$

where $\underline{u}$ is the basic imperturbed flow. The formalism, eqs. (2.15-2.18), is still valid within the assumption of slow distorsion. With the choice of (4.3) for $\mathcal{L}$, eqs. (2.19 - 2.21) now become

$$AW = \underline{u}.\nabla_y \underline{W} + \underline{W}.\nabla_y \underline{u} + \nabla_y p' + \bar{\underline{u}}.\nabla_y \underline{W} + \underline{W}.\nabla_y \bar{\underline{u}} + \nu \nabla_y^2 \underline{W} \tag{4.5}$$
$$BW = \underline{u}.\nabla_x \underline{W} + \underline{W}.\nabla_x \underline{u} + \nabla_x p' + \bar{\underline{u}}.\nabla_x \underline{W} + 2\nu \nabla_x.\nabla_y \underline{W} \tag{4.6}$$
$$CW = \nu \nabla_x^2 \underline{W} \tag{4.7}$$

(A 1) guaranties that we still have a null-space $\zeta = \partial_{y_1} u$ (with the associated pressure field $p' = \partial_{y_1} P$) for the operator A. Hence

$$\underline{W}_o = \lambda(\tau, x_1)\underline{\zeta} \tag{4.8}$$

Where λ is a scalor (the translational invariance is in only one direction). (A2) ensures that solvability conditions analog to (2.27) and (2.31) are still valid.Except that the angular brackets are now understood as an ensemble average combined with a space integration in those directions where the turbulence is not translation - invariant but decays at infinity. The solvability condition analog to (2.30) is now a <u>vector</u> equation. However the unknown λ is a scalor. Fortunately it may be checked that the parity assumption (A3) ensures the vanishing of the x_2 and x_3 components of the vector equation. The x_1 component produces again a wave-like equation

$$\partial_\tau^2 \lambda = c^2 \partial_{x_1}^2 \lambda \tag{4.9}$$

The shear modulus c^2 is also determined by an auxillary linear equation analog to (2.31), and in principle may be calculated numerically.

It may be now concluded that visco-elastic behaviour is also expected for Navier-Stokes turbulence, just like for the KS model.

5. CONCLUSION

Using the technique of homogenization, we have investigated the response of turbulent flows to large scale perturbations in the linear regime. Under large scales slow straining, visco-elastic behaviour is obtained. We would like to suggest that this is a fairly general property of turbulence which has often been ignored and may have important consequences. We expect the elastic effects to be particularly conspicuous in free turbulence away from rigid boundaries. Elastic oscillations may possibly be observed (e.g. by imaging techniques) in one or several of following flows:

i) Flame fronts with sufficiently mature cellular structures
ii) Solar granulation and super granulation
iii) Cloud streets and squall lines in the Atmosphere

We now briefly indicate how elastic effects may be included in practical turbulence modelling such as k-ε modelling. In the traditional approach the Reynolds stresses $< u_i'.u_j' >$ are modelled by a diffusive term proportional to the average strain $\nabla\underline{\chi} + (\nabla\underline{\chi})^T$. Here T means matrix transposition and χ is the average displacement field which satisfies

$$\partial_t\underline{\chi} + \underline{u}.\nabla\underline{\chi} = \underline{u} + \mu_t \nabla^2\underline{\chi} \qquad (5.1)$$

where $\underline{u}$ is the mean flow and μ_t is a turbulent diffusivity which accounts for the scrambling of displacements by the turbulent small scale motion. The determination of μ_t (which may be a tensor in anisotropic cases) can be done by homogenization techniques in the spirit of the work of Mc LAUGHLIN, PAPANICOLAOU and PIRONNEAU (1983). We would like to call the above modelling k - ε - χ modelling of turbulence.

Acknowledgements : We are very grateful to M.E. BRACHET, P. COULLET, P. DELACHE, S. FAUVE, R. GRAPPIN, D. GALLOWAY, J. LEORAT, K. MOFFATT, T. PASSOT, O. PIRONNEAU and E. SIREGAR, for many useful discussions.

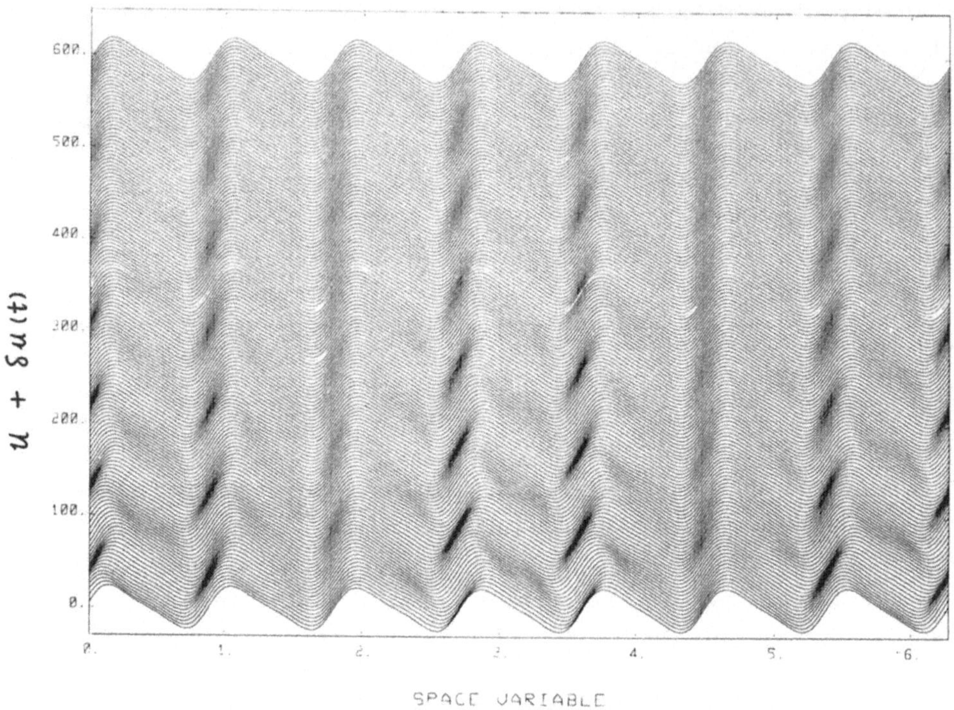

Fig.1 - Visco-elastic relaxation of weakly perturbed cellular solution. Basic wavenumber is k = 7 (ν = 0. 123, η = 0.397). Initial perturbation is 0,8 cos (x). Time step is 2.10^{-4}. Output is every loo time steps. Successive outputs are shifted by $\delta u = 4$

REFERENCES

Bensoussan, A., Lions, J.L. and Papanicolaou, G., 1078. Asymptotic Analysis for Periodic Structures, North Holland.
Chacon, J. and Pironneau, O., 1984. On the mathematical foundations of the K-ε turbulent model, Preprint, I.N.R.I.A.
Coullet, P. 1984. Private communication
Crow, 1968. Phys. Fluids 33, 1.
Faure, S., 1984. Private communication
Frisch, U., 1983. Turbulent transport of temperature, magnetic field and monumentum, preliminary notes available.
Frisch, U., She, Z.S. and Thual, O., 1984. On the elastic behaviour of turbulence, A case study of the Kuramoto- Sivashinsky model, preprint submitted to J. Fluid Mech.
Gottlieb, P. and Orszag, S., 1977. Numerical Analysis of Spectral Methods, SIAM, Philadelphia.
Kuramoto, Y., 1978. Prog. Theor. Suppl. 64, 346.
Larsen, E., 1980. Nucl. Sci. Eng. 73, 274.
Mc Laughlin, D., Papanicolaou, G. and Pironneau, O., 1983. Simulation numérique de la turbulence par homogénization des structures de sous maille, preprint, I.N.R.I.A.
Moffatt, H.K., 1967. Eds. Yaglom and Tatarsky, Nanka, Moscow, p. 139
Papnicolaou, G. and Pironneau, O., 1981. in "Stochastic Nonlinear Systems" Eds. Arnold and Lefever, Springer, p.
Rivlin, 1957. Q. Appl. Math. 15, 212.
Sivashinsky, G.I., 1977. Acta Astraunaut, 4, 1177.
Townsend, A.A., 1956. "The Structure of Turbulent Shear Flow" 2nd Ed. 1976 Cambridge University Press.

Sedimentation of a Random Dilute Suspension [1]

Russel E. Caflisch
Courant Institute of Mathematical Sciences

ABSTRACT

Recently Batchelor has calculated the average sedimentation speed of a random dilute suspension. A description of that work is presented here, along with a computation of the variance in the speed. It is shown that under Batchelor's assumptions and in the appropriate continuum limit, the variance is infinite.

1. Introduction

The motion of a solid spherical particle through a viscous fluid under the effect of gravity can be described by Stokes equations if the Reynolds number $\mathrm{Re} = a|\bar{v}_{ST}|\,\rho_P/\mu$ is sufficiently small. In this formula the Stokes speed is

$$\bar{v}_{ST} = -\hat{e} m \bar{g} / 6\pi\mu a \tag{1}$$

and a is the particle radius, m is the particle mass, ρ_P and ρ_f are the particle and fluid densities, μ is the fluid viscosity, $\bar{g} = g(1 - \rho_f/\rho_P)$ is the bouyancy - reduced gravitational acceleration and $\hat{e}$ is the unit vertical vector. Since $\rho_P < \rho_f$, the Reynolds number based on particle density ρ_P has been used so that particle inertia and fluid inertia are both negligible.

Consider a suspension of N of these identical particles which are randomly distributed throughout a fixed container of volume V. The average number density of particles is $n = N/|V|$ and the volume fraction is $\beta = \frac{4}{3}\pi a^3 n$. The suspension may be expected to behave as a continuum if n is large, i.e. in the limit

$$N \to \infty \quad \textit{with } \beta \textit{ constant.} \tag{2}$$

For a dilute suspension with $\beta \ll 1$, the dependence of the sedimentation speed $\bar{v}_S$ has been analyzed in several different cases. For a periodic array of spheres, Hasimoto [4] found $\bar{v}_S$ to be

[1] This research was partially supported by the Office of Naval Research under contract # N00014-81-K-0002 and by the National Science Foundation under contract # NSF-DMS-83-1229.

$$\vec{v}_S = \vec{v}_{ST}\,(1 - c\beta^{1/3} + O(\beta)) \tag{3}$$

in which the positive constant c depends on the type of periodic lattice. For a uniformly distributed random suspension Batchelor [1] showed that

$$\vec{v}_S = \vec{v}_{ST}\,(1 - 6.55\,\beta + O(\beta^2)). \tag{4}$$

These results are important for prediction of the behavior of dilute suspensions, as well as for indication of the effect of higher volume fractions. However the correct particle distribution - uniformly random, periodic, or something in between - is uncertain.

In this paper the results of Batchelor for a random suspension are derived and discussed and the variance in the sedimentation speed is calculated for such a suspension.

2. Stokes Equations.

Stokes equations, for the motion of a fluid in which inertial effects are negligible, are

$$\mu\,\nabla \vec{u} - \vec{\nabla} p = 0 \tag{5}$$

$$\nabla \cdot \vec{u} = 0. \tag{6}$$

If the fluid contains N particles with centers $\vec{x}_1,\ldots,\vec{x}_N$ each of radius a, then equations (5) and (6) hold for $\vec{x} \in V$ with $|\vec{x} - \vec{x}_i| > a$ for all i. In addition the conditions of no slip on the particle surface and balance of forces on the particle (neglecting inertia) are (for $i = 1,\ldots,N$)

$$\vec{u}(x) = \vec{v}_i \quad on \quad |\vec{x} - \vec{x}_i| = a \tag{7}$$

$$\vec{F}_i = \int_{|\vec{x} - \vec{x}_i| = a} \sigma \cdot \vec{n}\, d\,s_x \tag{8}$$

in which $\vec{v}_i$ is the particle velocity and is independent of $\vec{x}$, F_i is the external force on the particle, $\hat{n}$ is the outward normal to the sphere and $\sigma_{ij} = -p\,\delta_{ij} + 2\mu(\partial u_i/\partial x_j + \partial u_j/\partial x_i)$ is the stress tensor in the fluid.

In these equations particle rotation has been artifically set to zero since its effect on sedimentation is negligible. For the case in which the forces $\vec{F}_i$ are prescribed, the velocities $\vec{v}_i$ must be determined as part of the problem. On the other hand if the velocities $\vec{v}_i$ are prescribed, equation (8) is superfluous.

The subsequent developments in this paper rely on Faxen's law, which relates the solution of Stokes equations with a single particle to the solution without any particles. Its statement is the following:

Suppose that $\vec{u}$ solves Stokes equations with a force $\mathcal{F}$ distributed throughout the fluid (or at ∞) but with no boundaries, i.e.

$$\mu \nabla \vec{u} - \vec{\nabla}\vec{p} = -\mathcal{F} \tag{9}$$

$$\vec{\nabla} \cdot \vec{u} = 0 \tag{10}$$

for all $\vec{x} \in R^3$ and with $\mathcal{F} = 0$ for $|\vec{x} - \vec{x}_o| < a$. Let $\vec{w}$ solve Stokes equations with the same distributed force $\mathcal{F}$ and with a particle of radius a centered at $\vec{x}_o$, subjected to an external force $\vec{F}$, i.e.

$$\mu \nabla^2\vec{w} - \vec{\nabla}\pi = -\mathcal{F} \tag{11}$$

$$\vec{\nabla} \cdot \vec{w} = 0 \tag{12}$$

for $|\vec{x} - \vec{x}_o| > a$ and

$$\vec{w} = \vec{v}_P \quad \text{for} \quad |\vec{x} - \vec{x}_o| = a \tag{13}$$

$$\vec{F} = \int_{|x - x_0| = a} \hat{n} \cdot \vec{\sigma} ds_x . \tag{14}$$

Then $\vec{v}_P$ and $\vec{F}$ are related by

$$\vec{v}_P = \vec{F}/(6\pi\mu a) + (1 + \frac{1}{6} a^2\nabla_{\vec{x}}^2)\, \vec{u}|_{\vec{x} = \vec{x}_0} . \tag{15}$$

This result is proved [1] by showing that

$$\vec{v}_P = \vec{F}/6\pi\mu a + (4\pi a^2)^{-1}\int_{|\vec{x} - \vec{x}_o| = a} \vec{u} ds_x \tag{16}$$

which is a consequence of the balance of forces. Since $\nabla^4\vec{u} = 0$ for $|\vec{x} - \vec{x}_o| \leq a$, the integral in (16) has the value $\vec{u}(\vec{x}_0) + (1/6)a^2 \nabla^2\vec{u}(\vec{x}_0)$.

Faxen's law (15) is altered if there are boundaries in the flow $\vec{u}$. Suppose that $\vec{u}$ is a solution of Stokes equations with a particle of radius a centered at $\vec{x}_1$. Then (15) becomes

$$\vec{v}_P = \vec{F}/(6\pi \mu a) + (1 + \frac{1}{6}a^2 \nabla_{\vec{x}}^2)\, \vec{u}|_{x = x_0} + O(|\vec{v}_{ST}|\, a^4\, |\vec{x}_0 - \vec{x}_1|^{-4}) . \tag{17}$$

3. Probabilistic Description of a Suspension

Consider an ensemble of systems each with N particles in a volume V and with particle centers on the random configuration $C^N = \{\vec{x}_1,\ldots,\vec{x}_N\}$. The probability density $P(x_1)$ for one particle is normalized so that

$$\int_V P(\vec{x}_1)\, d\,\vec{x}_1 = N . \tag{18}$$

If the distribution is uniform, then $P(\vec{x}_1) \equiv n$, the number density. The N-particle density $P(C^N)$ and the conditional density $P(C^{N-1}|\bar{x}_1)$ are normalized so that

$$\int_{V^N} P(C^N)\, d\, C^N = N! \tag{19}$$

$$\int_{V^N} P(C^{N-1} \mid \bar{x}_1) d\, C^N = (N-1)!$$

in which $C^{N-1} = \{\bar{x}_2, \ldots, \bar{x}_N\}$.

The average of any function $f(\bar{x},C^N)$ is then

$$<f>(\bar{x}) = (N!)^{-1} \int_{V^N} f(\bar{x},C^N)\, P(C^N) d\, C^N \tag{20}$$

$$<f>(\bar{x}|\bar{x}_1) = (N-1)^{-1} \int_{V^{N-1}} f(\bar{x},C^N) P\, (C^{N-1}|\bar{x}_1) d\, C^{N-1} .$$

We shall assume , as in [1], that the particles are uniformly, independently distributed except for volume exclusion. For small β this implies that

$$P(\bar{x}_1|\bar{x}_0) = \begin{cases} 0 & \text{if } |x_0 - x_1| < 2\,a \\ P(\bar{x}_1) & \text{if } |x_0 - x_1| < 2\,a \end{cases} \tag{21}$$

with a relative error of size β. This choice is motivated by two facts: First two particles in an unbounded fluid move together without changing their relative positions. Thus if only two particle interactions are included any two-particle distribution is an equilibrium. Second Brownian motion of the particles will tend to make the particle positions independent.

In the subsequent analysis, the following approximation rule is used: If $f = f(\bar{x},C^N)$ is consistently defined for any N, then as $N \to \infty$,

$$<f>(\bar{x}) = \int f(\bar{x},\bar{x}_1) P(\bar{x}_1) dx_1 + O(\beta^2) . \tag{22}$$

We are unable to make a precise statement of the conditions for validity of this approximation, but certainly one condition is that the integral on the right be absolutely convergent. For motivation this approximation will be demonstrated in a simple context.

Suppose that f solves

$$\begin{aligned} (\nabla^2 - \lambda) f &= 0 \quad \text{for } |\bar{x} - \bar{x}_i| > a \\ f &= \bar{f} \quad \text{for } |\bar{x} - \bar{x}_i| = a \\ f &\to 0 \quad as\ |\bar{x} - \bar{x}_i| \to \infty \end{aligned} \tag{23}$$

for $i = 1, \ldots, N$. Then f satisfies (22)

This is proved by introducing the Green's function $G(\bar{x},\bar{y})$ solving $(\nabla_x^2 - \lambda)\, G(\bar{x},\bar{y}) = \delta(\bar{x} - \bar{y})$ for all $\bar{x}$ and $\bar{y}$. Then

$$f(\bar{x},C^N) = \sum_{i=1}^{N} \int_{|\bar{x}_i - \bar{y}| = a} \{\bar{f}\frac{\partial}{\partial n_y}G(\bar{x},\bar{y}) - G(\bar{x},\bar{y})\frac{\partial}{\partial n_y} f(\bar{y})\} ds_y \,. \tag{24}$$

Using the fact that the particles are identical, the average is

$$<f>(\bar{x}) = (N!)^{-1} \int f(\bar{x},C^N)P(C^N)dC^N$$

$$= (N!)^{-1}N \int_V \int_{V^{N-1}} P_1 P(C^N|x_1)$$

$$\int_{|\bar{y} - x_1| = a} \{\bar{f}\frac{\partial}{\partial n_y}G(\bar{x},\bar{y}) - G(\bar{x},\bar{y})\frac{\partial}{\partial n_y} f(\bar{y})\} ds_y \, d\, C^{N-1} d\, \bar{x}_1 \tag{25}$$

$$= \int_V P(\bar{x}_1) \int_{|\bar{x}_1 - \bar{y}| = a} \{\bar{f}\frac{\partial}{\partial n_y} G(\bar{x},\bar{y}) - G(\bar{x},\bar{y}) \frac{\partial}{\partial n_y} <f>(\bar{y}|\bar{x}_1)\} ds_y \, d\, \bar{x}_1$$

Now to leading order $<f>(\bar{y}|\bar{x}_1) = f(\bar{y},\bar{x}_1)$, the solution of (23) with $N = 1$. Then the inner integral on the right of (25) is exactly $f(\bar{x},\bar{x}_1)$. This is the result (22).

In this model calculation the functions f and G decay exponentially to zero at ∞, so that all integrals are absolutely convergent. Note also the normalization of the integral in (22): If $f=1$, the value of the integral is N and the approximation rule is invalid. If however f decreases as $|\bar{x} - \bar{x}_1| \to \infty$ the integral may be finite. It represents the single-particle effects on f.

4. The Sedimentation Speed

The approximation rule (22) is now used to calculate the sedimentation speed. Let $\bar{v}_P(\bar{x}_0,C^N)$ be the velocity of a particle centered at $\bar{x}_0$, with N other particles positioned on the configuration C^N and with gravitational force applied to the particles. The sedimentation speed is just

$$\bar{v}_S = <\bar{v}_P>$$

$$= (N!)^{-1} \int \bar{v}_P(\bar{x}_0,C^N)P(C^N|\bar{x}_0)\, dC^N \,. \tag{26}$$

The simplest use of (22) is to approximate

$$<\bar{v}_P> = \int \bar{v}_P(\bar{x}_0,\bar{x}_1)\, P(\bar{x}_1|\bar{x}_0)\, d\bar{x}_1 \,. \tag{27}$$

Since $\bar{v}_P(\bar{x}_0,\bar{x}_1)$ is approximately $\bar{v}_{ST}$ and $P(\bar{x}_1|\bar{x}_0) = n$ (for $|\bar{x}_0 - \bar{x}_1| > 2a$) which goes to infinity, the right side of (27) diverges in the limit (2). A second attempt is made by first subtracting off $\bar{v}_{ST}$ to get

$$<\bar{v}_P> = \bar{v}_{ST} + \int \{\bar{v}_P(\bar{x}_0,\bar{x}_1) - \bar{v}_{ST}\} P(\bar{x}_1|\bar{x}_0)d\bar{x}_1 \,. \tag{28}$$

Again the integral diverges because the bracketed quantity is size $|\bar{x}_0 - \bar{x}_1|^{-1}$ at infinity. However this quantity varies in sign so that it may be conditionally convergent in some sense.

In order to find $<\bar{v}_P>$ some other quantities must be subtracted from v_P before making the approximation (22). Following Faxen's law write

$$\bar{v}_P(\bar{x}_0,C^N) = \bar{v}_{ST} + (1 + \frac{1}{6} a^2 \nabla_x^2) \bar{u}(\bar{x}_0,C^N) + \bar{w}(\bar{x}_0,C^N) \tag{29}$$

in which $\bar{u}(\bar{x},C^N)$ is the fluid velocity at $\bar{x}$ for particles at C^N but no particle at $\bar{x}_0$, and $\bar{w}$ is a remainder. According to (17), $\bar{w}(\bar{x}_0,\bar{x}_1) = O(\bar{v}_{ST}\, a^4|\bar{x}_0 - \bar{x}_1|^{-4})$ which is integrable. Thus the approximation (22) is valid for $\bar{w}$ and

$$<\bar{v}_P> = \bar{v}_{ST} + \bar{v}' + \bar{v}'' + \bar{\bar{w}} \tag{30}$$

in which

$$\bar{v}' = (N!)^{-1} \int \bar{u}(\bar{x}_0,C^N)\, P(C^N|\,\bar{x}_0) dC^N \tag{31}$$

$$\bar{v}'' = (N!)^{-1} \int \frac{1}{6} a^2 \nabla^2 \bar{u}(\bar{x}_0,C^N)\, P(C^N|\bar{x}_0) d\, C^N \tag{32}$$

$$\bar{\bar{w}} = \int \bar{w}(\bar{x}_0,\bar{x}_1)\, P(\bar{x}_1|\,\bar{x}_0)\, d\bar{x}_1 \,. \tag{33}$$

The last integral was calculated by Batchelor [1] by direct solution of the two particle problems and under assumption (21), as

$$\bar{\bar{w}} = -1.55\, \beta\, \bar{v}_{ST} \,. \tag{34}$$

He also calculated the non-absolutely convergent integrals (31) and (32) using two physical principles.

The first of these principles is that the average velocity of the fluid-particle mixture should be zero, i.e.

$$<\bar{u}>(\bar{x}_0) = (N!)^{-1} \int \bar{u}(\bar{x}_0,C^N)\, P(C^N) dC^N = 0 \,. \tag{35}$$

Note that in (35) there is no conditioning that $\bar{x}_0$ be in the fluid as it is in (31). Subtracting (35) from (31) and using approximation (22) results in

$$\begin{aligned} \bar{v}' &= (N!)^{-1} \int \bar{u}(\bar{x}_0,C^N)\, \{P(C^N|\bar{x}_0) - P(C^N)\}\, dC^N \\ &= \int \bar{u}(\bar{x}_0,\bar{x}_1)\, \{P(\bar{x}_1|\bar{x}_0) - P(\bar{x}_1)\}\, dx_1 \\ &= -\, n \int_{|\bar{x}_0 - \bar{x}_1| < 2a} \bar{u}(\bar{x}_0,\bar{x}_1) d\bar{x}_1 \\ &= -\, 5.5\, \beta\, \bar{v}_{ST} \,. \end{aligned} \tag{36}$$

The last computation is made using the explicit solution for Stokes flow around a single particle.

The second principle is that the effective pressure of the fluid-particle mixture is hydrostatic. Consider the isotropic part $\sigma_{ii}/3$ of the stress. In the fluid

$\vec{\nabla}(\sigma_{ii}/3) = \vec{\nabla} P = \mu \nabla^2 \vec{u}$; in the particles it is defined through some constitutive laws. The fluid-particle mixture is considered to have an effective density $\beta \rho_P$ (since the density of the fluid has been absorbed into the bouyancy term) and an effective pressure $\bar{p}$ satisfying

$$\vec{\nabla} \bar{p} = - \beta \rho_P \bar{g} \hat{z}$$
$$= - \beta \vec{v}_{ST} \frac{9}{2} \frac{\mu}{a^2} . \tag{37}$$

Thus as a physical principle, we assume that

$$<a^2 \vec{\nabla} \frac{1}{3} \delta_{ii}> = (N!) \int a^2 \frac{1}{3} \vec{\nabla} \sigma_{ii} (\bar{x}, C^N) P(C^N) dC^N$$
$$= - \beta \vec{v}_{ST} \frac{9}{2} \mu . \tag{38}$$

This assumption is somewhat different from that of Batchelor but has the same result.

Using (38), $\vec{v}''$ can be calculated. First (38) is rewritten as

$$- \frac{3}{4} \beta \vec{v}_{ST} = < \frac{a^2}{6\mu} \vec{\nabla} \frac{1}{3} \sigma_{ii}>$$
$$= (N!)^{-1} \frac{1}{6} a^2 \int_{\bar{x} \text{ in fluid}} \nabla^2 \vec{u}(\bar{x}_0, C^N) P(C^N) dC^N$$
$$+ (N!)^{-1} \frac{a^2}{6\mu} \int_{\bar{x} \text{ in particle}} \vec{\nabla} \frac{1}{3} \sigma_{ii}(\bar{x}_0, C^N) P(C^N) dC^N . \tag{39}$$

The second integral is approximately

$$\frac{a^2}{6\mu} \int_{|\bar{x}_0 - \bar{x}_1| < a} \frac{1}{3} \vec{\nabla} \sigma_{ii}(\bar{x}, \bar{x}_1) P(\bar{x}_1) d\bar{x}_1$$
$$= - \frac{a^2 n}{6\mu} \int_{|\bar{x} - \bar{x}_1| = a} \frac{1}{3} \hat{n} \sigma_{ii}(\bar{x}, \bar{x}_1) d\bar{x}_1 \tag{40}$$
$$= - \frac{1}{4} \beta v_{ST}$$

in which the last calculation uses the one-particle Stokes solution. Therefore

$$(N!)^{-1} \frac{a^2}{6} \int_{\bar{x} \text{ in fluid}} \nabla^2 \vec{u}(\bar{x}_0, C^N) P(C^N) dC^N = - \frac{1}{2} \beta \vec{V}_{ST} . \tag{41}$$

Subtracting this from (32) and using (22) yields

$$\vec{v}'' = (N!)^{-1} \int_{\bar{x}_0 \text{ in fluid}} \frac{a^2}{6} \nabla^2 \vec{u}(\bar{x}_0, C^N) \{P(C^N | \bar{x}_0) - P(C^N)\} dC^N + \frac{1}{2} \beta \vec{V}_{ST}$$

$$= \int_{|\bar{x}_0 - \bar{x}_1| > a} \frac{a^2}{6} \nabla^2 \bar{u}(\bar{x}_0,\bar{x}_1)\{P(\bar{x}_1|\bar{x}_0) - P(\bar{x}_0)\}d\bar{x}_1 + \frac{1}{2}\beta \bar{V}_{ST}$$

$$= \frac{1}{2}\beta \bar{V}_{ST} \tag{42}$$

using the one-particle Stokes solution.

In summary

$$\bar{v}_S = <\bar{v}_P> = \bar{v}_{ST} + \bar{v}' + \bar{v}'' + \bar{w}$$

$$= \bar{v}_{ST}(1 - 6.55\beta + 0(\beta^2) \tag{43}$$

as stated in (4).

5. Effect of Finite Container

Suppose the container V which holds the suspension has a characteristic length scale R with $V = R^3$. In the previous section the effect of finite R was neglected. In fact since R is finite all of the integrals in that section were actually convergent, but it should be shown that $\bar{v}_S$ does not depend on R (for R large).

Feuillebois [3] accomplished this by deriving the physical principles (35) and (37) for a suspension in a finite container. Here we present only the first step of his calculation, in which the particles are approximated by point forces which are uniformly, independently distributed. Let $\bar{u}$ solve

$$\mu \nabla^2\bar{u} + \bar{\nabla}p = -6\pi\mu a\bar{v}_{ST}\sum \delta(\bar{x} - \bar{x}_i) \tag{44}$$

$$\bar{\nabla}\cdot\bar{u} = 0$$

for $\bar{x} \in V$ with

$$\bar{u} = 0 \quad on \quad \partial V. \tag{45}$$

Since there are no particle boundaries, spatial derivatives commute with averaging. Thus

$$\mu \nabla^2 <\bar{u}> - \bar{\nabla} <p> = -6\pi\mu a\bar{v}_{ST}\, n \tag{46}$$

$$\bar{\nabla}\cdot<\bar{u}> = 0$$

for $\bar{x} \in \bar{V}$ and

$$<\bar{u}> = 0 \quad on \quad \partial V. \tag{47}$$

The solution of (46) - (47) is

$$<\bar{u}> = 0 \tag{48}$$

$$\vec{\nabla} <p> = -6\pi\ \mu\ a\ \vec{v}_{ST} n$$

$$= -\frac{9\mu}{6a^2}\ \beta\ \vec{v}_{ST} \tag{49}$$

which are just the physical principles (35) and (37). The other effects of finite particle size and volume exclusion can be included as shown in [3].

6. Variance of v^P

The variance in v^P was calculated in [2] using a scaling argument. Here we present the calculation only for the model of independent, uniformly distributed point forces as in the previous section. The solution $\vec{u}$ of (46), (47) can be written as the sum

$$\vec{u}(\vec{x}) = 6\ \pi\ a |v_{ST}| \sum_{i=1}^{N} \vec{z}(\vec{x}; \vec{x}_i, R) \tag{50}$$

in which $\vec{z}$ is the one particle Stokeslet in the box $V = V_R$, i.e. $(\vec{x}; \vec{y}, R)$ solves

$$\nabla_x^2 \vec{z} - \vec{\nabla}\, \Pi = -\hat{e}\ \delta(x-y) \tag{51}$$

$$\vec{\nabla} \cdot \vec{z} = 0$$

for $\vec{x} \in V_R$ with

$$\vec{z} = 0 \quad on \quad \partial\, V_R\ . \tag{52}$$

The velocity $\vec{z}$ satisfies the properties

$$<\vec{z}> = 0 \tag{53}$$

$$\vec{z}(\vec{x}; \vec{y}, R) = R^{-1}\ \vec{z}(\vec{x}/R; \vec{y}/R, 1) \tag{54}$$

$$<|\ \vec{z}(\vec{x}; \vec{y}, 1)|^2 > = \sigma\ . \tag{55}$$

In (54) $\vec{y}$ is distributed uniformly over V_R, while in (55) $\vec{y}$ is distributed uniformly over V_1, the box of lateral size 1. The quantity σ which depends on $\vec{x}$ is non-zero away from the boundary $\partial\ V_1$.

With the notation $\vec{z}_i = \vec{z}(\vec{x}\ ; \vec{x}_i, R)$, the variance in $\vec{u}$ is calculated as

$$<|\vec{u}|^2> = (6\pi\ a|\vec{v}_{ST}|)^2 <|\sum \vec{z}_i|^2> \tag{56}$$

$$= (6\pi\ a\ |\vec{v}_{ST}|)^2 \{N<|\vec{z}_1|^2> + N(N-1) <\vec{z}_1 \cdot \vec{z}_2>\}\ .$$

The second term vanishes since the $\vec{x}_i$ are independent and $<\vec{z}> = 0$. Using the scaling (54),

$$<|\vec{u}|^2> = (6\pi\ a|\vec{v}_{ST}|)^2\ N\ R^{-2} <|\vec{z}(\vec{x}/R; \vec{y}/R, 1)|^2 > \tag{57}$$

$$=(6\pi\, a|\vec{v}_{ST}|)^2 N\, R^{-2}\, \sigma\ .$$

Therefore

$$<(|\vec{u}|\,/|\vec{v}_{ST}|)^2> \ = N^{1/3}\, \beta^{2/3} c\, \sigma \qquad (58)$$

in which $c = (6\pi)^2(\frac{4}{3}\pi)^{-2/3}$.

This shows that the variance goes to ∞ as N goes to infinity. The effects of particle size and volume exclusion can be included as in [2] and the variance in $\vec{v}_P$ can be shown to have the same behavior.

The most likely interpretation of this result is that the two particle distribution is more complicated than that given in (21). That distribution may be accurate enough for computation of $<\vec{v}_P>$ but not for $<|\vec{v}_P|^2>$. One reason for the importance of velocity variance is that it enters the Kubo formula for calculation of the particle diffusion due to random interactions. Finally note that the infinite value of the variance is related to the infinite energy in the Stokes flow around a single particle, i.e.

$$\int |\vec{u}(\vec{x},\vec{x}_1)|^2 d\vec{x}_1 = \infty \qquad (59)$$

since $|\vec{u}| = O(|\vec{x} - \vec{x}_1|^{-1})$. For small, nonzero Reynolds number this will be finite but large.

7. References

1. G. K. Batchelor. "Sedimentation in a Dilute Suspension of Spheres". JFM (1972) **52**, 245-268.
2. R. E. Caflisch and J. H. C. Luke. "Variance in the Sedimentation Speed of a Suspension". Physics of Fluids, submitted.
3. F. Feuillebois. "Sedimentation in a Dispersion with Vertical Inhomogenities". JFM (1984) **139**, 145-171.
4. H. Hasimoto. "On the Periodic Fundamental Solutions of the Stokes Equations and Their Application to Viscous Flow Past a Cubic Array of Spheres". JFM (1958) **5**, 317-328.

REMARKS ON OSCILLATIONS AND STOKES' EQUATION

Luc Tartar
C.E.A. Limeil-Valenton and Ecole Polytechnique
FRANCE

Homogenization and compensated compactness, in the way I have developped them with François Murat are concerned with understanding oscillations in nonlinear partial differential equations and in more intuitive terms to understand what are the equations governing macroscopic quantities in presence of microscopic variations of physical quantities.

The applications to realistic problems have had some success but mainly the tools have to be improved. I present here two remarks that I found useful for questions related to turbulence ; I will certainly be told that specialists do not see how to use them : I hope to learn during the workshop more about turbulence so as to make comments more adequate.

I - An homogenization theorem on Stokes' equation

We consider the following problem in R^3 :

$$(1)\quad \begin{cases} \dfrac{\partial u_\varepsilon}{\partial t} - \nu\, \Delta u_\varepsilon + u_\varepsilon \times \operatorname{curl}(v_0 + \lambda v_\varepsilon) + \operatorname{grad} p_\varepsilon = f_\varepsilon \\ \operatorname{div} u_\varepsilon = 0 \end{cases}$$

and we assume that : $(T < +\infty)$

$$(2)\quad u_\varepsilon \rightharpoonup u_0 \text{ in } L^2(0,T;H^1(R^3)^3) \text{ weak and } L^\infty(0,T\,;\,L^2(R^3)^3) \text{ weak}^*$$

The object of this problem is to understand how the oscillations of the sequence v_ε create oscillations in grad u_ε which then dissipate energy by viscosity, an effect that should be seen in the equation satisfied by u_o ; $\lambda > 0$ is a strength parameter which is there in order to emphasize the quadratic effect of the oscillations.

The reason for working in all the space R^3 is in obtaining a good estimate for the pressure p_ε :

(3) $\quad p_\varepsilon \in$ bounded in $L^2(0,T\ ;\ L^2(R^3))$

The precise hypothesis on f_ε and v_ε will be

(4) $\quad f_\varepsilon = \sum_j \frac{\partial}{\partial x_j}(g_{\varepsilon j})$ with $g_{\varepsilon j} \to g_{oj}$ in $L^2(0,T;L^2(R^3))$ strong

(5) $\quad v_o \in L^2(0,T;L^\infty(R^3)^3) + L^\infty(0,T;L^3(R^3)^3)$

(6) $\quad \begin{cases} v_\varepsilon = v_{1\varepsilon} + v_{2\varepsilon} \text{ with } v_{1\varepsilon} \rightharpoonup 0 \text{ in } L^q(0,T;L^\infty(R^3)^3) \text{ weak * for} \\ \text{some } q > 2\ ;\ v_{2\varepsilon} \rightharpoonup 0 \text{ in } L^\infty(0,T;L^r(R^3)^3) \text{ weak * for some } r > 3 \end{cases}$

[if then the initial data was given in $L^2(R^3)^3$ with zero divergence, we will have, using Sobolev's imbedding theorem $H^1(R^3) \subset L^6(R^3)$ a unique solution u_ε bounded in $L^2(0,T;H^1(R^3)^3) \cap L^\infty(0,T;L^2(R^3)^3)$ and by using Fourier transform in x, p_ε satisfying (3)].

As in many other examples an homogenization theorem relies on the construction of suitable test functions and taking weak limits of suitable quadratic quantities, a form of a compensated compactness argument which in some instances like ours will reduce to a simple integration by parts.

In order to do this we construct, for exercy $k \in R^3$ a sequence w_ε satisfying.

$$(7)\quad \begin{cases} -\dfrac{\partial w_\varepsilon}{\partial t} - \nu\,\Delta w_\varepsilon + k \times \operatorname{curl} v_\varepsilon + \operatorname{grad} q_\varepsilon = 0 \\ \operatorname{div} w_\varepsilon = 0 \\ w_\varepsilon \rightharpoonup 0 \text{ in } L^2(0,T;H^1(\mathbb{R}^3)^3) \text{ weak and } L^\infty(0,T;L^2(\mathbb{R}^3)^3) \text{ weak } * \end{cases}$$

which is possible if we take zero data at time T and assume

(8) v_ε bounded in $L^2(0,T;L^2(\mathbb{R}^3)^3)$

and we will have

(9) q_ε bounded in $L^2(0,T;L^2(\mathbb{R}^3))$;

Hypothesis (6) ensures that $w_\varepsilon \times \operatorname{curl} v_\varepsilon$ is bounded in $L^2(0,T;H^{-1}(\mathbb{R}^3)^3)$ so we can conclude that

$$(10)\quad \begin{cases} \text{a subsequence satisfies} \\ w_\varepsilon \times \operatorname{curl} v_\varepsilon \quad Pk \text{ in } L^2(0,T;H^{-1}(\mathbb{R}\{)^3) \text{ weak.} \end{cases}$$

Let $\phi \in \mathcal{D}(\mathbb{R}^3 \times]0,T[)$, multiply (7) by $\phi\, w_\varepsilon$ and integrate by parts ; then, using the fact that $w_\varepsilon \to 0$ in $L^2(\Omega \times]0,T[)^3$ strong for every bounded set Ω of $\mathbb{R}^3$, we obtain

(11) $\nu\,|\operatorname{grad} w_\varepsilon|^2 \rightharpoonup (Pk,k)$ in $\mathcal{D}'(\mathbb{R}^3 \times]0,T[)$.

In order to obtain an equation satisfied by u_o we must find the limit of the term $u_\varepsilon \times \operatorname{curl} v_\varepsilon$ which stays bounded in $L^2(0,T;H^{-1}(\mathbb{R}^3)^3)$; so we can assume that a subsequence satisfies

(12) $u_\varepsilon \times \operatorname{curl} v_\varepsilon \rightharpoonup \ell$ in $L^2(0,T;H^{-1}(\mathbb{R}^3)^3)$ weak

and we have to identify ℓ.

We now multiply (1) by $\phi\, w_\varepsilon$, (7) by $-\phi\, u_\varepsilon$, add and integrate on $\mathbb{R}^3 \times]0,T[$. Most of the terms, suitably paired, converge to 0 (notice for example that $\dfrac{\partial}{\partial X_i}(u_{\varepsilon j}\, w_{\varepsilon k})$ converges weakly to 0 in $L^1(0,T;L^{3/2}(\mathbb{R}^3))$ or $L^2(0,T;L^1(\mathbb{R}^3)))$; it remains

(13) $\int_0^T \int_{R^3} (\phi\ \lambda(u_\varepsilon \times \mathrm{curl}\ v_\varepsilon, w_\varepsilon) - \phi(k \times \mathrm{curl}\ v_\varepsilon, u_\varepsilon))\ dx\ dt \to 0$

If we notice that

(14) $((u_\varepsilon - u_0) \times \mathrm{curl}\ v_\varepsilon\ , w_\varepsilon) \rightharpoonup 0$ in $\mathcal{D}'(R^3 \times]0,T[)$

we will deduce that

(15) $\int_0^T \int_{R^3} (\phi(\ell,k) - \lambda\ \phi(u_0, Pk))\ dx\ dt = 0$

But (14) follows from the fact that w_ε as well as $(u_\varepsilon - u_0)$ converge strongly to 0 in $L^2(0,T;L^a(R^3)^3)$ for $a < 6$ and Ω bounded, and also in $L^b(0,T;L^2(\Omega)^3)$ for $b < +\infty$, and the hypothesis (6).

If we do this for 3 independent vectors k (15) will prove that
(16) $\ell = \lambda\ P^*\ u_0$ in the sense of $L^2(0,T;H^{-1}(R^3)^3)$ and so u_0 will satisfy

(17) $\frac{\partial u_0}{\partial t} - \nu\ \Delta\ u_0 + u_0 \times \mathrm{curl}\ v_0 + \lambda^2\ P^*\ u_0 + \mathrm{grad}\ p_0 = f_0$

where p_0 is the weak limit of a subsequence and $f_0 = \sum_j \frac{\partial}{\partial x_j} g_{0j}$

If now we multiply (1) by $\phi\ u_\varepsilon$ and integrate by parts and compare to the result of multiplying (17) by $\phi\ u_0$ and integrating by parts we obtain

(18) $\nu\ |\mathrm{grad}\ u_\varepsilon|^2 \rightharpoonup \nu|\mathrm{grad}\ u_0|^2 + \lambda^2(P^*\ u_0, u_0)$ in $\mathcal{D}'(R^3\ V\ ]0,T[)$

So we can state a theorem, noting $M = P^*$:

<u>Theorem</u> : Assume f_ε satisfies (4), v_0 satisfies (5) and v_ε satisfies (6), (8) ; then there is a subsequence and a matrix M satisfying

(19) $\begin{cases} M \in L^2(0,T;H^{-1}(R^3)^9) \\ (Mk,k) \geq 0 \text{ in } \mathcal{D}'(R^3 \times]0,T[)\ \forall k \in R^3 \end{cases}$

(M and the subsequence depending only on the sequence v_ε) such that : if a solution u_ε of (1) satisfies (2) with p_ε satisfing (3), the weak limit of u_ε satisfies

$$(20)\quad \begin{cases} \frac{\partial u_o}{\partial t} - \nu\, \Delta\, u_o + u_o \times \text{curl } v_o + \lambda^2 M\, u_o + \text{grad } p_o = f_o \\ \text{div } u_o = 0 \end{cases}$$

$$(21)\quad \nu\, |\text{grad } u_\varepsilon|^2 \rightharpoonup \nu\, |\text{grad } u_o|^2 + \lambda^2 (M\, u_o, u_o) \text{ in } \mathcal{D}'(\ R^3 \times]0,T[)$$

The functional space for M in (19) is certainly not optimal, because we also have $M\, u_o \in L^2(0,T;H^{-1}(\ R^3)^3)$; some technical improvements have to be made on this result.

The main defect is that this analysis does not give (at least, I do not see how) the same result in a bounded domain of R^3 : the crucial difficulty lies in the estimate on the pressure.

(1) was modelled on Navier Stokes equation :

$$\frac{\partial u}{\partial t} - \nu\, \Delta\, u + u \times \text{curl } (-u) + \text{grad } (p + \frac{u}{2}^2) = f$$

but I am no longer certain that the above analysis is of any use in understanding some aspects of turbulence.

II - Remarks on special quadratic quantities

When sequences of functions converge weakly and satisfy some differential equations, something can be said about weak limits of quadratic quantities : this is the basis of the compensated compactness method which I have developped with François Murat.

These quantities being more robust than others with respect to oscillations it is natural to try to use them in situations where oscillations are expected ; I do not yet know of any important applications to turbulence problems but the idea seems promising.

These special quadratic quantities can often be handled with a classical compactness argument through a factorisation property like

$\frac{\partial u}{\partial x}\frac{\partial v}{\partial y} - \frac{\partial u}{\partial y}\frac{\partial v}{\partial x} = \frac{\partial}{\partial x}(u\frac{\partial v}{\partial y}) - \frac{\partial}{\partial y}(u\frac{\partial v}{\partial x})$ (more general properties of this kind have been investigated by Hanouzet-Joly, with the wrong idea that the compensated compactness method was only making a list of sequentially weakly continuous quadratic forms).

The estimates are then obtained by using Fourier transform ; in some cases a little more care is necessary and the use of special functional spaces helps (if some Sobolev's spaces have a natural physical meaning, some are certainly used as technical tools).

We work here in R^2 and consider the equation

$$(22) \qquad -\Delta p = \sum_{i,j} \frac{\partial u_i}{\partial x_j}\frac{\partial u_j}{\partial x_i}$$

which appear for the pressure in Navier Stokes equation without exterior forces (or with a divergence free forcefield).

Lemme : If $u \in H^1(R^2)^2$ with div $u = 0$ then $p \in C^0(R^2)$, grad $p \in L^2(R^2)^2$ with

$$(23) \qquad \|p\|_{L^\infty} + \|\text{grad } p\|_{L^2} \leq c\,\|\text{grad } u\|^2_{L^2}$$

Proof : Of course by (22) we wean the only continuous solution converging to 0 at infinity (which does exist by the following argument) By Fourier transform we have

$$(24) \qquad \hat{p}(\xi) = -\frac{1}{|\xi|^2}\int_{R^2} \sum_{i,j} \hat{u}_i(\xi-\eta)\,(\xi-\eta)_j\,\hat{u}_j(\eta)\,\eta_i\,d\eta$$

The integrand $\sum_{i,j} \hat{u}_i(\xi-\eta)\,\hat{u}_j(\eta)\,(\xi_j-\eta_j)\,\eta_i$ has a modulus bounded by

$|\hat{u}(\xi-\eta)|\,|\hat{u}(\eta)|\,|\xi-\eta|\,|\eta|$; but because div $u = 0$ it is also

$\sum_{i,j} \hat{u}_i(\xi-\eta)\,\hat{u}_j(\eta)\,\xi_j\,\eta_i$ which is bounded by $|\hat{u}(\xi-\eta)|\,|\hat{u}(\eta)|\,|\xi|\,|\eta|$ and similarly by $|\hat{u}(\xi-\eta)|\ |\hat{u}(\eta)|\ |\xi|\ |\xi-\eta|$; the integrand is then bounded by $|\hat{u}(\xi-\eta)|\ |\hat{u}(\eta)|\ |\xi|\ |\eta|^{\frac{1}{2}}\ |\xi-\eta|^{\frac{1}{2}}$, so

$$|\hat{p}(\xi)| < \frac{1}{|\xi|} \int_{R^2} |\xi-\eta|^{\frac{1}{2}}\ |\hat{u}(\xi-\eta)|\ |\eta|^{\frac{1}{2}}\ |\hat{u}(\eta)|\ d\eta$$

We see a convolution product of $|\eta|^{\frac{1}{2}}|\hat{u}|$ by itself ; the use of Young's inequality in L^p spaces is not adequate here and we need some properties of Lorentz spaces $L^{p,q}(R^2)$ here. [Lorentz spaces are interpolation spaces between L^1 and L^∞ and can be defined using decreasing rearrangements of functions ; the basic properties which follow from general theorems of J.L. Lions and Peetre are :

$L^{p,p} = L^p$; $L^{p,q_1} \subset L^{p,q_2}$ if $q_1 < q_2$; multiplication acts from

$L^{p_1,q_1} \times L^{p_2,q_2}$ into $L^{p,q}$ $\quad \frac{1}{p} = \frac{1}{p_1} + \frac{1}{p_2}$, $\frac{1}{q} = \frac{1}{q_1} + \frac{1}{q_2}$;

convolution acts from $L^{p_1,q_1} \times L^{p_2,q_2}$ into $L^{r,q}$ $\quad \frac{1}{r} = \frac{1}{r_1} + \frac{1}{r_2} - 1$,

all the values used being in $[1,+\infty]$;

The dual of $L^{p,q}$ is $L^{p',q'}$ if $1 < p,q < +\infty$; $\frac{1}{r^\lambda} \in L^{N/\lambda,\infty}(R^N)$]

with this is mind we have $|\eta|\ |\hat{u}| \in L^2 = L^{2,2}$ and $|\eta|^{-\frac{1}{2}} \in L^{4,\infty}$ so

$|\eta|^{\frac{1}{2}}|\hat{u}| \in L^{4/3,2}$; then $|\eta|^{\frac{1}{2}}\ |\hat{u}| * |\eta|^{\frac{1}{2}}\ |\hat{u}| \in L^{2,1} \subset L^2$ so

grad $p \in L^2$; then as $\frac{1}{|\xi|} \in L^{2,\infty}$ we get

$\hat{p} \in L^1$ so $p \in \mathcal{F} L^1 \subset C^0(R^2)$.

Continuity of the pressure was noticed before by L.E. Fraenkel ; similar results were noticed by Wente with a right hand side

$\frac{\partial u}{\partial x}\frac{\partial v}{\partial y} - \frac{\partial u}{\partial y}\frac{\partial v}{\partial x}$: it was in order to understand this result that I derived a proof using interpolation theorems and bilinear mappings which I think has more general possibilities.

Although these functional spaces may frighten some, they could be more useful in order to understand the exchanges between low and high frequencies by introducing adapted functional spaces were the Fourier transform will be.

It is important to notice that the above analysis requires more than a simple argument of scaling : a precise algebraic property has to be satisfied by the quadratic quantity involved ; playing with these special quadratic quantities is an interesting game thay may lead on a good track : it was by this way that I discovered some (apparently not known before) invariants for Euler equation in even dimension generalizing the vorticity, a result that helped D. Serre to generalize the helicity to any odd dimension. It has not given any important result for Euler of Navier Stokes equation in realistic 2 or 3 dimensions but I still believe it is a useful point of view.

Bigliography

I have written two survey articles on the methods of homogenization and compensated compactness ; they contain more references.

[1] Etudes des oscillations dans les équations aux dérivées partielles non linéaires. Trends and Applications of Pure Mathematics to Mechanics, Ciarlet-Roseau ed. Lecture Notes in Physics, Springer, 195 (1984) p. 384-412.

[2] Oscillations in non linear partial differential equations : compensated compactness and homogenization. Lecture Notes in Applied Mathematics, vol. 23. American Mathematical Society.

LARGE AND SMALL STRUCTURES IN THE COMPUTATION OF TRANSITION TO FULLY DEVELOPED TURBULENT FLOWS

P. PERRIER
AMD/BA, B.P. 300, 92214 St Cloud, France

ABSTRACT :

We shall discuss how the large structures of turbulent flows have to be modellized especially from the beginning of their onset, without any hypothesis based on fully developed turbulence. A completely mathematical derived model is presented, given by homogenization theory. That model would be able to evaluate the size and behaviour of large structures in presence of statistical equilibrium of small structures.

0. INTRODUCTION

Two minimal conditions are required in the modelizations of a given experimental turbulent flow field, with fully developed turbulence :

1) The statistical quantities of the flow can be defined in each point of the field.

2) A separation can be clearly defined between mean and fluctuating quantities. It is equivalent to the capability of separating with a sufficient accuracy the first moment of the velocity in each point as a fonction of time.

In the more simple case of a steady flow, with no clear dependance of the time coordinate, one can be less stringent on the definition of statistical properties at a given time because a good approximation can be obtained for them if a sufficiently long interval of time is selected.

In practical computations, the main interest is however on unsteady flows but with slow variation of statistical properties of the flow in a given point of the field compared to the derivatives of instantaneous turbulent quantities.

The incompressible Navier-Stokes equations can be expressed in the following form.

$$u_{,t} + u\cdot\nabla u + \nabla p = \nu\Delta u$$

$$\nabla\cdot u = 0$$

+ boundary conditions

+ initial conditions

If we retain an expression for the difference to the mean velocity u_o and pressure p_o, the classical equivalent equation to solve is.

$$u_{o,t} + u_o \cdot \nabla u_o + \nabla p_o - \nu \Delta u_o = \nabla \cdot a + b$$

$$\nabla \cdot u_o = 0$$

If the evaluation of a and b terms is sufficiently well approximated, a good approximation, at the same level as a and b approximations, could be given by solving that system of partial differential equations with a convenient set of boundary conditions. From an industrial point of view, two types of quantities of the turbulent flows are of major interest.

1) Mean quantities

Current interest is first in mean and root mean square quantities of mean and statistical velocity and pressure distribution. Those data give sufficient information for restitution of mean flow fields and some information on the intensity of turbulence at each time. Secondary quantitites as the turbulent stress tensor or the higher order moment of probability distribution of turbulent part of the flow field are more important for checking the validity of the modeling than from a practical point of view.

2. Extreme quantitites

The industrial concern with mean quantities is only a part of the definition of the performances required for the design of a system using turbulent flows. If it is required to know the capability of such systems to work whatever will be external conditions; it is necessary to know extreme quantities of the flow field for a given probability level. The knowledge of the extreme quantitites, and a realization of the complete flow field corresponding to extreme case, are required as unique feature of the studied systems, for the design of a system able to fulfill special requirements given by extreme flow field conditions. For example in meteorology the extreme rainfalls define architecture norms of water evacuation, the extreme gust to be encountered in flight will define the majorant test case of structure calculation for aircraft design with an acceptable level of probability compared to life cycle of the aircraft. It can be seen however that the knowledge of the extreme value of gust interacting with an aircraft is not sufficient for having an evaluation of unsteady loads given on the aircraft structure. It is necessary also to have a realization of the flow field so that the aircraft response to the entrance and exit of a real gust gradient can be evaluate. If a realization cannot be evaluated, something equivalent can be deduced from the minimal knowledge of externe local flow field value, and that correspond to the minimum industrial requirement. The probability distribution of current interest covers the 10^{-3} to 10^{-6} interval, so that the selection of direct numerical simulation amongst many such numerical simulation of flows is a procedure of unacceptable cost.

On the figure 2, is given a typical evalution of maximal dispersion of a criteria relative to the surge of a turbojet engine mounted in the air inlet of an aircraft versus probability of encounter such extreme turbulent realization in critical flight regime. The industrial target is to predict a priori that law of probability distribution, from the resolution of one to several unsteady turbulent flow fields inside the air intake of the aircraft.

A typical acceleration curve measured at the passenger seat of a commercial aircraft flying in rough atmosphere is given figure 3. The mean value of acceleration is obviously 1 g but the r.m.s. of the fluctuation of the acceleration is to be caracterized. In addition to r.m.s. band of fluctuation, that are to be reduced to passenger tolerable level, it seems important to know the acceleration peaks level if such peaks are frequent.

1. PHYSICS OF LARGE TURBULENT STRUCTURES

The detailed study of a fully developped turbulence can be pushed to the detailed analysis of spectral distribution (in wave number tridimensional coordinates) in one point and to the analysis of space and time correlations. If it was possible to made a modeling of statistical quantities at a sufficient level of accuracy (the 3^{d} moment at minimum) it will be possible to solve correctly the Navier-Stokes equation expressed in fluctuating quantities ; however it would not be probably possible to make evaluation of exact level of turbulent quantities, for example a correlation length or a dissipation length ; but, for a given set of initial such quantities, a turbulent closure more or less sophisticated, give the capacity of complete modelling of more or less turbulent flows. However the characteristic lengths and/or the level of energy contained in turbulent motion at input of the computation, are strongly a function of boundary conditions of the problem : it is not correct to solve the averaged Navier-Stokes equation without taking in account the correct initial data or initial boundary conditions : the interest for fully developed turbulence without any memory of the boundaries of the problem cannot be fruitful without the specification a priori or from experimental data of the energy inside the turbulent part of flows and of their spectral distributions.

In many practical applications, the main interest is related to turbulent flows far from the bodies that generate the turbulence inside the flowfield. In that case the progress in modelling capabilities have improved significantly the quality of computations. It is specifically true for all the thin layers or boundary layers or wakes or parietal jets.

On the contrary the major part of useful flows to be considered are free flows near the boundaries that have generated their shear layer and with just spread out turbulence in a more larger wake or flow : near "potential" core of jets, cavity flows, flow around bodies near base, flow inside nuclear reactor or on the rearpart of a car or truck, flow inside a combustion chamber. In all these cases the current hypothesis is to apply to the small scales the fully turbulent equilibrium properties (Smagorinsky model) and use the average Navier-Stokes equation for large scales with the simple cutting between mean and fluctuating quantities ; usually the Navier-Stokes formulation is retained by putting in the second hand the additional turbulent stress terms to be modellized.

A detailed analysis of the flows in the immediate neighbour of boundaries shows evidence that the larger part of turbulent flow is generated by unsteadiness of very thin turbulent layers, boundary layers becoming wakes at separation of body. So two significantly different scales of turbulence ar present : one for the small turbulent flow, before their mixing in one well-mixed turbulent wake and second the large scale given by maximum size of the body or cavity that turbulence finally fill up completely or spread fast inside it because of large unsteadiness of the flow. The unsteadiness is more characteristic of such a turbulence at large size that microstructure of turbulent initial eddies because the energy had not usually the time to be transferred effectively from large to small size of structures in a continuous manner ; the large structures are deterministic and distort themselves to destroy ramdomly themselves in a turbulent wake in statistically stable quasi equilibrium unsteadiness of the flow due to the effect of turbulence contained in small scales.

That transition to turbulence is of main practical interest in all the cases given before and for which the characteristic length is of the same order as the dimension of the flowfield to be computed. That difficult requirement is the main origin of the onset of new turbulence model able to take in account two scale lengths well separated (e.g. 1 to 2 order of magnitude range) so that the trace of deterministic unsteady flow, with their own bifurcation processes, be used as initial conditions for the downstream turbulence and give the access to knowledge of detailed flows in the transition zone to statistically classical turbulent flows.

However an analysis of fully developed turbulent flows with conditional sampling can show the existence inside the turbulent homogenous flow of coherent structures of large size. The probability level of such large coherent structure of significant size is in the range (10^{-13} to 10^{-6}) of extreme practical applications as we have defined them above. At some time, between time intervalls not too large, depending on the size of coherent structure studied, it is possible to have realization of large flat eddies or shear layers, with caracteristics feuillet like shape

Analysis of direct simulation of turbulent flows by Mc Williams (ref) shows clearly that the inverse cascade can create such large structure from initially homogeneous turbulent flows for a limited time life but with a well organised feuillet like shape without any forcing influence of boundary conditions. With forcing conditions such large structure are so frequent that it can be well checked experimentally in many cases for example in meteorological structures downwind of an island in 10^{8} - 10^{10} Reynolds number range.

In practical cases it is of main interest to have a correct modeling of the coupling procedure between small and large scales of turbulence, the later ones given by boundary conditions. For example in the very simple flow behind a bidimensionnal step it is well known that coherent bidimensional rolls-like vortices are present on the jet-like line generated by upstream boundary layer and that the reattachment area of the flow downstream of the step is particularly rich in large structures giving very large fluctuations of pressures near the reattachment point.

If the curvature of the plate downstream of the step is increased until the reattachment can no more be realized, it can be checked that the limit of the attached flow varies a lot with the energy generated inside the rolls-like large vortex structure in the wake generated at the corner of the step.

A careful analysis of that effect of great practical importance requires a very good modelling of the turbulence, much better than currently available because it is necessary to handle the growth and the dynamic of such large structures and of their interaction with the smaller scales of turbulence near the wall.

2. LARGE PULSED UNSTEADY STRUCTURES

It is necessary to recall that the Navier-Stokes equations can have quite different solutions corresponding to very small perturbations in the initial conditions of generation of large viscous flows but also in the case where the turbulent viscous flows are limited to boundary layers and thin wakes downstrean of the bodies in the flow.

For having instability it is sufficient that several solutions of the inviscid equations can exist without any viscous effect or entropy larger generation (potential flow equations or Euler equations). For example a transonic wing section with a large leading edge radius can present, in a specific angle of attack and Mach number range, triple exact solutions of inviscid equations with different position of main shock wave near the leading edge or near the crest at mid section.

The law of lift variation with angle of incidence and the schematic equivalent flow are given from a computation on figure 4 below, and experimentally in wind tunnel such a behaviour have been also obtained. However in the double regime between the 2 extreme stable solutions, the flow is oscillating in a random way from one to the another regime with a distribution of probability of one state that can be deduced from experimental analysis. The unstable solution is given by lift incidence negative slope and is physically unattainable but can be determined accurately with a continuation mathematical method. On the contrary the probability law that gives, at each point in the unsteadiness domain, the probability for having one of the stable solutions cannot actually be predicted and is probably dependant of external conditions, and of general turbulence level of external flows.

In the same way for high angles of attack, separated or non separated double solutions give some hysteresis on a profile or some instabilities with unsteady separation (periodic on random) in nozzles. These cases of instability is directly connected to the boundary conditions effects and occur in many cases of unsteady flows and every time one has multiple solution of Navier-Stokes equations or potential flow or Euler's equations.

When the flow is varying in a piped flow, one notes (detailed experience of SARPKAYA) that the generation of turbulent structures is extremely unsteady and strongly depends on the coupling of the prescribed unsteady behaviour and small structures in the pipe flow.

In all that cases, the current closure are not known to be abble to compute correctly a turbulent flow due to very different sizes of structures.

At last a coupling can exist between large structures and unsteady behaviour because many large structures do not come from the initial homogeneous turbulent flow as these studied by Mc Williams, but from the boundary conditions ; for instance there is an exponential process of generation of large structures from an unsteady mode of the solutions of the Navier-Stokes equations, and one has to be able to compute this mode when modelling the Navier-Stokes equations. Usually it is the case for the generation of the Karman vortex streets in the locally 2-D wakes for high Reynolds number ; an other case is the Goërtler eddies: in boundary layers or wakes curved by the effects of the concavity (after 2-D concavity) generation of eddies colinear with the velocity the large structures are generally more stable than vortices normal to velocity.

So it is important that the modeling of the Navier-Stokes equations retain the natural low frequency unsteady solutions given by the boundary conditions, in order to select the good solutions : the requirements of the computed results require the modelling of transition to large turbulence structure near the boundaries.

On the figures 6 et 7 below, the specific place of such a model is given with respect of the domain of validity of usual models in the case of a strongly separated wake after a body (car or aircraft with high angle of attack) and the distribution of energy versus the wave numbers is given for specification of the limits of validity of each modelling.

On the figure 8 some measures of spectral distribution of energy, made in the entry of a pipe with an important separating area, show that the steady turbulence is progressively reached in the pipe, after the separating point. This initial transient phasis has to be modelled. For this experimental case the result of a direct simulation of the Navier-Stokes equations is given, a simulation with a coarse turbulent model, ($k-\varepsilon$) which allows to see the separated zone and the corresponding vorticity. New computation models must be improved for the zone after the first half-diameter (that initial part being wall computed by a direct solution of Navier-Stokes equations) and before the 3 or 4 diameters needed to have a reasonable statistical equilibrium of the turbulence. Such a model is presented below, it relies on the theory of homogeneization, in order to allow extreme variations of mesh density, which is needed to catch correctly the unsteady boundaries of turbulent wakes and the thin layers of the wakes after the edges of a body. This model has been derived by MM. MACLAUGHIN, PAPANICOLAOU, PIRONNEAU from a first approach of M. PIRONNEAU and the author.

3. CONVECTION OF MICROSTRUCTURES BY EULER'S EQUATIONS : A TURBULENT MODEL BY HOMOGENEIZATION

Convection of microstrutures is obtained by the following equations where u is the "homogenized" quantity with respect to the length scale ε , ε small, and where u is the velocity and p the pressure, solution of the second equation, R a tensor function of the lagrangian coordinate a and the turbulent energy q, turbulent "helicity" r :

(1)
$$\begin{cases} u_{,t} + u\cdot\nabla u + \nabla p - \nu\Delta u - \nabla\cdot R = O(\varepsilon) \\ \nabla\cdot u = 0 \\ R = \langle w\otimes w\rangle \end{cases}$$

(2)
$$\begin{cases} q_{,t} + u\cdot\nabla q + \nabla\cdot B\nabla q - R(\nabla a\nabla a^T) : \nabla u + b(q,r) = 0 \\ r_{,t} + u\cdot\nabla r + \nabla\cdot C\ \nabla r - S(\nabla a\nabla a^T) : \nabla u + \phi(q,r) = 0 \\ \langle u'\otimes u'\rangle = -\mu\nabla a\nabla a^T \end{cases}$$

R, S, b, ϕ are computed by solving

(3)
$$\begin{cases} u'\cdot\nabla_y u' + M\ \nabla_y p' = 0 \\ \nabla_y\cdot u' = 0 \\ u' - y \text{ periodic} \end{cases} \qquad \begin{array}{l} M = \nabla a^T\nabla a \\ u' = \nabla a^T w \end{array}$$

as a function of a solution of

(4)
$$\begin{cases} a_{,t} + u\cdot\nabla a = 0 \\ a(x,t=0) = x \end{cases}$$

For the present the model is only valid for a flow with two well separated scales for the initial conditions and with equilibrium for the microstructures. Without these hypothesies, the model has to be improved.

The equations (2) (3) (4) are obtained by an asymptotic expansion of :

$$u^{\varepsilon}_{,t} + u^{\varepsilon}\cdot\nabla u^{\varepsilon} + \nabla p^{\varepsilon} - \nu\Delta u^{\varepsilon} = 0$$

$$\nabla\cdot u^{\varepsilon} = 0$$

+ boundary conditions

$$u^{\varepsilon}(x,t=0) = u_o(x) + w_o(\frac{x}{\varepsilon})$$

where

$$u^{\varepsilon}(x,t) = u(x,t) + w(\frac{a(x,t)}{\varepsilon},x,t) + \varepsilon u^1(\frac{a(x,t)}{\varepsilon},x,t) + 0(\varepsilon^2)$$

$$p^{\varepsilon}(x,t) = p(x,t) + \pi(\frac{a(x,t)}{\varepsilon},x,t) + \varepsilon p^1(\frac{a(x,t)}{\varepsilon},x,t) + 0(\varepsilon^2)$$

A two equations model is found (k-ε type model), with a "elastic" main term (cf. FRISCH) ; here the power of such an analysis by homogenization is demonstrated.

The interest of this model is the purely theoretical determination of its constants and its ability to work in the transition to the fully developped turbulence. Also it has not to be used alone in the zone of turbulence generation where the stress tensor and the strain tensor have not the same eigen values, as it is underlying in k-ε model.

On the figure 9, one can see a result of a computation of flow behind a cylinder where the turbulent production in the high main gradient zones is visible, though the flow velocity has been computed at each time with large unsteady structures, without any information on mean shear stress of the flow.

4. CONCLUSION

This paper is a review of the complicated cases of transition to a large homogeneous turbulent zone of the flow after bodies where the initial turbulence is confined in small wakes or jets and where the scales of the final large structures and of the inital turbulence are really separated; a modelisation of such a complex flow is required by the actual practical industrial objectives. Comparison of computational requirements and of the complexity of physics to be modelized suggests to retain a new turbulent model by homogeneization which should permit to compute the delicate phasis of transition to a classical full developped turbulence.

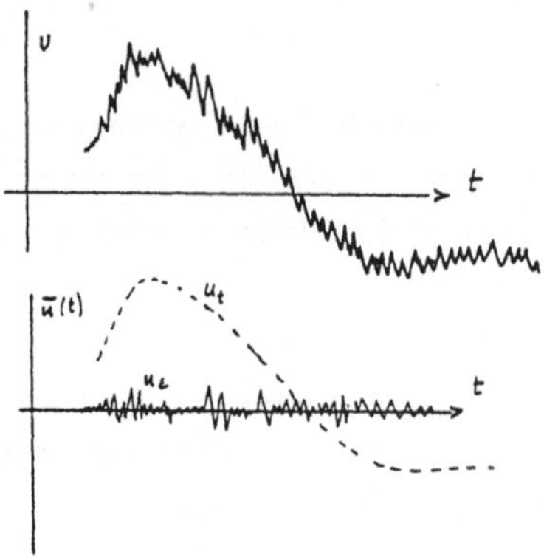

Figure 1

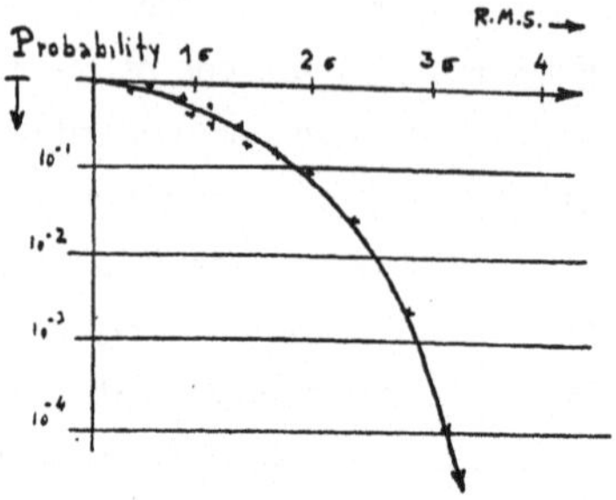

Figure 2

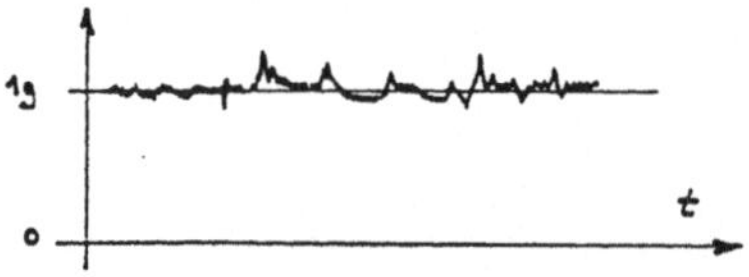

Figure 3

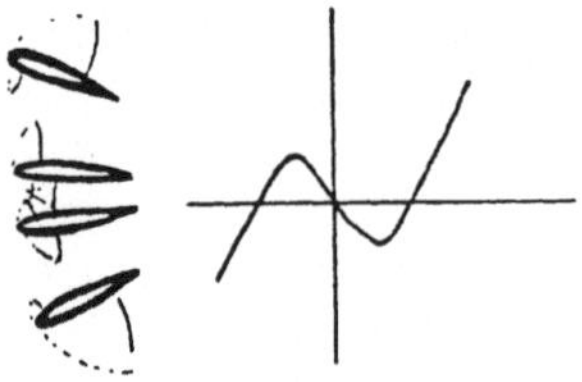

Figure 4

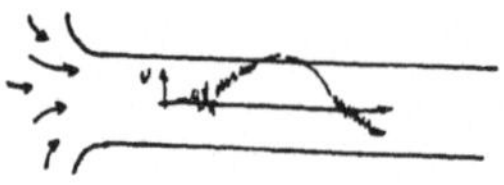

Figure 5

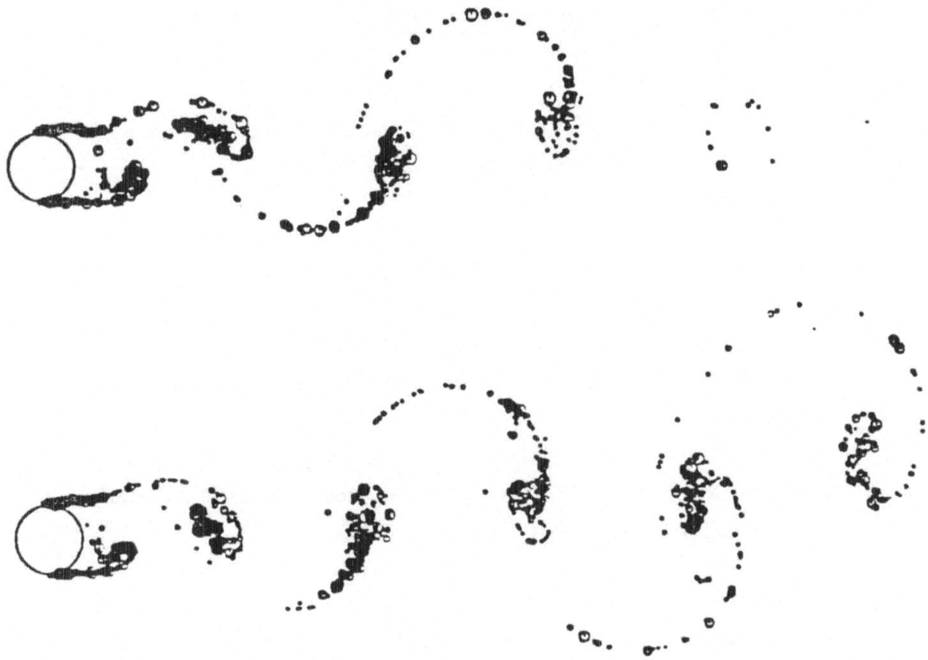

Figure 9

Flow behind a cylinder. Display of the kinetic turbulent energy q(x,t) at two different times.

Figure 6

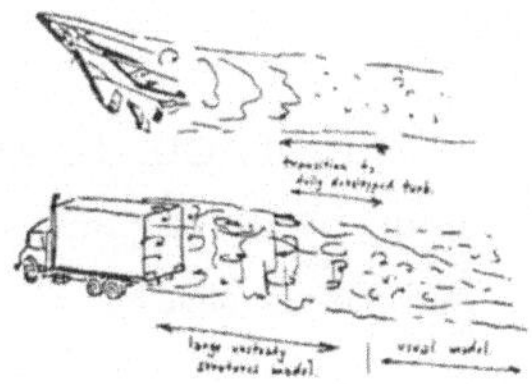

Figure 7

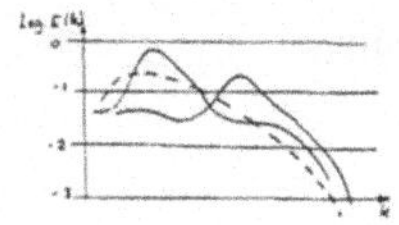

Figure 8

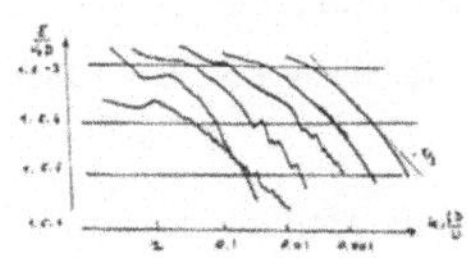

REFERENCES

(1) Mc WILLIAMS J.C. The emergence of isolated coherent vortices in turbulent flow. J. Fluid Mech. (1984) vol. 146 pp. 21-43.

(2) A. BENSOUSSAN, J.L. LIONS and G. PAPANICOLAOU, Asymptotic Methods for Periodic Structures, North-Holland, Amsterdam (1978).

(3) J.S. SMAGORINSKY, Mon. Weather Rev. 91, 99-164.
U. FRISCH, Z.S. SHE, O. THUAL, On the elastic behaviour of turbulence.

(4) P. PERRIER, O. PIRONNEAU, Couplage des grosses et petites structures turbulentes par l'homogénéisation, CRAS. 13 Février 1978.

(5) P. PERRIER, O. PIRONNEAU, Subgrid turbulence modelling by homogenization, Math. Modelling Vol. 2, 295-317 (1981).

(6) O. PIRONNEAU, Homogenization transport processes and turbulence modelling. Proc. INRIA-Novossibirsk, Dec. 1978 (to appear).

(7) G. PAPANICOLAOU, O. PIRONNEAU, On the asymptotic behavior of motion in random flow in "Stochastic non linear systems" Arnold-Lefever eds. Springer (1981).

(8) D. Mc. LAUGHIN, G. PAPANICOLAOU, O. PIRONNEAU, Non linear evolution equations with rapidly oscillating initial data.
Lecture Note in Physics 154 Springer (1981).

(9) D. Mc. LAUGHIN, G. PAPANICOLAOU, O. PIRONNEAU, Convection of micro-structures. Proc. INRIA Conf. Dec. 1981, North-Holland (Glowinski ed.)

(10) C. BEGUE, "Simulation Numérique de la turbulence pour méthode d'homogénéisation". Thèse de 3ème cycle - 1983 - Paris VI.

(11) T. CHACON, Contribucion al estudio del modelo m.p.p. de turbulencia - These doctoral - Université de Séville - Septembre 1984.

EDDY VISCOSITY SUBGRID SCALE MODELS FOR HOMOGENEOUS TURBULENCE

B. AUPOIX
ONERA/CERT
2 avenue Edouard Belin
31055 TOULOUSE Cedex - FRANCE

1 - INTRODUCTION

A complete simulation of the NAVIER equations can be performed only at very low Reynolds numbers, due to computers limitations. As the Reynolds number is increased, the range of scales in the flow extends and only a part of it can be captured by the computation mesh. The large eddy simulation technique (LES) proposed by LEONARD (1973) solves filtered NAVIER equations for the large eddies while the small eddies, which are less flow-dependent, are filtered out. The filtering operation gives rise to new terms, the subgrid scale terms, which account for the non-linear interactions between large and small eddies.

This paper will be devoted to simple subgrid scale modelling for homogeneous turbulence. Subgrid scale terms are analysed in Chapter 2 in the simple case of homogeneous turbulence without mean velocity gradient. In Chapter 3, a cascade of subgrid scale models for isotropic turbulence is investigated, with reference and comparison with previous models. The proposed models are then extended to the cases of rotating turbulence in Chapter 4 and strained turbulence in Chapter 5.

2 - SUBGRID SCALE ANALYSIS

For the sake of simplicity, we are only considering here homogeneous isotropic turbulence without mean velocity gradients. For homogeneous turbulence submitted to mean velocity gradients, the numerical method generally used (i.e. FOURIFR method) requires the use of a new set of variables linked to the distorsion of the mesh by the mean flow (ROGALLO, 1981). The subgrid scale analysis is similar but the connection between computational variables and physical ones is much more intricate.

2.1. Filtering of the NAVIER equations

Incompressible, isotropic turbulence satisfies both the continuity equation :

$$\frac{\partial u_i}{\partial x_i} = 0$$

and the momentum equation :

$$\frac{\partial u_i}{\partial t} + \frac{\partial}{\partial x_j} u_i u_j = -\frac{1}{\rho}\frac{\partial p}{\partial x_i} + \nu \frac{\partial^2 u_i}{\partial x_\ell \partial x_\ell}$$

For homogeneous turbulence, it is often much more convenient to deal with FOURIER transform of the NAVIER equation which reads :

$$k_i \hat{u}_i(\underline{k}) = 0$$

$$\frac{\partial}{\partial t} \hat{u}_i(\underline{k}) + \nu k^2 \hat{u}_i(\underline{k}) = - i k_j (\delta_{i\ell} - \frac{k_i k_\ell}{k^2}) \iint \delta(\underline{k} - \underline{p} - \underline{q}) \; u_j(\underline{p}) \; u_\ell(\underline{q}) \; d^3\underline{p} \; d^3\underline{q}$$

where $\hat{u}(\underline{k})$ is the FOURIER transform velocity field and $\delta(\underline{k}) = 0$ if $k \neq 0$ is the DIRAC function while δ_{ij} is the KRONECKER tensor. The right hand side of the momentum equation stands for both the advection and the pressure term. It points out the fact that each wave vector $\underline{k}$ interacts with all wave vectors $\underline{p}$ and $\underline{q}$ such as $\underline{k} = \underline{p} + \underline{q}$, i.e. $\underline{k}$, $\underline{p}$ and $\underline{q}$ form a triangle.

Velocities and pressure can be decomposed into two terms, a large-scale component (e.g. $\overline{u}$) and a small-scale component ($u' = u - \overline{u}$) with the help of a convolution filter. The filtered value of a variable f reads :

$$\overline{f}(\underline{x}) = f * G = \int f(\underline{x}') \; G(\underline{x} - \underline{x}') \; d^3\underline{x}$$

or, in FOURIER space :

$$\hat{\overline{f}}(\underline{k}) = \hat{f}(\underline{k}) \; \hat{G}(\underline{k})$$

Both to capture the maximum of energy in the simulation and to easily model the subgrid scales with the help of spectral turbulence closure, an isotropic, low-pass filter has been selected. This filter is defined as :

$$\hat{G}(\underline{k}) = 1 \quad ||\underline{k}|| \leq k_c$$

$$\hat{G}(\underline{k}) = 0 \quad ||\underline{k}|| > k_c$$

In what follows, the modes selected by the filter will improperly be named "large eddies", the modes discarded, "small eddies". The governing equations for the large eddies then read :

$$\frac{\partial \overline{u}_i}{\partial x_i} = 0$$

$$\frac{\partial}{\partial t} \overline{u}_i + \frac{\partial}{\partial x_j} (\overline{\overline{u}_i \overline{u}_j} + R_{ij}) = - \frac{1}{\rho} \frac{\partial \overline{p}}{\partial x_i} + \nu \frac{\partial^2 \overline{u}_i}{\partial x_j \partial x_j}$$

with $R_{ij} = \overline{\overline{u}_i u'_j} + \overline{u'_i \overline{u}_j} + \overline{u'_i u'_j}$

or, in FOURIER space : $k_i \hat{\overline{u}}_i(\underline{k}) = 0$

$$\frac{\partial \hat{\overline{u}}_i}{\partial t}(\underline{k}) + \nu k^2 \hat{\overline{u}}_i(\underline{k}) = - i k_j (\delta_{i\ell} - \frac{k_i k_\ell}{k^2}) \iint_{\substack{p \leq k_c \\ q \leq k_c}} \delta(\underline{k} - \underline{p} - \underline{q}) \; \hat{u}_j(\underline{p}) \; \hat{u}_\ell(\underline{q}) \; d^3\underline{p} \; d^3\underline{q}$$

$$- i k_j (\delta_{i\ell} - \frac{k_i k_\ell}{k^2}) \iint_{p \text{ or } q > k_c} \delta(\underline{k} - \underline{p} - \underline{q}) \; \hat{u}_j(\underline{p}) \; \hat{u}_\ell(\underline{q}) \; d^3\underline{p} \; d^3\underline{q}$$

The first term of the RHS represents the resolvable part of the advection i.e. $\bar{u}_i\bar{u}_j$ and its counterpart in the pressure term. The second term represents the subgrid scale term R_{ij} and its pressure counterpart. The first term can be computed as it contains only filtered velocity as $\hat{u}_j(\underline{p}) = \hat{\bar{u}}_j(\underline{p})$ for $p < k_c$ while the second term has to be modelled.

2.2. Role of the subgrid scale

This subgrid scale term represents the interactions between wave numbers above the filter cut and wave numbers below the cut. It is well known that small eddies extract energy from the large eddies and scramble the motion of the large eddies. These two actions can be brought into evidence while looking at the equation for the large scale kinetic energy at a given point :

$$\frac{\partial}{\partial t}\frac{1}{2}\bar{u}_i\bar{u}_i + (\overline{\bar{u}_i\bar{u}_j} + R_{ij})\frac{\partial \bar{u}_i}{\partial x_j} + \nu\frac{\partial \bar{u}_i}{\partial x_\ell}\frac{\partial \bar{u}_i}{\partial x_\ell} =$$

$$\frac{\partial}{\partial x_\ell}\left[\nu\frac{\partial}{\partial x_\ell}\frac{1}{2}\bar{u}_i\bar{u}_i - \bar{u}_i(\overline{\bar{u}_i\bar{u}_\ell} + R_i\ \ ) - \frac{\bar{p}\ \bar{u}_\ell}{\rho}\right]$$

The second term of the LHS is the energy transfer due to vortex stretching. This energy transfer redistributes energy between the large scales and exchanges energy with the small scales. The RHS represents the diffusion of energy. Statistically, this term is null for homogeneous turbulence, but locally the small scales play a mixing role upon the large scales.

2.3. Subgrid scale modelling with two-point closures

Several models such as TFM or EDQNM give expressions for the detailed energy transfer S(k, p, q) at wave number k due to interactions with wave numbers p and q. With these models, the LIN equation for the energy spectrum reads :

$$(\frac{\partial}{\partial t} + 2\nu k^2)\ E(k) = T(k) = \iint S(k,p,q)\ d^3p\ d^3q$$

where E(k) is the energy spectrum :

$$E(k) = 2\pi\ k^2\ \ \langle \hat{u}_i(\underline{k})\ \hat{u}_i(-\underline{k})\rangle$$

The energy transfer can be splitted into two parts :

$$T(k) = \iint\limits_{\substack{p\leq k_c\\ q\leq k_c}} S(k,p,q)\ d^3p\ d^3q + \iint\limits_{\substack{p>k_c\\ q>k_c}} S(k,p,q)\ d^3p\ d^3q$$

and   or

The first term represents interactions with modes below the cut, while the second term corresponds to interactions with the modes above the cut.

On the other hand, the filtered NAVIER equation reads :

$$\left(\frac{\partial}{\partial t} + \nu k^2\right) \hat{\bar{u}}_i(\underline{k}) = T_i^R(\underline{k}) + T_i^S(\underline{k})$$

where T_i^R is the resolvable term and T_i^S the subgrid scale term. The energy spectrum equation can be deduced from the above equation as :

$$\left(\frac{\partial}{\partial t} + \nu k^2\right) 2\pi k^2 \langle\hat{\bar{u}}_i(\underline{k})\ \hat{\bar{u}}_i(-\underline{k})\rangle = 2\pi k^2 \langle T_i^R(\underline{k})\ \hat{\bar{u}}_i(-\underline{k}) + T_i^R(-\underline{k})\ \hat{\bar{u}}_i(\underline{k})\rangle$$

$$+ 2\pi k^2 \langle T_i^S(\underline{k})\ \hat{\bar{u}}_i(-\underline{k}) + T_i^S(-k)\ \hat{\bar{u}}_i(+\underline{k})\rangle$$

The first term of the RHS corresponds to interactions between large eddies while the second term reflects interactions between large and small eddies.

As previously quoted, the resolvable term T^R needs no modelling. The two above forms of the LIN equation can be compared to model the subgrid scale term. The subgrid scale term must satisfy :

$$2\pi k^2 \langle T_i^S(\underline{k})\ \hat{\bar{u}}_i(-\underline{k}) + T_i^S(-\underline{k})\ \hat{\bar{u}}_i(\underline{k})\rangle = \iint_{p \text{ or } q > k_C} S(k,p,q)\ d^3p\ d^3q$$

As two-point closures are dealing with ensemble-average, this equation just imposes the equality of ensemble-averaged energy transfer between wave number k and the small scales. The ensemble-averaged energy transfer given by the subgrid scale model for $T^S(\underline{k})$ will be consistent with the two-point closure used to model $S(k,p,q)$. The above equation has an infinity of solutions as information about the individuality of the computed flow realization have been discarded by taking ensemble averages. BARDINA (1983) and BERTOGLIO (1984) have proposed models to restore information about the computed realization.

Nevertheless, the use of two-point closure modelling is an improvement over the first subgrid scale models which only verify the energy flux across the filter cut and not the above transfer equation for each wave number.

2.4. Effective viscosity and eddy viscosity

An interesting case occurs when the wave number k is small compared to the cut k_c. Because of the triangle relation $\underline{k} = \underline{p} + \underline{q}$, wave numbers which contribute to subgrid scale terms must be close and great compared to k, i.e. $k \ll k_c \leqslant p \sim q$. A spectral gap exists between the considered large scales and the small scales. LESIEUR and CHOLLET (1979) have shown that the small scales then behave like a brownian motion superimposed on the large scales.With analogy to the kinetic theory of gases, the subgrid scale term can be modelled by an effective viscosity hypothesis :

$$k \ll k_c \quad T_i^S(\underline{k}) = -\nu_e k^2 \hat{\bar{u}}_i(\underline{k})$$

where the effective viscosity ν_e is linked to the small eddies.

For all wave numbers, a solution of the transfer equation :

$$2\pi k^2 \langle T_i^S(\underline{k})\, \hat{\bar{u}}_i(-\underline{k}) + T_i^S(-\underline{k})\, \hat{\bar{u}}_i(\underline{k})\rangle = \iint_{p \text{ or } q > k_c} S(k,p,q)\, d^3p\, d^3q$$

can be obtained in a wave number dependent eddy viscosity form. The eddy viscosity is defined as :

$$\iint_{p \text{ or } q > k_c} S(k,p,q)\, d^3p\, d^3q = -2\nu_t(k)\, k^2\, E(k)$$

and a model for the subgrid scale term is :

$$T_i^S(\underline{k}) = -\nu_t(k)\, k^2\, \hat{\bar{u}}_i(\underline{k})$$

The eddy viscosity $\nu_t(k)$ is given by the two-point closure. As its definition involves an integral over the small scales, this eddy viscosity is linked to the small scales, at variance with standard subgrid scale models. At last, the effective viscosity ν_e can be interpreted as the limit of $\nu_t(k)$ as the ratio k/k_c tends towards zero.

3 - SUBGRID SCALE MODELLING FOR ISOTROPIC TURBULENCE

3.1. LES/EDQNM coupling

Among the various two-point closures, the EDQNM model (ORSZAG (1970), ANDRE-LESIEUR (1977)) has been selected as it is easy to implement and gives good results when compared with experiments (VIGNON et al. (1979)). Moreover, this model has been extended to anisotropic cases by CAMBON (1981) and BERTOGLIO (1981).

A method to compute a wave number dependent eddy viscosity has been developed in a way similar to the one used by CHOLLET (1983). Two computations are carried out together : a realization of a velocity field is calculated by the large eddy simulation technique for the large scales and the energy spectrum of the small scales is given by the EDQNM.

At each time step, the energy spectrum of the large scales is calculated. With the knowledge of the energy spectrum of both the large and the small scales, the EDQNM routine can then compute on one hand the subgrid scale transfer and the eddy viscosity $\nu_t(k)$ in the large scales and, on the other hand, the energy transfer $T(k)$ in the small scales. So, the evolution of all scales can be computed.

It must be pointed out that the evolution of the large scales and the small scales are linked. The large eddy velocity field evolution is governed by the resolvable term $\overline{\bar{u}\,\bar{u}}$ and by the subgrid scale in which the eddy viscosity is computed by the EDQNM as an integral over the small scales. The small eddies spectrum evolution is linked to the energy spectrum of the large scales with the EDQNM (fig. 1).

The celebrated experiments of isotropic decaying turbulence performed by COMTE-BELLOT and CORRSIN (1971) have been successfully simulated (fig. 2).

The time and wave number evolutions of the eddy viscosity coefficient are plotted on fig. 3. The eddy viscosity coefficient remains close to the effective viscosity for wave numbers far from the cut, but exhibits an increase in the vicinity of the cut.

3.2. Constant eddy viscosity approximation

KRAICHNAN (1976), LESLIE and QUARINI (1979) and CHOLLET and LESIEUR (1982) have studied the behaviour of the eddy viscosity coefficient. KRAICHNAN brought into evidence an important cusp near the filter cut in the case of an infinite inertial range ($E(k) \propto k^{-5/3}$). This cusp is due to non local interactions between two wave numbers on both sides of the cut and a wave number near zero as the energy spectrum is infinite at k = 0. LESLIE and QUARINI or CHOLLET and LESIEUR studied the evolution of the cusp according to spectrum shape and cut location. Fig. 4 shows the evolution of eddy viscosity as function of the cut location for a high Reynolds number case. As the cut is located away from the energy maximum, this energy maximum contributes to non local interactions near the cut and builds up the cusp.

Large eddy simulation is aimed at computing the large, energy-containing eddies while filtering out the small eddies. So, to reduce mesh size and computational time, it is convenient to locate the cut at the beginning of the inertial range as done in the COMTE-BELLOT simulation (fig. 2, 3). In such cases, the cusp remains moderate. A good approximation to model the eddy viscosity is to assume it constant (in wave number, not in time) and equal to the effective viscosity :

$$\nu_t(k) = \nu_e$$

3.3. Effective viscosity models

Following KRAICHNAN's ideas, the effective viscosity can be derived from the expression for the energy transfer $S(k,p,q)$ by expanding it in terms of k/p and retaining only the leading terms. The effective viscosity then reads :

$$\nu_e = \frac{1}{15} \int_{k_c}^{\infty} \theta_{oqq} \, (5\, E(q) + q \frac{dE(q)}{dq}) \, dq$$

where θ_{kpq} is the relaxation time linked to the eddy damping :

$$\theta_{kpq} = \frac{1}{\mu(k) + \mu(p) + \mu(q)}$$

$$\mu(k) = \nu\, k^2 + \lambda \left[\int_o^k p^2\, E(p)\, dp \right]^{1/2}$$

where ν is the kinematic viscosity and λ the only adjustable constant in the EDQNM model related to the KOLMOGOROV constant.

Let us notice again that ν_e is linked to the small eddies.

An hypothesis about the shape of the energy spectrum is needed to evaluate the effective viscosity. If the energy spectrum is assumed an inertial range shape

$$E(k) = K_o\, \varepsilon^{2/3}\, k^{-5/3} \qquad K_o \sim 1.4$$

the effective viscosity reads :

$$\nu_e = .315\, \varepsilon^{1/3}\, k_c^{-4/3}$$

SMAGORINSKY proposed the first and still popular eddy viscosity model expressing the eddy viscosity in terms of the large eddies in a mixing length form :

$$\nu_t(x) = (c\Delta)^2\, (\bar{S}_{ij}\, \bar{S}_{ij})^{1/2}$$

with $\Delta = \pi/k_c$

$$\bar{S}_{ij} = \frac{1}{2} \left(\frac{\partial \bar{u}_i}{\partial x_j} + \frac{\partial \bar{u}_j}{\partial x_i}\right)$$

LILLY (1966) and LESLIE and QUARINI linked this model to two-point closures by assuming a space-averaged eddy viscosity :

$$\nu_e = (c\Delta)^2 \left[\int_o^{k_c} 2\, k^2\, E(k)\, dk \right]^{1/2}$$

Then, LILLY evaluated c by assuming the energy flux through the filter cut to be equal to the dissipation rate ε. This led him to $c = 0.1825$.

AUPOIX and COUSTEIX (1982) have shown that the effective viscosity is consistent with SMAGORINSKY model with $c = 0.148$. The difference between the two constants is due to the cusp of the eddy viscosity $\nu_t(k)$ near k_c. For infinite inertial range, this cusp is very important and contributes to the energy flux through the cut. LILLY's constant tries to integrate the cut while AUPOIX and COUSTEIX neglect it.

As the effective viscosity is linked to the small scales, CHOLLET and LESIEUR (1982) expressed it in terms of the energy spectrum at the cut. The effective viscosity then reads :

$$\nu_e = 0.267 \quad \left(\frac{E(k_c)}{k_c}\right)^{1/2}$$

A third approach has been proposed by AUPOIX and COUSTEIX (1982). As the effective viscosity is linked to the small scales, they expressed the effective viscosity in terms of these small scales. The kinetic energy of the small scales can be expressed as :

$$\frac{1}{2} q'^2 = \int_{k_c}^{\infty} K_o \, \varepsilon^{2/3} \, k^{-5/3} \, dk = \frac{3}{2} K_o \, \varepsilon^{2/3} \, k_c^{-2/3}$$

so that the effective viscosity can take the following forms :

$$\nu_e = .0685 \, \varepsilon^{1/3} \, \Delta^{4/3} \qquad \Delta = \pi/k_c$$

$$\nu_e = .0819 \quad \frac{1}{2} q'^2 \, \Delta$$

$$\nu_e = .0714 \, \frac{(1/2 \, q'^2)^2}{\varepsilon}$$

The third expression has been selected by AUPOIX and COUSTEIX (1982).

In addition, they extended this model to low Reynolds number cases by assu ming that the energy spectrum above the cut does not satisfy KOLMOGOROV's law but more intricate laws taking viscous cut-off into account such as :

SAFFMAN $\qquad E(k) = K_o \, \varepsilon^{2/3} \, k^{-5/3} \, \exp\left[- 2(k/k_k)^2\right]$

PAO $\qquad E(k) = K_o \, \varepsilon^{2/3} \, k^{-5/3} \, \exp\left[- \frac{3}{2} K_o (k/k_k)^{4/3}\right]$

$\qquad$ where $k_k = \left(\frac{\varepsilon}{\nu^3}\right)^{1/4}$

Finally, as in the case of the LES/EDQNM coupling, the evolution of the small scales has to be linked to the large scales. The kinetic energy of the small scales is deduced from the difference of the total kinetic energy and of the kinetic energy of the large scales. The total kinetic energy evolution could be prescribed from experimental data or an independent computation. If so, AUPOIX and COUSTEIX (1982) have demonstrated that the evolution of the kinetic energy partition diverges. The global kinetic energy $1/2 \, q^2$ is computed with the simulation, using the usual transport equation :

$$\frac{\partial}{\partial t} \frac{1}{2} q^2 = - \varepsilon$$

where the kinetic energy dissipation rate ε is linked to the spectrum shape selected, the kinetic energy of the small scales $1/2 \, q'^2$ and the filter cut k_c.

3.4. Comparison with experiments

Fig. 5 to 7 show large eddy simulation results for the COMTE-BELLOT experiment. Fig. 5 and 6 are comparisons of energy spectra at the last station with various subgrid scale models. When LILLY's constant is used in the SMAGORINSKY model, the effective viscosity and the energy flux to the small scales are overestimated so the energy spectrum tends towards a $k^{-5/3}$ law. With our constant, agreement with experiment is improved. CHOLLET's model (fig. 5) and various forms of our (q'^2, ε) model (fig. 6) give fair predictions of the energy spectrum at the final station. These last models give a slight accumulation of energy near the filter cut and an overestimation of the energy spectrum in this region. This is due to the neglect of the increase in the eddy viscosity $\nu_t(k)$ in the vicinity of the cut. Our model gives access to the total kinetic energy which is better predicted (fig. 7) with PAO's spectrum shape as it is the spectrum shape the closer to the experimental one.

4 - SUBGRID SCALE MODELLING FOR ROTATING TURBULENCE

4.1. EDQNM model for rotating turbulence

Various experiments show different effects of rotation upon turbulence ; in some cases, the energy decay is faster, in others, slower. However, experiments nearly satisfying homogeneity such as the one performed by WIGELAND (1978) or direct simulations of rotating turbulence performed by BARDINA (1983), both indicate that rotation decreases the rate of decay of turbulence.

Such an effect is not accounted for by the standard EDQNM model. CAMBON (1982) noticed that the damping

$$\mu(k) = \nu k^2 + \lambda \left[\int_0^k p^2 \, E(p) \, dp\right]^{1/2}$$

takes into account the vorticity of all scales up to wave number k. He proposed to include the mean rotation which corresponds to k = 0, so the damping reads :

$$\mu(k) = \nu k^2 + \lambda \left[\int_0^k p^2 \, E(p) \, dp + 2\omega^2\right]^{1/2}$$

This extra term increases the damping and so decreases energy transfer from large scales to small scales, so that the dissipation rate of the kinetic energy of turbulence is decreased.

This model is consistent with earlier observations and gives good predictions of the WIGELAND and NAGIB experiments (CAMBON et al. (1982)).

4.2. Validation of the EDQNM model

Direct simulations of homogeneous isotropic turbulence submitted to rotation have been performed to check the EDQNM model. The initial energy spectrum is :

$$E(k) = k^4 \exp(2(1 - k^2)) \qquad (E(k) \text{ in } cm^3/s^2 \quad k \text{ in } cm^{-1})$$

and the Reynolds number $\frac{(q^2)^2}{9\nu\varepsilon}$ is about 80.

Fig. 8 shows the energy spectra at the final station, i.e. after seven eddy turnover times. As the rotation rate is increased, the energy is "trapped" in the large scales as energy transfer from the large scales to the small scales is inhibited. These results are consistent with BARDINA's previous simulations. The modified EDQNM model predictions are in very good agreement with the direct resolution of the NAVIER equations for every rotation rate.

4.3. Subgrid scale models

There are no reliable experiments of homogeneous rotating turbulence at high Reynolds number. WIGELAND's experiment can be computed with the direct simulation technique as its Reynolds number is very low. So, there is a lack of experiments to check a subgrid scale model for rotating turbulence.

We performed computation of rotating turbulence, using COMTE-BELLOT experimental spectrum as initial spectrum, for various rotation rate. The energy spectrum computed with the EDQNM model are compared with results of large eddy simulations with different subgrid scale models. This test is a check for both the subgrid scale model and the EDQNM. The Reynolds number is now 3 200 and the computation extends over thirteen eddy turnover times.

The first subgrid scale model is a wave number dependent eddy viscosity $\nu_t(k)$ obtained by coupling the large eddy simulation with an EDQNM routine. Fig. 9 shows excellent agreement between the large eddy simulation and the EDQNM, whatever the rotation rate. Furthermore, the eddy viscosity evolution versus wave number is very similar to the no-rotation case ; the cusp of eddy viscosity remains moderate.

So, the eddy viscosity can still be assumed to be wave number independent and equal to the effective viscosity. If the energy spectrum satisfies KOLMOGOROV's law in the small scales, the effective viscosity reads (AUPOIX et al. (1983)) :

$$\nu_e = 0.0714 \frac{(1/2\, q'^2)^2}{\varepsilon} \frac{2}{1 + (1 + 0.0216 \cdot \frac{\omega\, 1/2\, q'^2}{\varepsilon})^{1/2}}$$

Fig. 10 shows comparison between EDQNM computations and large eddy simulations using a wave number independent eddy viscosity. To take low Reynolds number effects into account, a PAO's spectrum is assumed for the small scales. Two subgrid scale models are tested, one taking rotation effects into account, the other not. When rotation effects are taken into account, the energy spectra agree fairly well with some overestimation of the LES near the filter cut due to the neglect of the eddy viscosity cusp. When rotation effects are not included in the subgrid scale model, the eddy viscosity and the energy transfer to the small scales are overestimated ; the LES energy spectrum then lies under the EDQNM one and the difference increases with the rotation rate.

Finally, it must be noticed that, according to PROUDMAN's theorem, isotropic turbulence submitted to rotation should evolve towards an axisymmetric state or towards two-dimensional turbulence if the rotation rate is high enough. This trend towards axisymmetry is not clearly visible in our direct simulation, perhaps because of the low Reynolds number and is not accounted for in the EDQNM model which is isotropic, nor consequently in the subgrid scale model.

5 - APPLICATION TO STRAINED TURBULENCE

5.1. Validity of the model

Previous studies (e.g. BERTOGLIO (1983)) have shown that eddy viscosity models are no longer valid for anisotropic turbulence and especially sheared flows where the energy backscatter from the small scales to the large scales is important. However, our model has been extended to cases of relatively small anisotropy to compare LES predictions with experiments. Only strained turbulence cases have been tried at the present time.

5.2. Extension of the subgrid scale model

To perform direct simulation or large eddy simulations of homogeneous turbulence submitted to mean velocity gradients, ROGALLO (1981) proposed a method in which the computational domain is convected by the mean flow. Such a method simplifies the numerical approach as it allows the use of FOURIER transform but makes subgrid scale modelling more intricate. The filter can remain the usual sphere ($||\underline{k}|| < k_c$) and, in such a case, a large computational mesh is required to fill in this sphere at each time step because of the distorsion of the computational mesh. The other choice is to allow the filter to evolve with the grid, but in this case, the limiting volume becomes a prolate spheroid.

Even in the simplest case of the effective viscosity model, the integral has now to be evaluated outside of a prolate spheroid :

$$\nu_e = \frac{1}{15} \int \theta_{oqq} \left(5\ U(q) + q\ \frac{dU(q)}{dq}\right)\ d^3\underline{q}$$

where $U(q) = \frac{E(q)}{4\pi q^2}$ is the energy density which is a function of the wave number as the small scales are assumed isotropic to give an isotropic effective viscosity. The evaluation of the kinetic energy of the small scales also requires an integral outside of the filter domain.

First trials with that model have shown that, while the hypothesis of isotropy was necessary to derive an isotropic subgrid scale model, the subgrid scales were not really isotropic because of the location of the filter cut in the beginning of the inertial range. The neglect of the anisotropy of the small scales would lead to an important underestimation of the total Reynolds stresses and of the kinetic energy as the evolution of the global kinetic energy then reads :

$$\frac{\partial}{\partial t} \frac{1}{2} q^2 = - \langle u_i u_j \rangle \frac{\partial u_i}{\partial x_j} - \varepsilon$$

and $\frac{1}{2} q^2$ is still required to evaluate the small scales.

CAMBON (1981) has proposed a model to link the spherically-averaged anisotropy spectrum to the global anisotropy for low anisotropy cases. This model has been used to calculate the global Reynolds stresses.

5.3. Comparisons with experiments

Experiments performed by GENCE (1979) of homogeneous turbulence submitted to two successive plane strains of different axes have been selected to check our subgrid scale model. Evolution of the kinetic energy of turbulence and of the second invariant of the anisotropy tensor are plotted on fig. 11. The agreement with experiment is fairly poor when the two strains have different directions. The same phenomenon is observed either in one-point closure computations of these experiments or in direct simulations such as the one performed by DANG (1983) of analogeous experiment at a lower Reynolds number. Our LES simulation is in good agreement with both one-point closure and direct simulations. The discrepancy with experiments seems to be partly due to differences between theoretica and experimental strains.

It can however be concluded that the eddy viscosity model still behaves fairly well for moderate anisotropy.

6 - CONCLUSIONS

New subgrid scale models have been developed. The first one is based upon coupling the LES to an EDQNM routine and leads to a wave number dependent eddy viscosity. For the sake of simplicity, this model can be approximated by a wave number independent eddy viscosity. This constant eddy viscosity is expressed in terms of the small eddies, so that both the large scales and the small scales are known during the simulation for better comparison with experiments. Furthermore, this model has been extended to low Reynolds number turbulence by assuming various energy spectrum shapes, to rotating turbulence and to anisotropic filtering. By comparison with experiments or other simulations, these models have demonstrated their validity for moderate anisotropy.

We wish to thank Dr. BERTOGLIO, CAMBON and CHOLLET and Pr. LESIEUR and MATHIEU for helpful discussions on two-point closures and subgrid scale modelling. We are very grateful to Dr. J. COUSTEIX for his aid in the development of the (q'^2, ε) model and his remarks in the elaboration of this paper.

REFERENCES

ANDRE J.C., LESIEUR M. (1977) - "Influence of helicity on the evolution of isotropic turbulence at high Reynolds number" - JFM Vol. 81, pp. 187-207

AUPOIX B., COUSTEIX J. (1982) - "Subgrid scale model for isotropic turbulence" - Proceedings of the Symposium of Refined modelling of flows, PARIS, Sept. 7-10 1982

AUPOIX B., COUSTEIX J. (1982) - "Simple subgrid scale stresses models for homogeneous isotropic turbulence" - La Recherche Aérospatiale 1982-4

BARDINA J., FERZIGER J.H., ROGALLO R.S. (1982) - "Effects of rotation on isotropic turbulence : computation and modelling" - STANFORD University

BARDINA J., FERZIGER J.H., REYNOLDS W.C. (1983) - "Improved turbulence models based on large eddy simulation of homogeneous, incompressible, turbulent flows" - Report n° TF-19 - STANFORD University

BERTOGLIO J.P. (1981) - "A model of three-dimensional transfer in non isotropic homogeneous turbulence" - Turbulent Shear Flow 3, DAVIS

BERTOGLIO J.P., MATHIEU J. (1983) - "Study of subgrid models for sheared turbulence" - Turbulent Shear Flow 4, KARLSRUHE

BERTOGLIO J.P. (1984) - "Subgrid modelling for sheared turbulence" - Macroscopic Modelling of Turbulent Flows and Fluid Mixtures, NICE, Déc. 10-14, 1984

CAMBON C., JEANDEL D., MATHIEU J. (1981) - "Spectral modelling of homogeneous non isotropic turbulence" - JFM Vol. 104, pp. 247-262

CAMBON C., BERTOGLIO J.P., JEANDEL D. (1981) -"Comparison of computation with experiments" - The 1980-81 AFOSR-HTTM STANFORD Conference on Complex Turbulent Flows, Vol. III, pp. 1307-1311

CHOLLET J.P., LESIEUR M. (1981) - "Parametrization of small scales of three-dimensional isotropic turbulence utilizing spectral closures" - JAS, Vol. 38, N° 12, pp. 2747-2757

CHOLLET J.P. (1983) - "Two-point closures as a subgrid scale modelling for large eddy simulations" - Turbulent Shear Flow 4, KARLSRUHE

COMTE-BELLOT G., CORRSIN S. (1971) - "Simple Eulerian time correlations for full- and narrow-band velocity signals in grid-generated, "isotropic" turbulence" - JFM, Vol. 48-2, pp. 273-337

DANG K. (1983) - "Evaluation of simple subgrid scale models for the numerical simulation of homogeneous isotropic and anisotropic turbulence" - AIAA 16th Fluid and Plasma Dynamics Conference, DANVERS (MASSACHUSSETTS), July 12-14, 1983

GENCE J.N., MATHIEU J. (1979) - "On the application of successive plane strains to grid-generated turbulence" - JFM, Vol. 93-3, pp. 501-513

KRAICHNAN R.H. (1976) - "Eddy viscosity in two and three dimensions" - JAS, Vol. 33, pp. 1521-1536

LEONARD A. (1973) - "Energy cascade in large eddy simulations of turbulent fluid flow" - Advances in Geophysics, Vol. 18 A, pp. 237-248

LESIEUR M., CHOLLET J.P. (1979) - "Introduction aux théories statistiques de la turbulence pleinement développée" in Bifurcation et problèmes non linéaires, C.BARDOS , C. SCHATZMANN, J.M. LASRY, Springer Verlag

LESLIE D.C., QUARINI G.L. (1979) - "The application of turbulence theory to the formulation of subgrid modelling procedures" - JFM, Vol. 91-1, pp. 65-91

LILLY D.K. (1966) - "On the application of the eddy viscosity concept in the inertial subrange of turbulence" - NCAR Manuscript N° 123

ORSZAG S. (1970) - "Analytical theories of turbulence" - JFM, Vol. 41-2, pp. 363-386

ROGALLO R.S. (1981) - "Numerical experiments in homogeneous turbulence" - NASA TM 81315

SMAGORINSKY J. (1963) - "General circulation experiment with the primitive equations" - Monthly Weather Review, Vol. 91, N° 3, pp. 99-164

VIGNON J.M., CAMBON C., LESIEUR M., JEANDEL D. (1979) - "Confrontation aux expériences de turbulence thermique homogène et isotrope de calculs basés sur la théorie EDQNM" - CRAS PARIS t 288 Série B 335

AUPOIX B., COUSTEIX J., LIANDRAT J. (1983) - "Effect of rotation on isotropic turbulence" - Turbulent Shear Flow 4, KARLSRUHE

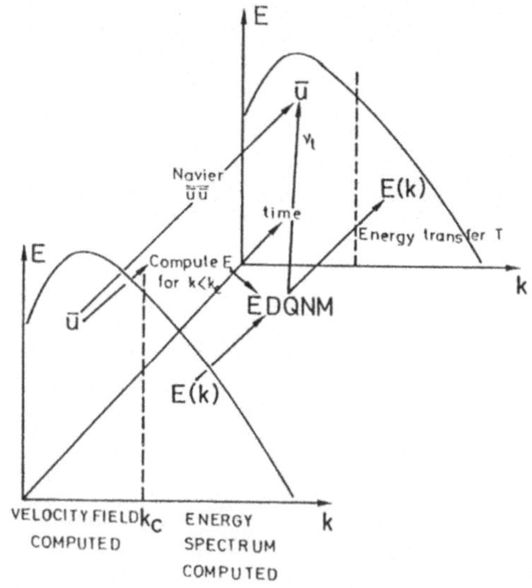

Fig. 1

Schematic of the coupling of the LES with the EDQNM

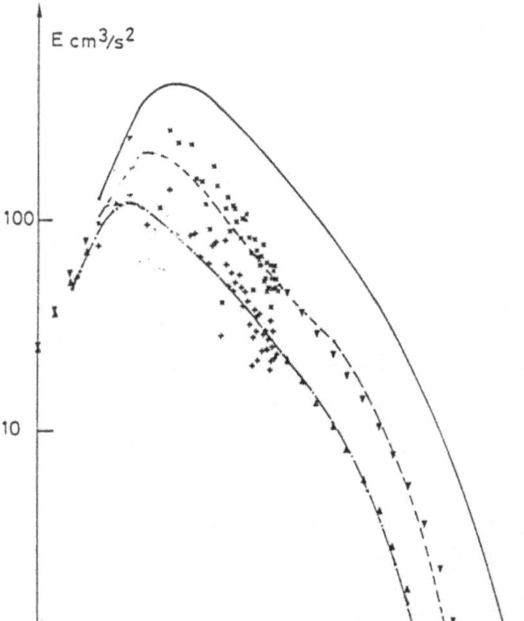

Fig. 2

COMTE-BELLOT experiment

Experimental spectra at

$\frac{U_o t}{M}$ = 42 ——
98 ----
171 —.—.

Large scales computed with LES coupled with the EDQNM

x $\frac{U_o t}{M}$ = 98

\+ 171

Small scales computed with EDQNM coupled with LES

▼ $\frac{U_o t}{M}$ = 98

▲ 171

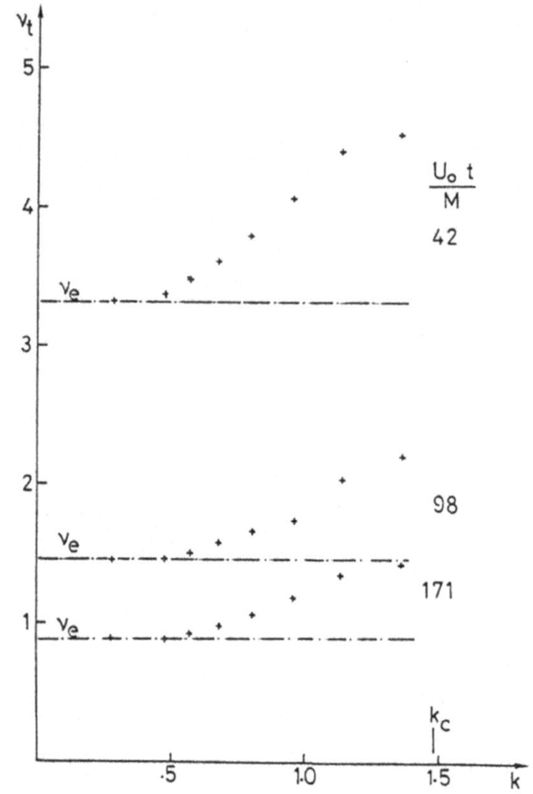

Fig. 3

LES/EDQNM coupling

Evolution of the eddy viscosity ν_t versus time and wave number. Comparison with the effective viscosity ν_e

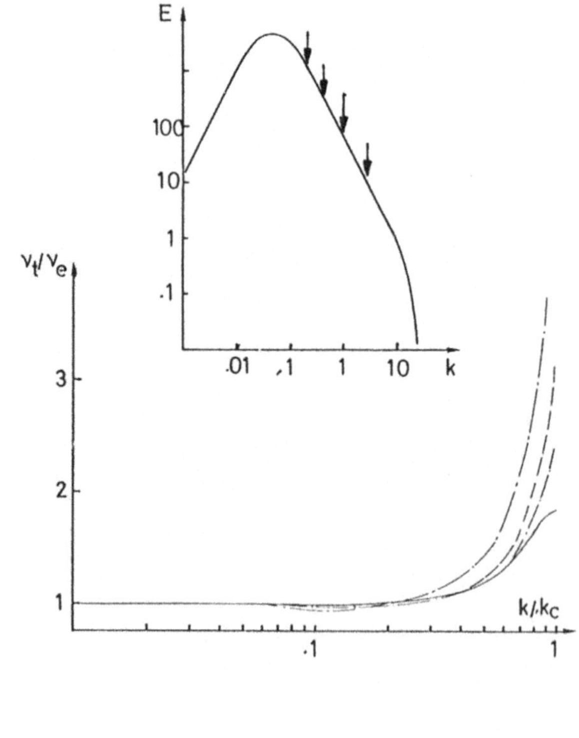

Fig. 4

Influence of the location of the filter cut on the eddy viscosity distribution at high Reynolds number

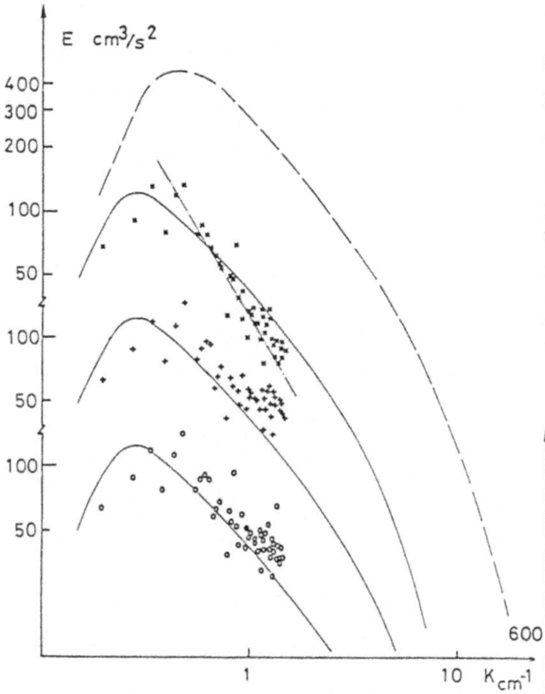

Fig. 5

COMTE-BELLOT experiment

Experimental energy spectrum

– – – initial spectrum

——— final spectrum

LES with various subgrid scale models

x SMAGORINSKY c = 0.1825
+ SMAGORINSKY c = 0.1478
o CHOLLET and LESIEUR
–.– $k^{-5/3}$ spectrum

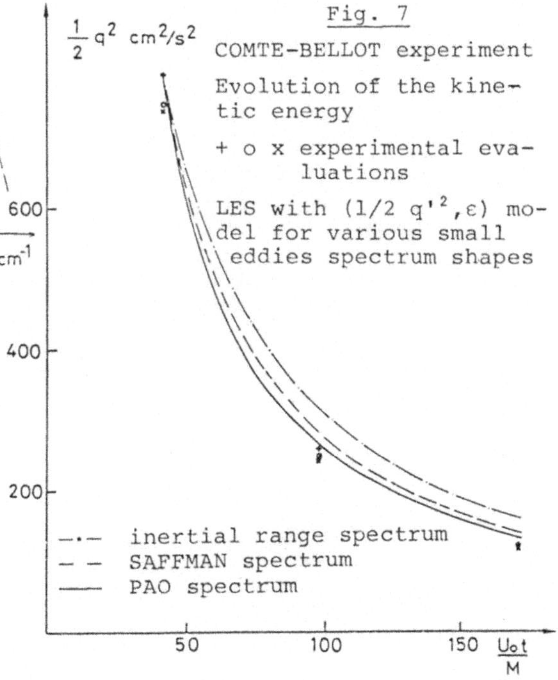

Fig. 7

COMTE-BELLOT experiment

Evolution of the kinetic energy

+ o x experimental evaluations

LES with ($1/2\ q'^2, \varepsilon$) model for various small eddies spectrum shapes

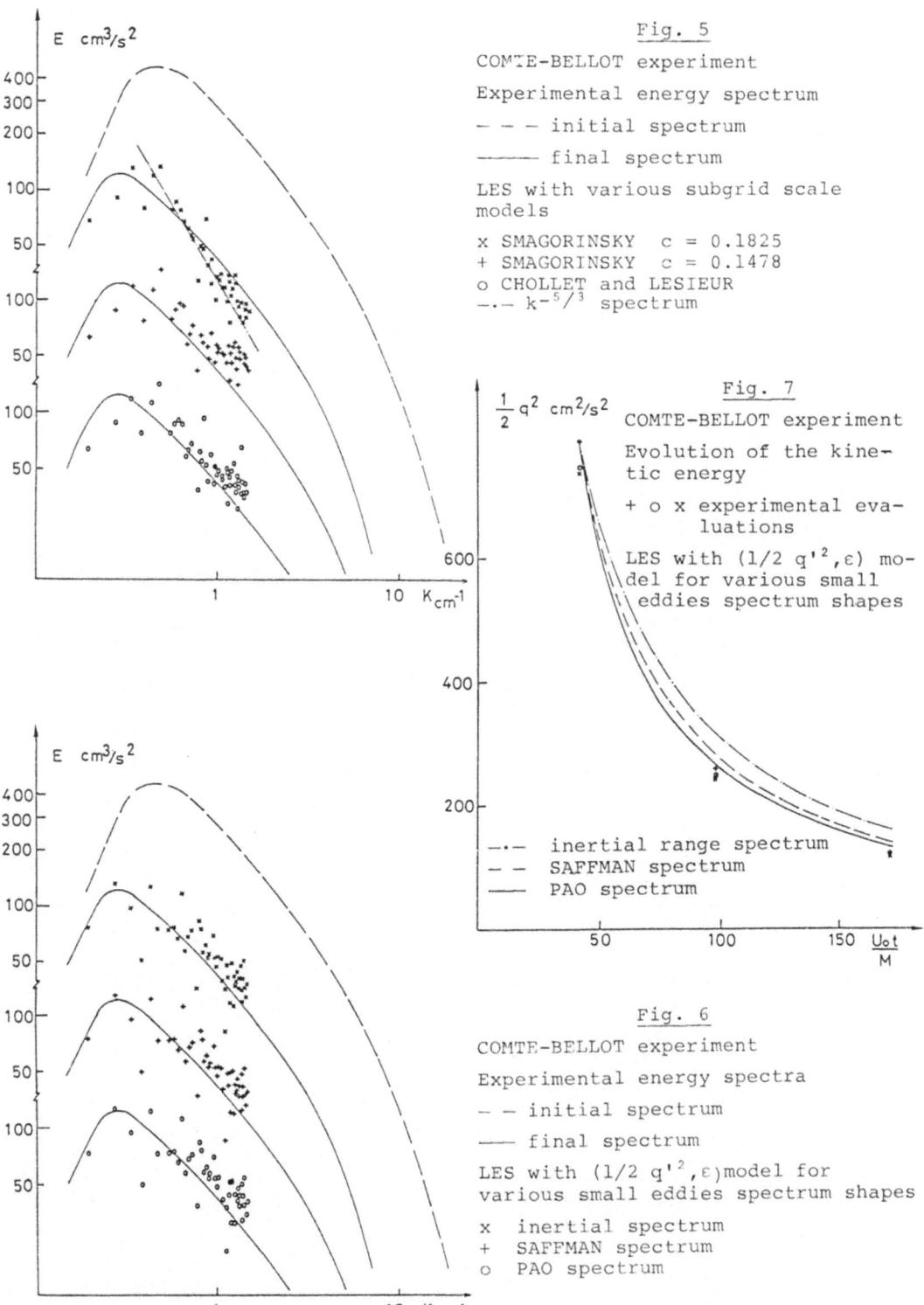

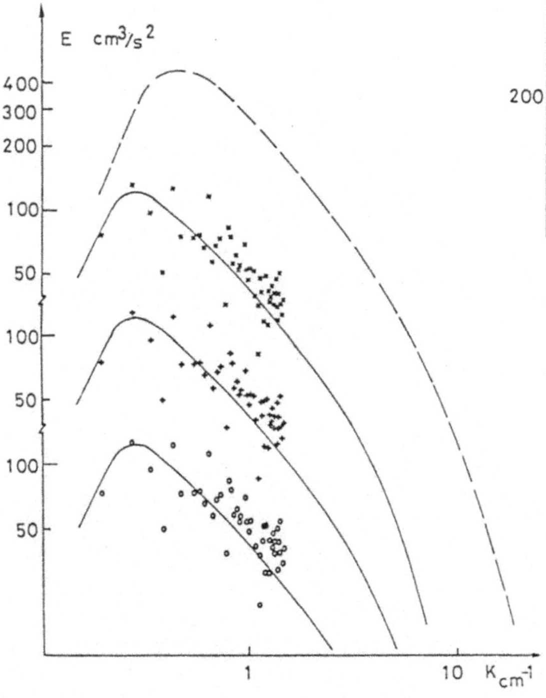

Fig. 6

COMTE-BELLOT experiment

Experimental energy spectra

– – initial spectrum

—— final spectrum

LES with ($1/2\ q'^2, \varepsilon$)model for various small eddies spectrum shapes

x inertial spectrum
+ SAFFMAN spectrum
o PAO spectrum

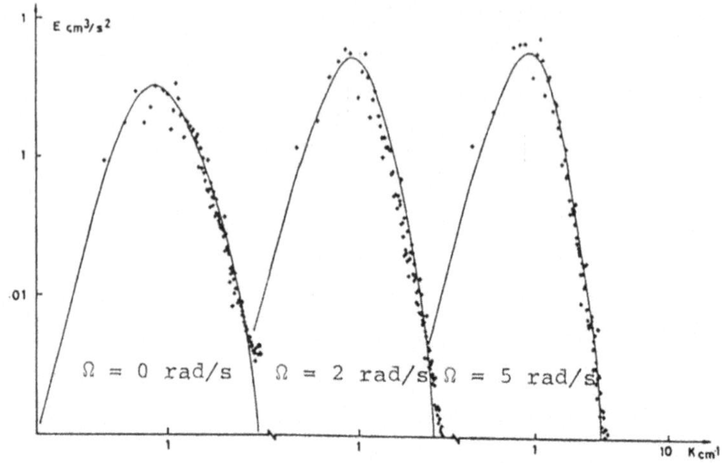

Fig. 8

Comparison between direct simulations and EDQNM computations of low Reynolds number, rotating turbulence

+ direct simulations
- EDQNM computations

Fig. 9

Rotating turbulence - Initial energy spectrum : COMTE-BELLOT
LES/EDQNM coupling - —— energy spectrum computed with EDQNM
\+ large scales computed with LES
▲ small scales computed with EDQNM

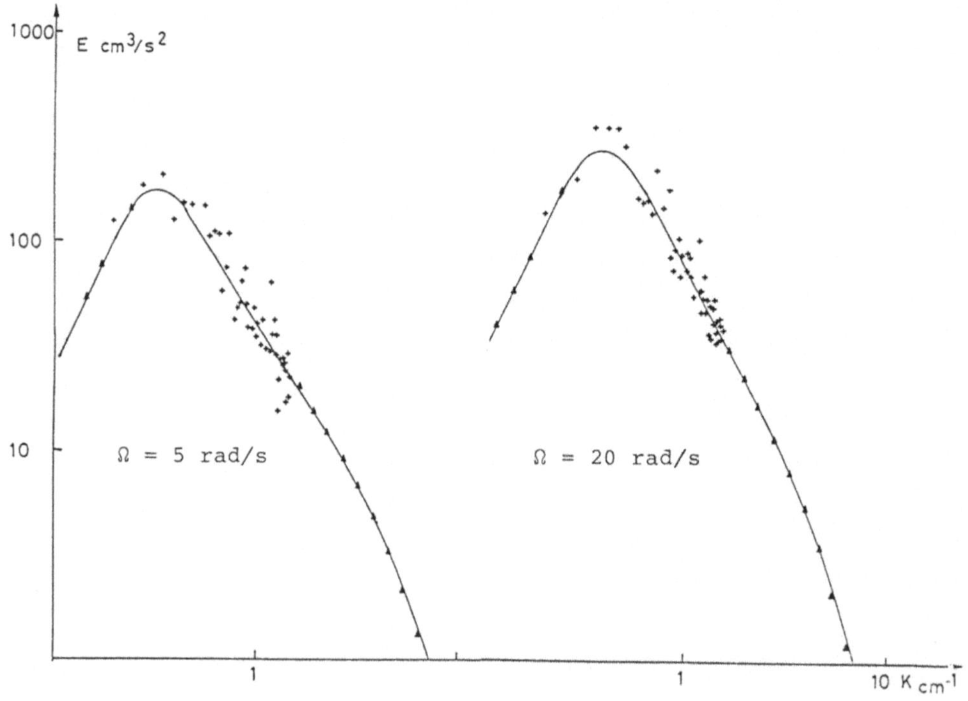

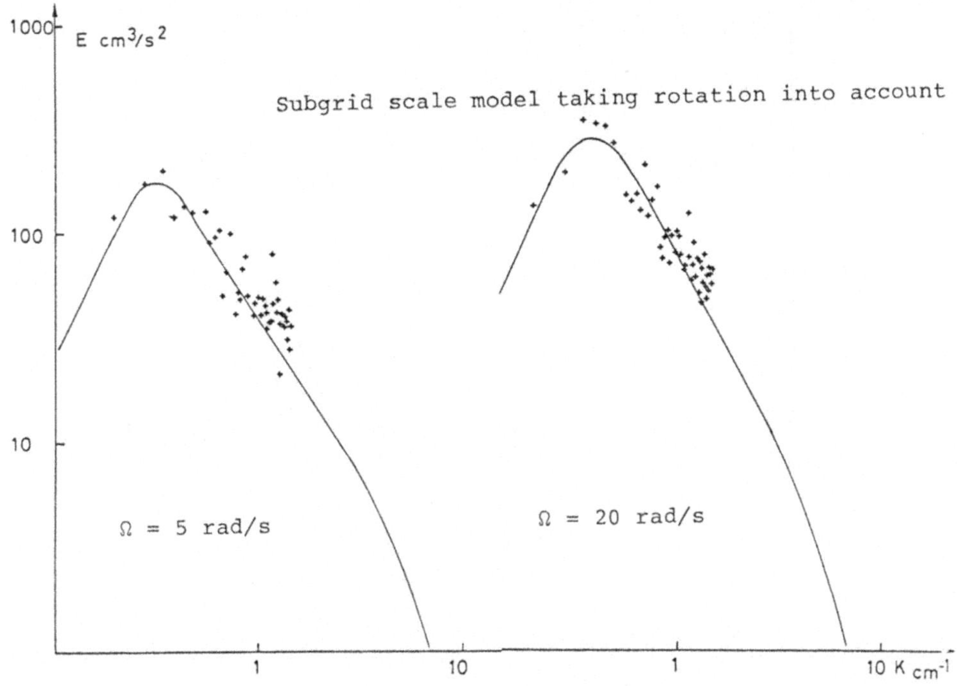

Fig. 10 - Rotating turbulence - Initial energy spectrum : COMTE-BELLOT
— energy spectrum computed with EDQNM
+ large scales computed with LES

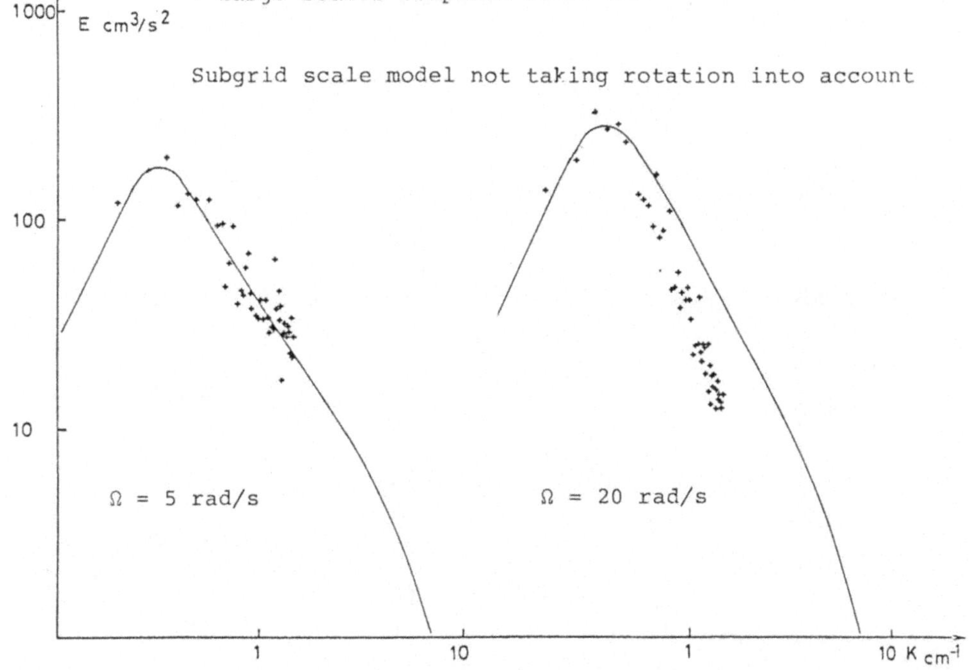

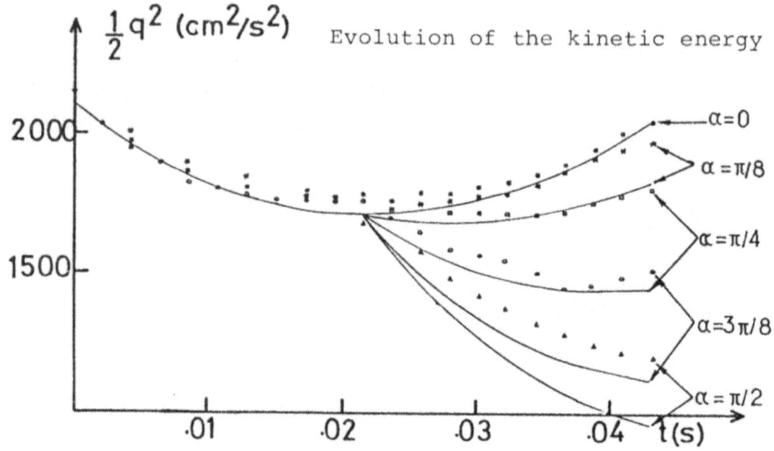

Fig. 11

Experiment of GENCE - Turbulence submitted to two plane strains of relative angle α

□ o Δ Experiment

—— LES

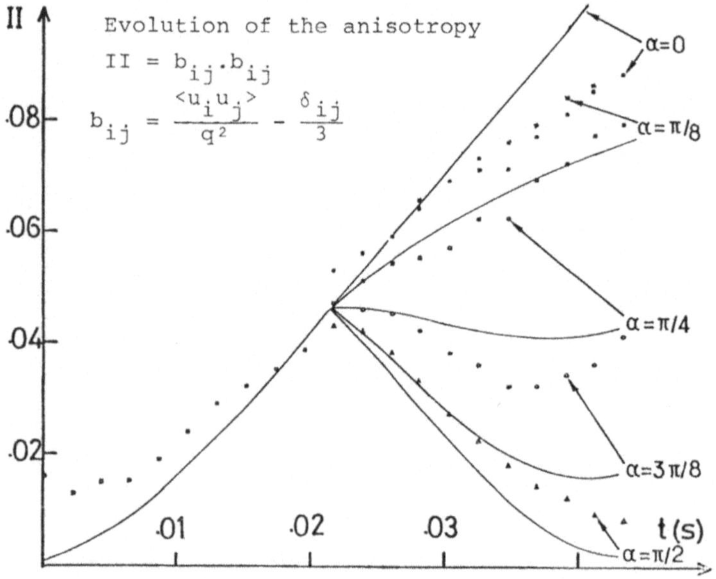

BLOW-UP IN THE NAVIER-STOKES AND EULER EQUATIONS

Eric D. Siggia
Laboratory of Atomic and Solid State Physics
Cornell University
Ithaca, N.Y. 14853

A series of calculations have been undertaken in which a single vortex filament with a variable core parameter is used to model how solutions to the Euler equations diverge. Mono-filament calculations suppress any instablities that might develop within the core. Nevertheless, as will be seen, we will still generate information about real solutions to Euler's equations which may not be stable but do exist. Technically, working with only one or two filaments permits one to interact with the computational process in a manner similar to investigations of dynamical systems.

Obtained seemingly for free from our calculations has been siginificant insight into how the Navier-Stokes equations might blow up. In fact, for the types of solutions we obtain, the viscosity may plausibly be added by hand and is counter balanced by the stretching. Viscosity we find is at best only marginally (i.e. to within logs) able to control the singularity that probably develops in the Euler equations. It is just conceivable that the sort of vortex filament pairing we see in our initial value problem is relevant to the bursting process one sees in wall bounded shear flows. Fluid uplifted from near the wall, has greater vorticity than its surroundings and numerical and laboratory experiment suggest it is pulled into a stream wise "hair pin" vortex. This configuration would then behave very much like our simulations.

A complete picture of the blow up process must include large spectral calculations to rigorously examine the core instabilities as well as much coarser calculations which exploit the vortex pairing we observe within Biot-Savart to derive a local P.D.E. for the filament pair. Both investigations are in progress and any definitive statement about whether the Euler equations to blow up must await the solution of our "local" model.

The equations we treat numerically are simply

$$v(r(\theta)) = \frac{\Gamma}{4\pi} \nabla \times \int \frac{(dr'/d\theta')d\theta'}{((r-r')^2 + \sigma^2(\theta) + \sigma^2(\theta'))^{1/2}} \tag{1}$$

where θ is a Lagrangian parameter and the core size σ obeys either

$$\sigma^2 L = \text{cst} \qquad \text{or} \tag{2a}$$

$$\sigma^2(\theta)\left|dr/d\theta\right| = \text{cst} \tag{2b}$$

In the former instance σ is independent of θ.

We typically begin with a single ellipse with a 4:1 axis ratio which is then twisted around the short axis to form in projection, a figure eight. Opposite sides of the ellipse pair together forming a sort of ribbon in which the two filaments are oppositly directed and separated by a distance of order σ. The initial stages of this pairing were examined previously.[1]

Detailed examination of the numerical output reveals that the vortex pairing persists i.e., the minimum spacing, d, between oppositly directed pieces of filament scales with the local σ as it evolves. In addition, the local radius of curvature r_c always exceeds σ, and by a factor of 4-5 when (2b) is used. Since σ decreases in time, the crinkling of the filament as measured by r_c proceeds from large to small scales i.e., little folds form on larger folds. This property is by no means obvious since within a linearized model all modes are unstable. Crow for instance has found for two straight antiparallel filament a dispersion relation of the form

$$\omega^2 = \Gamma^2(d^{-2}k^2 - k^4)$$

where $\mathrm{Re}(\omega) > 0$ implies instability. The constant axial elongation of the filaments must have a role in stabilizing the smaller scales.

The property $r_c/\sigma \gtrsim 1$ suggests that, even though for $d \sim \sigma$ the Biot-Savart law can not quantitatively represent the Euler equations, there is nevertheless a solution to the Euler equations nearby. The reason is simply a belief that for a nearly two dimensional structure (recall $r_c/\sigma \gtrsim$ 4-5 for (2b)) one can reasonably perturb the vortex dipole solutions that many authors have simulated in two dimensions. These have distributed vortex cores and are very stable with no tendency for plus and minus vorticity to mix.[2] By scale invariance there is always a one parameter family of dipole solutions for any core profile in which the velocity scales as Γ/σ which is all we require of our paired ribbon in three dimensions. Any singularity one might have thought was introduced by using a vortex filament as initial data rather than Fourier modes à la Taylor Green may be lumped into Γ/σ at $t = 0$ which merely sets the time scale.

The three dimensional vortex pairing also suggests an obvious and presumably quantitative local model for Biot-Savart equations. Namely expand the integrand in (1) about the point $\vec{r}(\theta)$ and about the point closest to $\vec{r}(\theta)$ of the other filament of the pair. Thus at each point along the pair there is naturally the position $\vec{r}(\theta)$ plus a separation vector $\vec{d}$; both evolve in time. We believe numerical solutions of a partial differential equation which is local in arclength should permit firm conclusions about whether finite time singularities exist for special solutions to the Euler equations.

For a nearly two dimensional vortex filament, perhaps also subject to uniform axisymmetric strain viscous effects can be sensibly included by letting the core size σ evolve according to

$$\frac{d\sigma^2}{dt} = \nu - \sigma^2(\frac{ds}{dt}\theta)/s_\Theta, \tag{3}$$

where ν is the kinematic viscosity and we used (2b) to describe the stretching (s=arclength). Now $d(\ln(s_\Theta))/dt$ is just the value of the rate of strain matrix, $\frac{1}{2}(\partial_i V_j + \partial_j V_i)$, projected onto the tangent and it can be estimated as

$$d(\ln s_\Theta)/dt \sim \Gamma/dr_c, \tag{4}$$

where Γ/d is just the velocity and r_c^{-1} comes from the gradient. Now if one recalls our solutions to (1) both r_c and d varied, but their distributions scaled with σ, thus (3) is dimensionally at least like

$$\frac{d\sigma^2}{dt} = \nu - \text{cst. } \Gamma \tag{5}$$

The stretching can balance the diffusion and lead to blow up. The Laplacian in Navier-Stokes is in a specific sense a marginal form of dissipation for if one had something like $\tilde{\nu}\ \nabla^{2+\alpha}$ then the analogue of (5) would always have a stable solution with $\sigma^\alpha \sim \tilde{\nu}/\Gamma$ and no blow up could occur. In a different sense it is just an accident that the viscosity and circulation have the same units.

The arguments above are necessary and not sufficient conditions for the Navier-Stokes equations to blow up. If one examines whether the circulation, Γ, of each filament is conserved in the presence of viscosity one finds that it is sufficient that $\alpha < 1/2$ in the scaling expression

$$s_\Theta \sim (t^*-t)^{-\alpha}$$

while if $\alpha = 1/2$ Γ may decrease to zero logarithmically.

Leray and later workers have shown that $\alpha \geq 1/2$ (e.g. $\sup_x(v) \gtrsim (\nu/(t^*-t))^{1/2}$, $\sup_x(\partial_i V_j) \sim 1/(t^*-t)$ etc.). For equations (1) and (2b) and $\nu=0$ we find an α measured from sup(vorticity) $\sim (\min_x(\sigma^2))^{-1}$ of order 1/2 or larger but the scaling is not convincing. We suggest whatever relation we find would reduce to $\alpha = 1/2$ due to the diffusion of vorticity when ν is included.

We, of course, can say nothing about the stability of a vortex pair in the presence of viscosity and in the absence of any experimental indications of singularities one presumes they don't exist. Of course, the logarithms could also account for their nonappearance in real flows. In either case we can claim to understand why it is so difficult to mathematically prove the regularity of Navier-Stokes; it requires getting the logarithms straight which is a difficult task even at the heuristic level.

Our research was supported in part by the Department of Energy under grant #DE AC02 83 ER 13004.

1. E.D. Siggia, Phys. Fluids, to appear. This article contains a complete set of references and figures.
2. H. Aref and E.D. Siggia, J. Fluid Mech. 100, 705 (1980).

LARGE EDDY SIMULATIONS OF TURBULENCE IN PHYSICAL SPACE. ANALYSIS OF SPECTRAL ENERGY TRANSFER

J.P. Benqué, A. Hauguel and D. Laurence
Electricité de France
Laboratoire National d'Hydraulique
6, quai Watier - 78400 Chatou (France)

INTRODUCTION

The basic idea in LES lies in the decomposition of the flow variables into large and small scale components through a spacial filter G :

$$v(x,t) = \bar{v}(x,t) + v'(x,t) \qquad \bar{v}(x,t) = \int G(x-y;\Delta f).v(y,t)\,dy$$

In practice, a second decomposition is introduced by substraction of the very large scale component (i.e. time averaged) $\mathcal{V}$. The total variable V is thus split as follow :

$$V(x,t) = \mathcal{V}(x) + \bar{v}(x,t) + v'(x,t)$$

$\mathcal{V}$ can be called the supporting field, $\bar{v}$ the large scale turbulent field, v' the sub-grid scale turbulent field, and $\bar{u}$ the computed field.

For homogeneous turbulent flows, $\mathcal{V}$ is known so the actual computation is restricted to $\bar{v}$ while the effect of $\mathcal{V}$ is analytical (i.e. : $\mathcal{V} = \mathcal{V}_0$ constant for grid turbulence, or $\mathcal{V}_1 = S.X_3$, with $S = d\mathcal{V}_1/dX_3$ constant for homogeneous shear). Our long term goal being LES of complex flows ($\mathcal{V}$ unknown), the computed variable remains $\bar{u} = \mathcal{V} + \bar{v}$ even for homogeneous flows which are only used as a bench mark for the code. Therefore, spectral methods are prohibited. The code is written in physical space and our task is then to make it as precise as possible, even for the higher frequencies and wave numbers where fluctuations are usually damped to ensure numerical stability of classical Navier Stokes codes. Only the results are transfered in spectral space for analysis.

The chronological order of our work is reversed with respect to the complexity of the simulated flows. The first LES attempted at EDF were performed on a confined jet flow [1], and channel flow. The scale of the turbulent structures appears very large when compared to thoses which can be observed in sheared turbulence simulations (fig.13, 14), although the same code is used. We concluded that the presence of a supporting field $\mathcal{V} \neq 0$ significantly increases the implicit filtering of the code, particularly through advection. This is consistant with the fact that for the turbulent jet, good agreement with experiment, of the turbulence intensity was found in the production region, but it was under-estimated in the lower half of the box where advection is dominant. Another fact to mention is that the non dimensional Reynolds stress in the shear layer was very large ($\langle U_1 \cdot U_3 \rangle = 0{,}75$), which means that the band width of the turbulent spectra of the LES is very narrow.

LES of homogeneous flows enables us to analyse the spectral behaviour of our code, and improve it before going back to more complex flows.

I - STANDARD NUMERICAL CODE

$(n+1)^{th}$ time step is computed in three sub-steps [6]: $t^n \rightarrow t^{n+1} = t^n + \delta t$

$U^n \rightarrow \tilde{\tilde{U}}^{n+1}$

$\tilde{\tilde{U}}^{n+1} \rightarrow \tilde{U}^{n+1}$

$\tilde{U}^{n+1} \rightarrow U^{n+1}$

. Advection of momentum :

This is resolved by a three dimensional curvilinear characteristics method :

$$\begin{cases} d\, C_M/dt = -U^n \\ C_M(t^{n+1}) = M \end{cases}$$

then : $\tilde{\tilde{U}}^{n+1}(M) = U^n(N)$

with : $N = C_M(t^n)$

Example of 1D characteristic

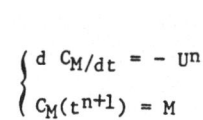

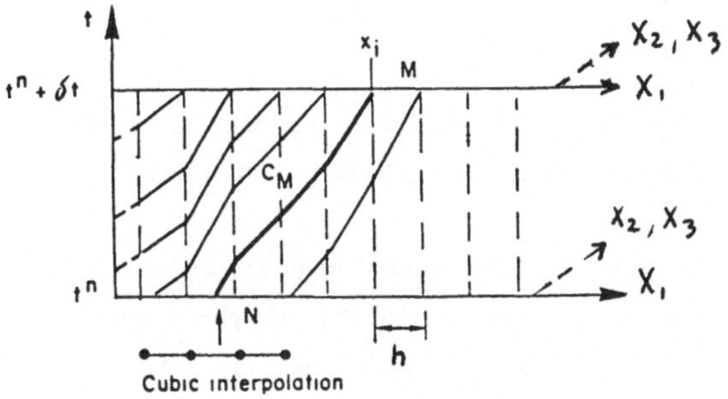

(projection of 4D space $\{t, X_1, X_2, X_3\}$ onto 2 D plane $\{t, X_1\}$)

. Diffusion of momentum (split into 3 directions) :

$$(\tilde{U}_i^{n+1} - \tilde{\tilde{U}}_i^{n+1})/\delta t = \frac{\partial}{\partial x_j}\left\{(\nu + \nu_T)\, \tilde{\tilde{D}}_{ij}\right\}$$ (Gauss elimination method)

where ν_T is SMAGORINSKY's subgrid viscosity model.

$$\nu_T = C_S^2 h^2 (D_{ij} D_{ij})^{1/2} , \quad D_{ij} = \frac{1}{2}\left(\frac{\partial U_i}{\partial x_j} + \frac{\partial U_j}{\partial x_i}\right)$$

. <u>Continuity and pressure</u> :

P^{n+1} is resolved by a GAUSS SEIDEL iterative method with over-relaxation under the form :

$$\frac{\partial \tilde{U}_i^{n+1}}{\partial x_i} = \frac{\delta t}{\rho} \frac{\partial^2 P^{n+1}}{\partial x_i \, \partial x_i} , \quad (U_i^{n+1} - \tilde{U}_i^{n+1}) / \delta t = -\frac{1}{\rho} \frac{\partial P^{n+1}}{\partial x_i}$$

Pressure is defined on staggered grid ("Pressure points" are at the center of "velocity cells").

II - BENCH TEST

Comte-Bellot and Corsin's experiments on grid generated turbulence [2] is used as bench-mark. Computation of the large scale field $\bar{u}$ is carried out in physical space. From the results, the energy spectrum is computed :

$$E(k) = \int_{\|\vec{P}\| = k} \underset{\sim}{\vec{U}}(\vec{P}) \cdot \underset{\sim}{\vec{U}}(-\vec{P}) \, dS \qquad \underset{\sim}{U} = \text{Fourier transform of } \bar{U}$$

Integration over the spherical shell removes random phases included in $\vec{U}(\vec{P})$. Variations of E(k) over each substeps yields information on the numerical filtering of the operators.

Simulation of homogeneous shear is also tested. Champagne, Harris & Corsin's experiment [3] is used because the shear is "slow" with respect to the non linear action of turbulence on itself.

Presently, grid turbulence is simulated in a computational domain carried by the supporting field $\mathcal{V}$, so the computed variable is $\bar{u} = \bar{v}$. Periodicity is used as boundary condition. On the contrary, for homogeneous shear, $\bar{u} = \bar{v} + \mathcal{V}$ and thus mesh deformation is avoided. In the shear direction X3, spacial periodicity has to be combined with a shift along the mean flow direction X1. To deal with this problem, the neighbourghing vertical domains in the X3 direction are staggered at each time step in the X1 direction. Computations are performed on CRAY 1 computer with $(32)^3$ mesh points.

III - ANALYSIS OF THE SUB-TIME-STEPS OPERATORS

. Advection :

Let $\bar{u} = \mathcal{V} + \bar{v}$. The time scale related to the resolution of the non linear interaction of may be large with respect to the Courant Friedrich Lewy number related to $\bar{v}$ (if $|\mathcal{V}| \gg |\bar{v}|$).

Therefore, we use a characteristics method for the resolution of advection (which has no CFL number restriction). The curvilinear characteristic CM is computed by a 2nd order Runge-Kutta method. The foot of this characteristic, N = CM (t^n), usually does not belong to the mesh, so a high order interpolation is required for computing $U^n(N)$.

This is where most of the filtering occures, instead of : $\hat{U}^{n+1}(M,K) = U^n(N,K)$

we have : $\hat{U}^{n+1}(M,K) = G_I(K) \cdot U^n(N,K)$

where K is the wave number, and $G_I(K)$ the filter introduced by interpolation. Various interpolation formula have been tested by homogeneous shifting of a turbulent field on the mesh. $G_I(K)$ is measured by comparing initial and final spectra. $G_I(K)$ can be fitted with the function : $G_\Delta(K) = e^{-\frac{\Delta^2 K^2}{24}}$

thus defining the filter width Δ_G, or rather, the non-dimensional width δ_G :

$$\Delta_G = \delta \cdot h \qquad \text{(h mesh step)}$$

For example, a linear interpolation yields $\delta_G = \sqrt{6}$, for CFL = 0.5.

This enabled us to tune the coefficient of the standard interpolation : a weighted average of Taylor developments located at the 8 node points of the cell containing N (32 points interpolation). It performes better than the $(4)^3$ point Lagrange polynominal as can be seen on array I.

For low CFL numbers, splitting the advection step in 3 directions allows us to use the "weak formulation" of the transport equation (advection of test functions, Ψ instead of the variable $\bar{U}$, as in finite element formulation) [5]. This formulation rejects the numerical diffusion into the dual space (Ψ) and is thus very conservative on the advected variable ($\bar{U}$) (see array I and fig. 1).

. <u>Continuity and pressure</u> :

The Poisson equation for pressure is obtained by applying the divergence operator to the filtered NS equation after discretisation in 3 sub-time steps. This leads to a 27 points operator, $\Delta_{27} = BB^T$, after the space discretisation of the Grad and Div operators have been defined (B for Div and B^T for Grad). Comparison of this compatible operator, Δ 27, with the classical 7 points "star" operator Δ 7 shows that in Fourier space, the error of the later is restricted to higher wave numbers. It nevertheless under-estimates the "return to isotropy" action of the pressure (fig. 8).

The splitting in 3 sub-time steps brings a net energy drain in the pressure steps :

$$2(dE/dt)_p = \| U^{n+1} \|^2 - \| \tilde{U}^{n+1} \|^2 = -\frac{\delta t^2}{\rho} (P^T, \Delta_{27} P) < 0$$

It can be reduced by looking for the second order in time, which means impliciting the pressure in the advection step :

$$\hat{U}^{n+1}(M) = U^n(N) - \frac{1}{\rho} \int_{t^n}^{t^{n+1}} grad\, P\, dt$$

The integration being carried out along the characteristic from N to M, it can be approximated by :

$$\hat{U}^{n+1}(M) = U^n(N) - \frac{1}{2\rho} \left[grad\, P^n|_N + grad\, P^{n+1}|_M \right] . \delta t$$

. <u>Smagorinsky model</u> :

In the litterature, the constant C_s of the Smagorinsky model ranges from 0.065 to 0.200, the later value being used by authors working with most accurate pseudo-spectral codes and thus, is probably the exact value for grid turbulence. From this value, C_s is decreased to account for numerical diffusion, and for our standard scheme is set to $C_s = 0.120$. This is consistant with the fact that most of the numerical diffusion takes place during the interpolation.

Consequently, when using the weak formulation of advection, one must set $C_s = 0.150$.

IV - INITIAL CONDITIONS

A velocity field $\underset{\sim}{U}^{\circ}(K)$ is computed from an experimental energy spectrum E (|k|), so that each Fourier space point K has the correct energy, random angles being chosen to share this energy between the 3 velocity components.

Care must the taken that this field is divergence free in the discreet sense, i.e. ; divergence equation : $K.\ \underset{\sim}{U}^{\circ}\ (K) = 0.$

is replaced by : $(K.\exp i\,\phi\,)\,.\,\underset{\sim}{U}^{\circ}\ (K) = 0$

where the phase shift ϕ results from the Fourier transform of the discreet divergence equation used in the LES code : B. U = 0.

Such a field is Gaussian, and thus has no skewness ($B'''_{LL,L}(0) = \langle (\frac{\partial U_1}{\partial x_1})^3 \rangle$) Skewness must then be allowed to build up by going through a few time iterations with a "frozen" spectrum E (|K|) = E° (|K|), lest the Taylor microscale grows to fast, B'''LL,L(0) being the only negative terme in it's balance equation :

$$\frac{1}{2}\frac{d}{dt} B''_{L,L}(0) = \frac{7}{6} B'''_{LL,L}(0) + \text{(positive molecular diffusion termes)}$$

V - RESULTS AND FUTHER DEVELOPMENTS

Decrease of grid turbulence is simulated between experimental stations : $tU_0/M = 42$ and $tU_0/M = 98$ [2].

Energy spectra at the latter are compared with measurements and filtered measurements ($\delta = 2.$). The initial spectrum is not pre-filtered (except for the sharp cut-off at kc $= 2\pi/2h$) because the implicit numerical filtering G_{Num} will always occure at every time step. If N is the total number of time steps, the total filtering will be $(G_{Num})^N$. Hereafter, the CFL number is computed with the r.m.s. of $\vec{U}$, so the maximum CFL is approximately for this field (CFL)max = 5 x CFL.

Spectrum on fig. 2 was obtained with CFL = 0.1 and is already satisfactory. Increasing CFL to 0.5 (fig. 3) costs us an important loss of energy at high wave numbers which is due to the 2nd order in time terms, and not interpolation as can be seen on array I. The weak formulation of advection performes remarkably well (fig. 4) but is restricted to low CFL (0.1) since characteristics are 1D (splitted in 3 directions).

The characteristics routine advects any variable f by a field U. it thus leads to a natural separation between advected field f = Ued, and advecting field U = Uing.

The standard scheme is Ued = Uing = $\overline{U}$. But the choice is open for new models, and could be more promising than diffusion type terms. Centering the pressure term in time simply leads to Ued = $\overline{U} - \frac{\delta t}{\rho}$ grad P. We can also modify the advecting field : for instance, let α be a purely random variable, then setting Uing = $\overline{U} + \alpha$, enables us to include the diffusion of Ued by the turbulent viscosity νT, if the mean square of α is $< \alpha^2 > = 2 \nu_T / \delta t$

This model has been tried for grid turbulence, and of course is equivalent, but no better than "turbulent diffusion" models, but shows that the procedure opens a new path for SGS modelling, which at least contains the previous models.

Bardina, Ferziger & Reynolds [4] found significan improvement of the model by adding a scale similarity term in relation with the smaller computed structures Ukc (near and left to kc), which can be extracted by a double filtering :

$$M_{ij} = C_r \left(\bar{u}_i \bar{u}_j - \bar{\bar{u}}_i \bar{\bar{u}}_j \right) \quad , \quad C_r \sim 1.$$

The B, F & R scale similarity model can also be re-written under the advective form (α being now strongly correlated with Ued). The model term Mij goes into the filtered Navier Stokes equation through its divergence, and due to incompressibility can be expressed as :

$$\frac{\partial M_{ij}}{\partial x_j} = C_r \left(\bar{u}_j \frac{\partial \bar{u}_i}{\partial x_j} - \bar{\bar{u}}_j \frac{\partial \bar{\bar{u}}_i}{\partial x_j} \right) \tag{1}$$

The L.H.S. of the filtered N.S. equation is then :

$$\frac{\partial \bar{u}_i}{\partial t} + \bar{\bar{u}}_j \frac{\partial \bar{\bar{u}}_i}{\partial x_j} + (1 + C_r) \left(\bar{u}_j \frac{\partial \bar{u}_i}{\partial x_j} - \bar{\bar{u}}_j \frac{\partial \bar{\bar{u}}_i}{\partial x_j} \right) = \ldots \tag{2}$$

With our choice of $\overline{U} = \mathcal{V} + \bar{v}$, Gallilean invariability is absolutly necessary, and so the model is modified as follows :

$$\frac{\partial M'_{ij}}{\partial x_j} = \frac{\partial M_{ij}}{\partial x_j} - \bar{\bar{u}}_j \frac{\partial (\bar{u}_i - \bar{\bar{u}}_i)}{\partial x_j} = (\bar{u}_j - \bar{\bar{u}}_j) \frac{\partial \bar{u}_i}{\partial x_j}$$

With this model, and Cr = 1, equation (2) may be written as :

$$\frac{\partial \bar{u}_i}{\partial t} + \bar{u}_j \frac{\partial \bar{u}_i}{\partial x_j} + (\bar{u}_j - \bar{\bar{u}}_j) \frac{\partial \bar{u}_i}{\partial x_j} = \ldots \tag{3}$$

This means again perturbating the advecting field by a fluctuation : $\alpha = \bar{u}_j - \bar{\bar{u}}_j$. There is no artificial imput of energy, Ukc is simply enhanced in the advecting field which is discarded after advection. The effect of the model is to increase energy transfer towards smaller computed scales. It has been used in combination with the previous for grid turbulence and gives good results although CFL = 0.5 (fig. 5).

On fig. 9-12, one can see that the scale similarity term corrects the excessive energy on the largest scale, correcting in turn the excessive anisotropy. Beyond St = 3., the numerical filtering is becoming large and so is the departure from experimental results.

CONCLUSION

Performances of our code have thus been improved for homogeneous turbulence, with the advantage of not being restricted to such flows. Perturbating the advecting field yields new opportunities for sub-grid stress modeling and could even lead to back scatter effects if the global energy input is known, for instance through a coupled EDQNM computation.

Much research has been done on the $\bar{v} \leftrightarrow v'$ interaction, but if LES is to escape the homogeneous field, some numerical developments will be necessary to deal with the $\mathcal{V} \leftrightarrow \bar{v}$ interaction. If the scales are well separated, this should be easier task since it means a long distance transport of a small fluctuation with a nearly homogeneous non-linear interaction in a local domain related to the v scales. Here again, the characteristics point of view could be useful.

REFERENCES

[1] F. BARON, D. LAURENCE
"Large Eddy Simulation of a Confined Turbulent Jet Flow and Homogeneous Shear". Turbulent Shear Flow IV - pp. 4.7-4.12 (1983).

[2] COMPTE-BELLOT, CORRSIN
"Simple Eulerian Time Correlation of Full and Narrow-band Velocity Signals in Grid generated, Isotropic, Turbulence". J. Fluid Mech (1971), vol. 48.

[3] CHAMPAGNE, HARRIS, CORRSIN
"Experiments on Nearly Homogeneous Turbulent Shear Flow". J. Fluid Mech. (1970).

[4] BARDINA, FERZIGER, REYNOLDS
"Improved Turbulence Models based on Large Eddy Simulation of Homogeneous, Incompressible Turbulent Flows". Report n° TF-19, Stanford University Calif. (1983).

Array I : width of the interpolation filter GI , δ_G

scheme	for 1 1 time step	for total transport over distance k	for total simula- tion grid turbu- lence experiment
standard 1 scheme CFL = 0.1	0.484	1.530	3.346
standard scheme CFL = 0.5	1.164	1.646	3.492
Lagrange polynomial CFL = 0.5	1.302	1.841	
Weak formulation of advection, CFL = 0.5 CFL = 0.5	0.785	1.105	
Weak formulation of advection, CFL = 0.1	0.239	0.755	1.603

SLIP-PERIODIC BOUNDARY CONDITIONS FOR HOMOGENEOUS SHEAR

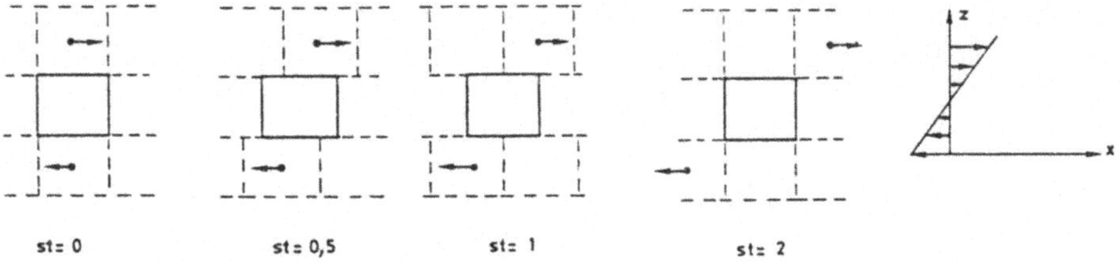

DISCRETISED LAPLACE OPERATORS

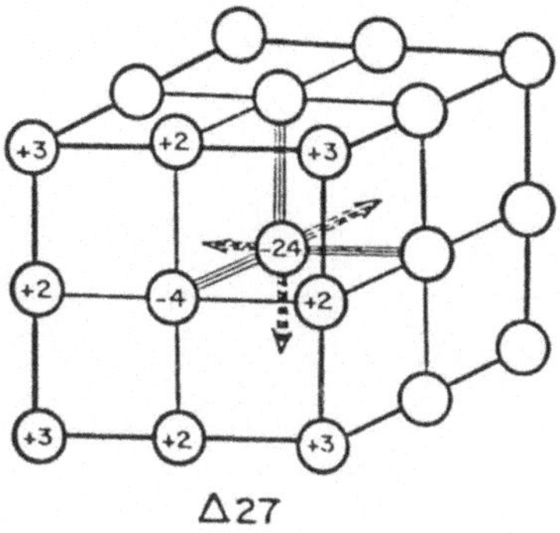

Compatible Δ

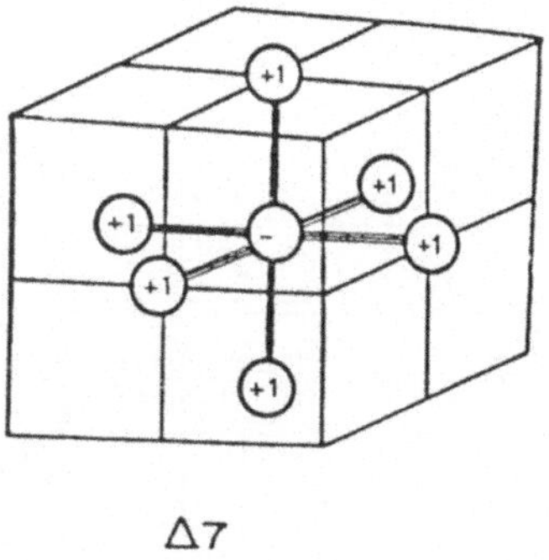

Simple 'star' Δ

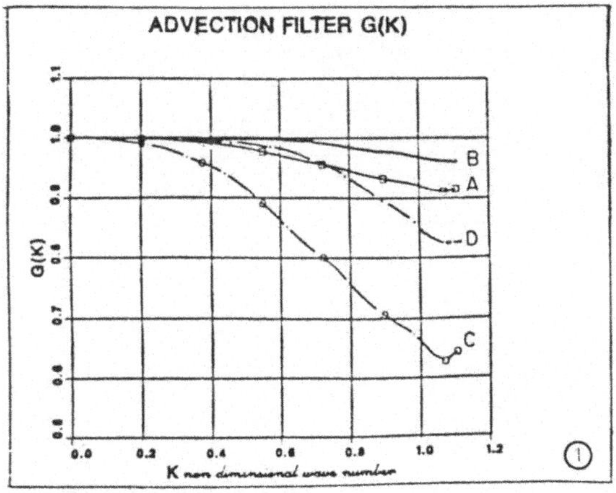

Numerical filter of the advection step

$G(k)$ = ratio of final to initial energy spectrum while advecting by a uniform field, for CFL = 0,1.

A : G_s for standard advection scheme

B : G_w for "weak formulation"

C : $(G_s)^5$ (five steps)

D : $(G_w)^5$

DECREASE OF GRID TURBULENCE

Spectra at station $tU^{\circ}/M = 98$

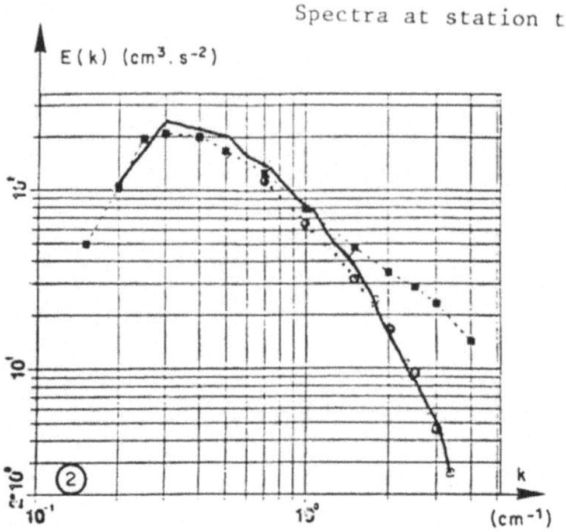

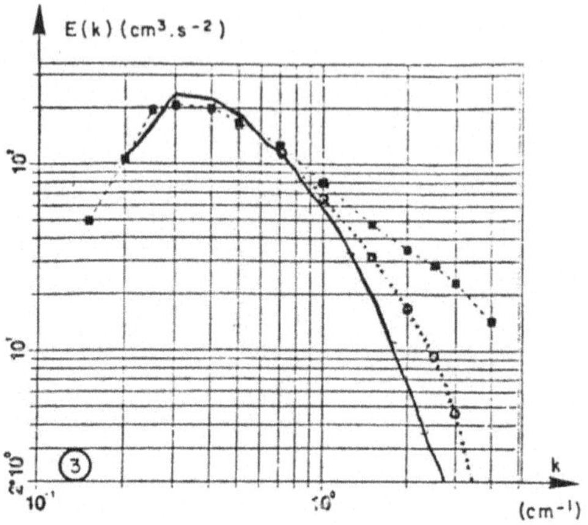

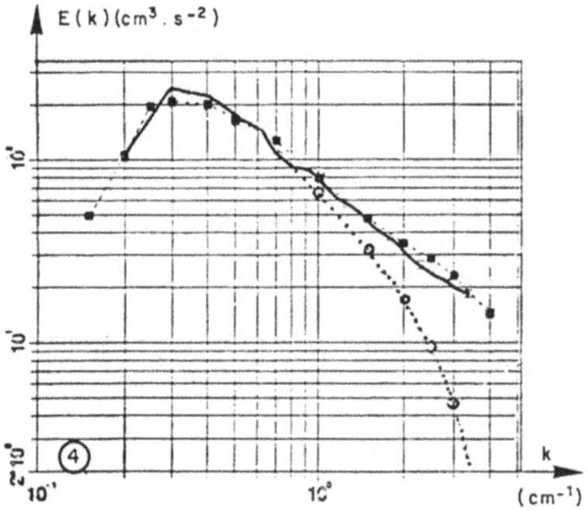

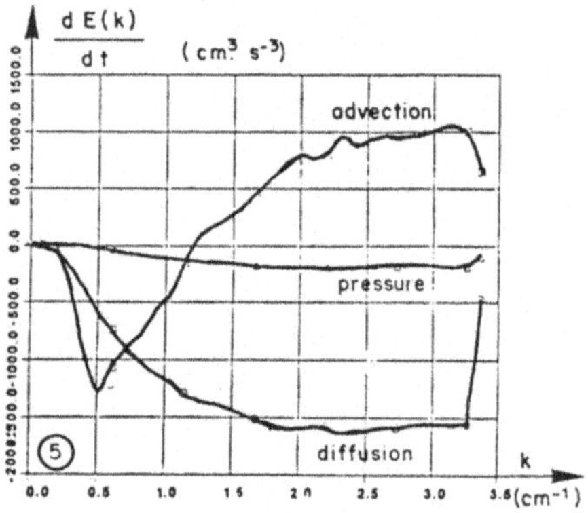

(2) Standard scheme ; final energy spectrum ('slow' run : CFL = .1).

(3) Standard scheme ; final energy spectrum ('fast' run : CFL = .5).

(4) Weak formulation of advection (CFL = .1) ; final energy spectrum.

(5) Weak formulation of advection (CFL = .1) ; energy variation per time step fraction.

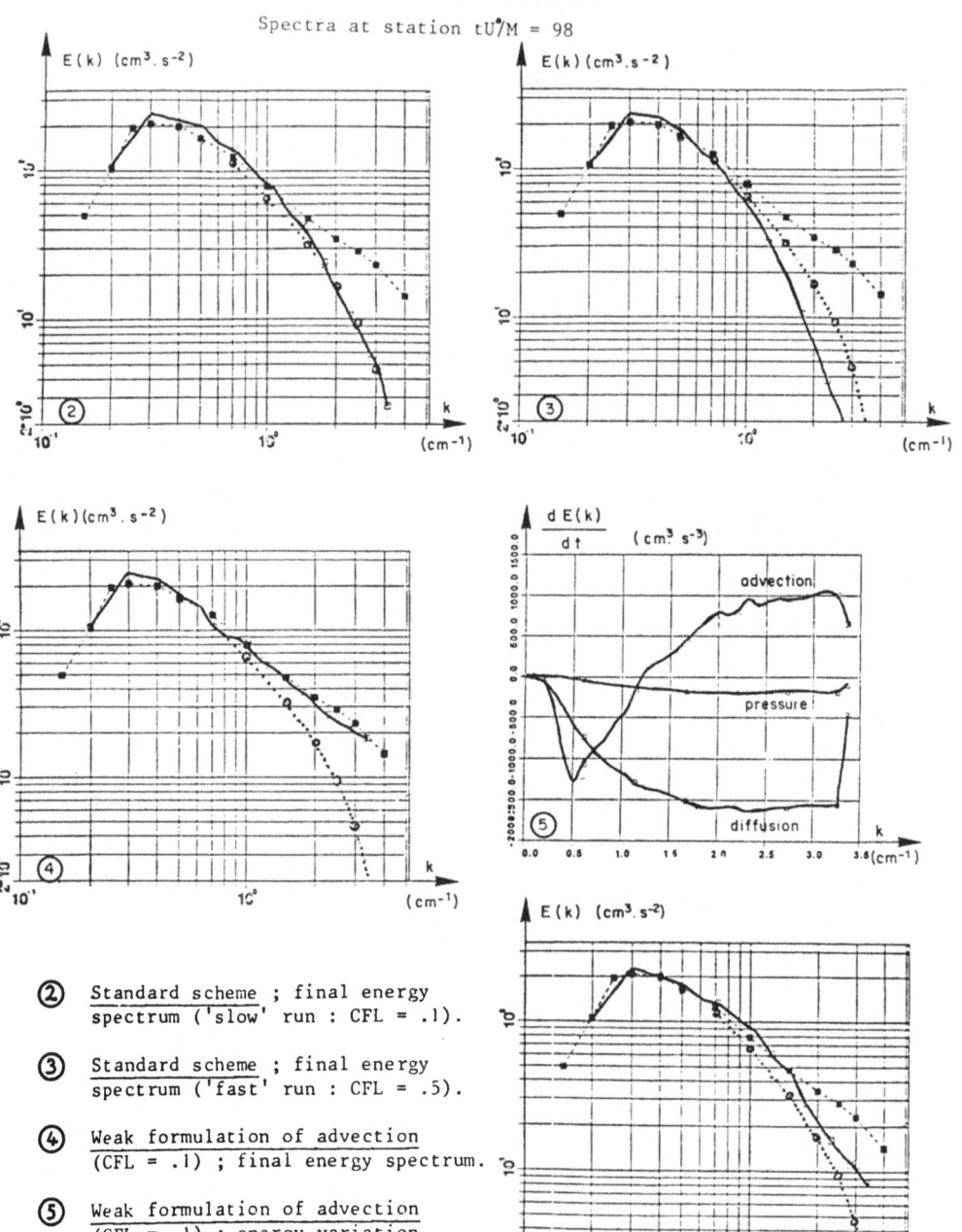

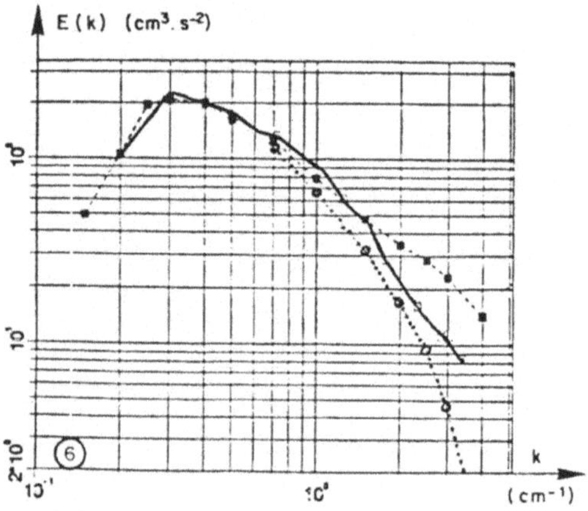

(6) Modified advection scheme (centered pressure gradient ; enhanced transfer Uk_c ; stochastic diffusion α) ; final energy spectrum (CFL = .5).

[All solid lines are L.E.S. results ; dashed lines are experimental results and dotted lines are filtered experimental results].

HOMOGENEOUS SHEAR

TURBULENCE INTENSITY

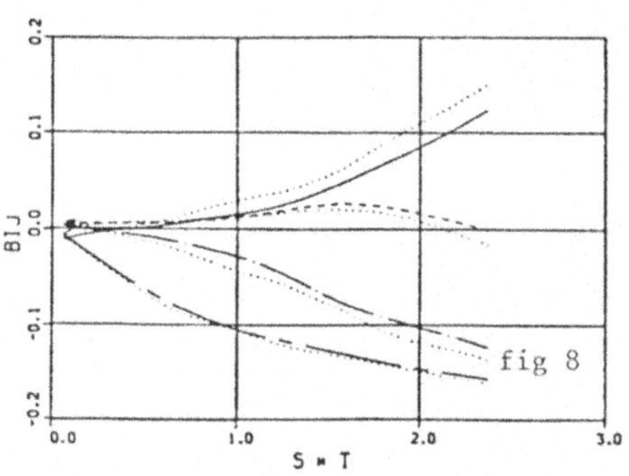

REYNOLDS STRESS ANISOTROPY

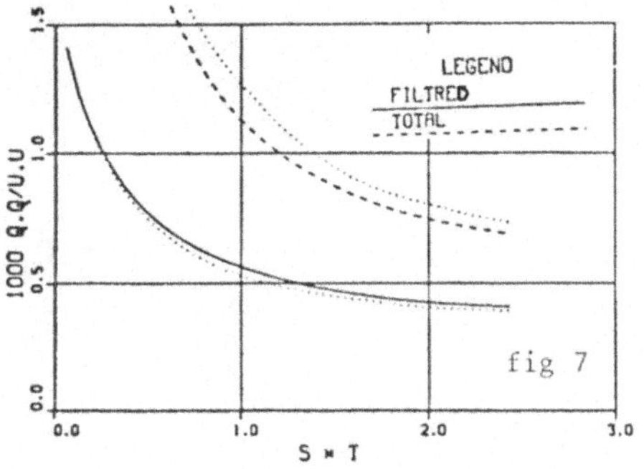

Standard run with Δ_{27}

(..... same run with Δ_7)

LEGEND
B11
B22
B33
B13

REYNOLDS STRESS ANISOTROPY

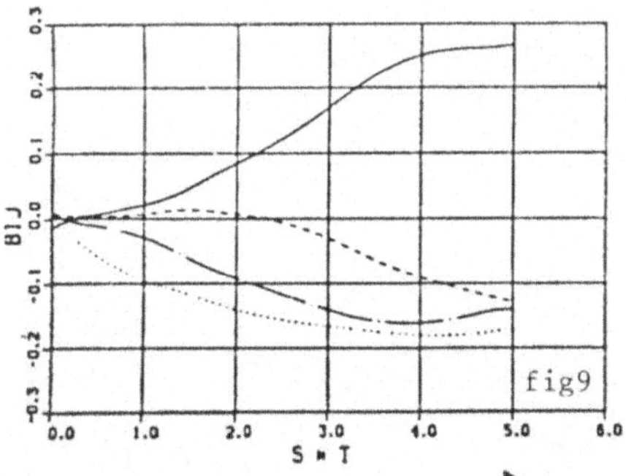

REYNOLDS STRESS ANISOTROPY

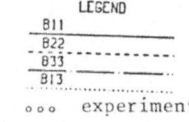

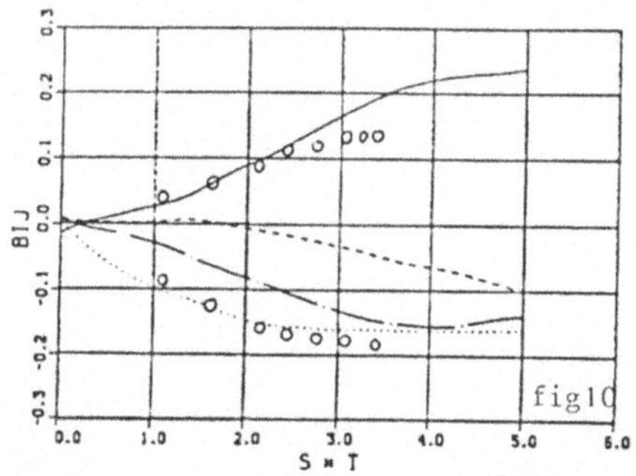

Smagorinsky model

Smagorinsky + scale-similarity model

LEGEND
B11
B22
B33
B13
ooo experiment

E(k) cm^3/s^2

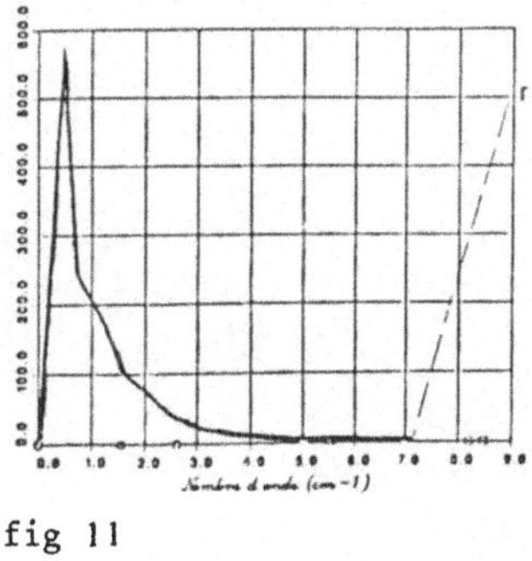

E(k)

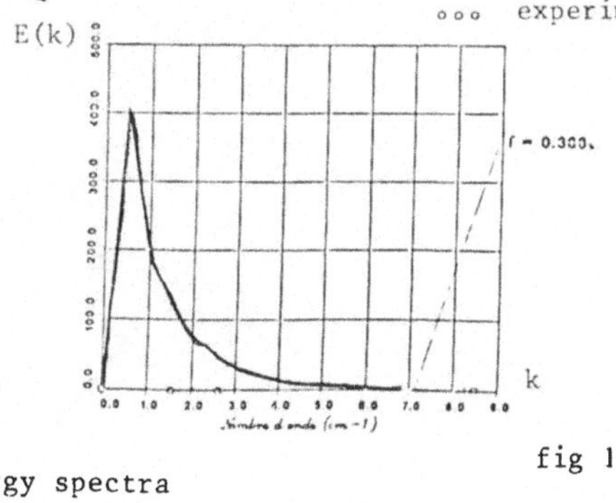

fig 11 Final energy spectra fig 12

CONFINED TURBULENT JET SIMULATION

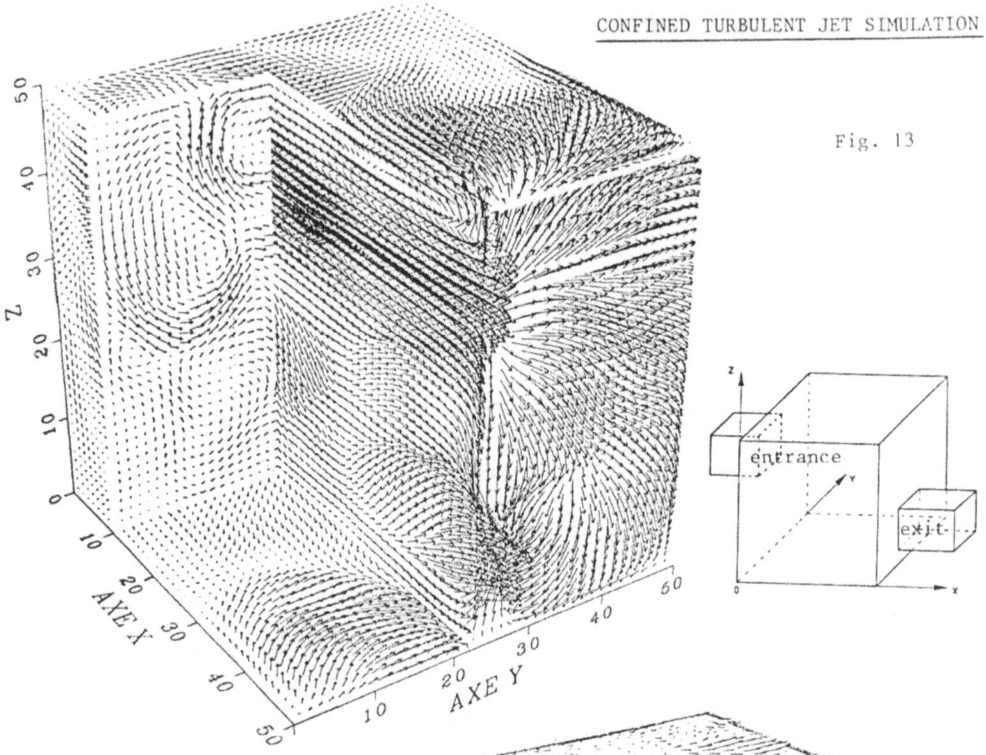

Fig. 13

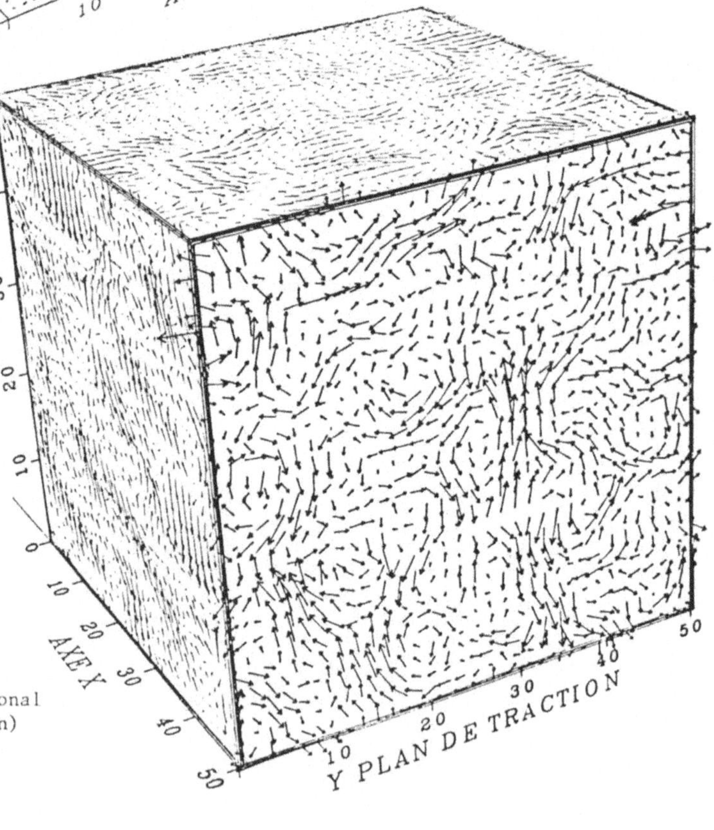

Fig 14

HOMOGENEOUS SHEAR SIMULATION

(front plane is orthogonal to stretching direction)

VORTEX STABILITY AND INERTIAL-RANGE CASCADES

Stephen Childress

Courant Institute of Mathematical Sciences
New York University
New York, N.Y. 10012

Abstract

We consider the inviscid stability and evolution of columnar vortices whose cores consist of a disc of constant vorticity ω_1 surrounded by an annular coat of constant vorticity ω_2. For core parameters in the range suggested by vortex-tube geometries of the inertial-range eddy structure, we find the breakup of unstable cores is largely two-dimensional, particularly if the vortex is being stretched. Tentative studies of the breakup of the core, using vortex methods, are described.

1. Introduction

The present paper is motivated by interest in the vortical structures which might adequately represent the energy cascade within inertial-range scales in fully-developed turbulence. Such structural models, usually based on vortex tubes or sheets, have often been used to give geometric meaning to the principal length scales of fully-developed turbulence [1-4]. To construct the inertial-range energy spectrum, however, structural models must admit an infinite hierarchy of scales. The simplest way to accomplish this is to introduce a statistical ensemble of structures of all sizes [5,6]. But it is physically more appealing to directly model the energy cascade as a deterministic event involving the evolution of vortical structures. Lundgren [7] accomplished this by considering stretched, spiral vortices which have a finite lifetime. Averaging over a great many events of this kind leads to a -5/3 Kolmogorov spectrum. In a further elaboration of these ideas we have recently proposed a realization of the β-model [8], in which the "inactive" and "active" eddies are represented by stable and unstable columnar vortices [9]. The resulting "γ-models" lead to cascades which conserve, in addition to energy, some of the Eulerian invariants implied by the evolution of circulation and of helicity in strictly inviscid flow. In Figure 1 we show the helical model $\gamma(2,2;H)$ developed in [9].

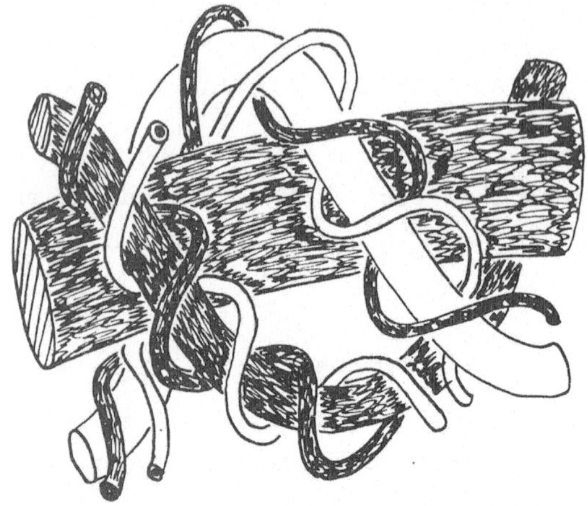

Figure 1. The helical model $\gamma(2,2:H)$. Active helices have a coated core, inactive helices a solid core. In a given step an active helix splits into two active daughter helices of opposite orientation, and an inactive tube carrying the original circulation. The two daughter helices are stretched to twice their length and wound around the inactive tube. The final result produces two new active helices geometrically similar to the starting helix. As the cascade continues, arbitrarily complex "coils upon coils" are produced as the residual inactive structures. At the termination of the cascade the active elements form a fractal set of dimension 13/5. Here we show inactive tubes from two generations together with currently active tubes.

In all γ-models the onset of dissipation in the inviscid limit can be studied by allowing the inactive structures to decay as straight vortex tubes. It then follows that the rate of dissipation obtained in the inviscid limit is zero only up to the moment of termination of the cascade, at which time finite dissipation ensues.

Our present intention is to study the inviscid stability of

columnar vortices and thereby to determine if the disintegration of cores postulated in the γ-models is consistent with high Reynolds number dynamics. It should be emphasized that the notion of "stability" is introduced here for the same reason that it is used in studies of transition to turbulence. Rather than try to describe exactly the evolution of a field of vorticity, we instead model a few features by perturbing simple steady flows. The results can provide a certain abstract bifurcation structure (cf. Bénard convection), but they cannot ever fully encompass the exact initial-value problem for the Navier-Stokes or Euler equations. Nevertheless it is in the solution of the initial-value problem which we have in mind when describing deterministic cascades.

2. Linear, 2-D stability of the coated vortex

The core of the coated vortex is shown in Figure 2. The "coat" of constant vorticity ω_2 surrounds the inner core of constant vorticity ω_1. A given core geometry is determined by the two dimensionless parameters $\omega = \omega_2/\omega_1$ and $\mu = \left(\frac{r_1}{r_2}\right)^2$. In order to produce self-similar cascades involving two daughter tubes of opposite orientation, we must take $\omega = -1$ and $\mu = 1 - 2^{-3/2}$ [9]. We refer to these values as the "standard case."

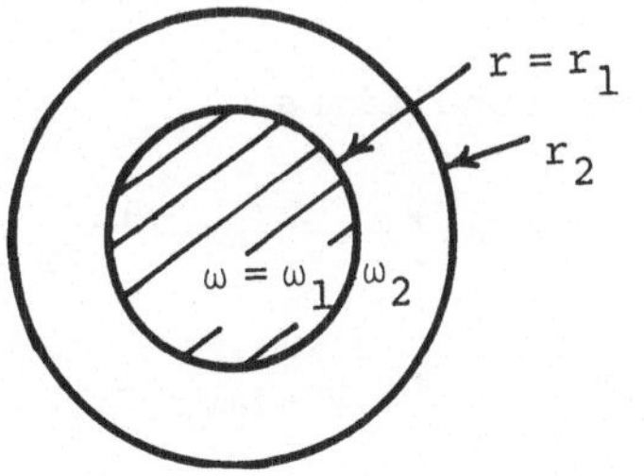

Figure 2. The coated vortex.

The equations to be solved are

(2.1a) $$\frac{\partial \omega}{\partial t} + \vec{u} \cdot \nabla \omega = 0 ,$$

(2.1b) $$\vec{u} = (u,v) = (\Psi_y, -\Psi_x) , \quad \omega = v_x - u_y = -\nabla^2 \Psi ,$$

where $\Psi(x,y,t)$ is the streamfunction. The unperturbed core is given by $\vec{u} = (u_r, u_\theta) = 0, V(r))$ where $V(r)$ is a continuous function of the form

$$(2.2) \qquad V(r) = \begin{cases} \omega_1 r/2 , & 0 \le r < r_1 , \\ \omega_2 r/2 + A/r , & r_1 \le r < r_2 , \\ B/r , & r \ge r_2 . \end{cases}$$

Writing, Ψ_0 as the streamfunction for V, we set

$$(2.3) \qquad \Psi = \Psi_0(r) + e^{\sigma t + im\theta}\psi(r) + \ldots$$

the linearized zero stability problem takes the form

$$(2.4) \quad (\sigma + imVr^{-1})[\psi'' + r^{-1}\psi' - r^{-2}m^2\psi] = \frac{im\psi}{r}[r^{-1}(rV)']' .$$

The last equation holds for any continuously differentiable V. For the special choice (2.2) the right-hand side of (2.4) reduces to a distribution and (2.4) is replaced by

$$(2.5a) \qquad (\sigma + imVr^{-1})(\psi'' + r^{-1}\psi' - r^{-2}m^2\psi) = 0$$

together with the jump conditions

$$(2.5b) \qquad (\sigma + imVr^{-1})[\psi']_-^+ = im\psi r^{-1}(\omega_2 - \omega_1) , \quad r = r_1 ,$$

$$(2.5c) \qquad (\sigma + imVr^{-1})[\psi']_-^+ = -im\psi r^{-1}\omega_2 , \quad r = r_2 .$$

In the last two equations the brackets denote the jump in the direction of increasing r. We seek solutions ψ of (2.5) which are piecewise continuously differentiable and vanish at $r = \infty$. When $\omega_1 = \omega_2$ this reduces to the constant vorticity core studied by Kelvin [10].

If we restrict attention to unstable modes, the first factor on the left of (2.5) cannot vanish and the problem is easily solved in terms of harmonic functions. After some manipulation, the solution yields the following equation for σ :

$$(2.6a) \qquad \sigma^2 + ib\sigma + c = 0 ,$$

$$(2.6b) \qquad 2b = m[\omega_2 + (\omega_1 - \omega_2)\mu] + (m-1)\omega_1 ,$$

$$(2.6c) \qquad 4c = m[\omega_2 + (\omega_1 - \omega_2)\mu][m\omega_1 + \omega_2 - \omega_1] - m\omega_1\omega_2 - \omega_2(\omega_1 - \omega_2)(\mu^m - 1) .$$

With $\omega = \omega_2/\omega_1$ and $\Omega = \omega + (1-\omega)\mu$, the necessary and sufficient condition for instability becomes

$$(2.7) \quad K \equiv (2/\omega_1)^2(4c - b^2) = [(2\omega - 1)^2 - 1]\mu^m - [m\Omega + 1 - m - 2\omega]^2 > 0 .$$

From (2.6) and (2.7) all of the relevant stability theory can be carried out explicitly. The $m = 1$ mode is seen to be neutrally stable with $\sigma = 0$ or $-i|K|^{\frac{1}{2}}\omega_1$. The stability boundary $\mathrm{Re}(\sigma) = 0$ can be derived for $m \geq 2$ as a quadratic equation in ω . For $m = 2$ the coefficient of ω^2 vanishes and there is the single boundary

$$(2.8) \qquad \omega = (2\mu - 1)^2/4\mu(\mu-1) ,$$

while if $m \geq 3$ the boundary has two branches

$$(2.8b) \qquad \omega = \frac{2\Gamma - 3m - m^2(\mu-1) \pm 2\mu^{m/2}(\mu^{m-2} + 2\mu^{m-3} + \ldots + m-1)^{\frac{1}{2}}}{4\Gamma - 4m - m^2(\mu-1)}$$

where

$$(2.8c) \qquad \Gamma = \mu^{m-1} + \mu^{m-2} + \ldots + 1 .$$

From (2.8) we then obtain Figure 3. The denominator in (2.8b) vanishes at $\omega = \mu_m$ and 1 , where $\{\mu_m\}$ is an increasing sequence with $\mu_2 = 0$. To the left of μ_m , the boundary consists of two branches in the domain $\omega > 1$, which meet at the point $(\mu,\omega) = (0, (m-1)/(m-2))$. In $\mu_m < \mu < 1$ the "+" sign in (2.8b)

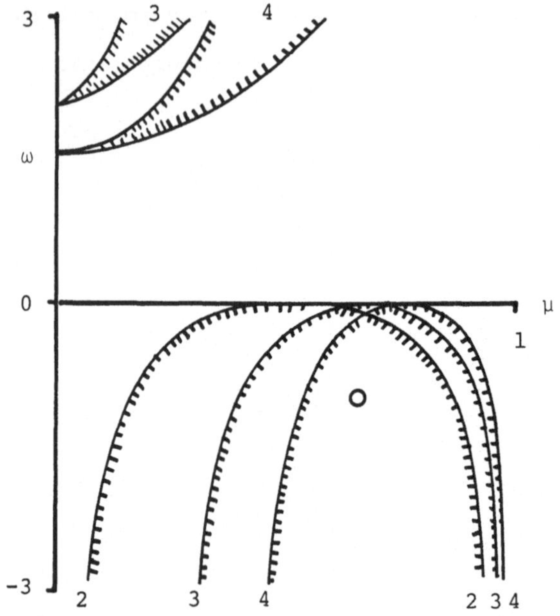

Figure 3. Instability domains for the coated vortex in two dimensions, indicated by cross hatching. The upper regions are the first two of an infinite sequence of shear-layer instabilities. The lower curves define the first three splitting instabilities. The corresponding values of m are indicated. The open circle represents the standard case.

gives a branch in $\omega < 0$, while the "-" sign gives the continuation of the lower curve in $\omega > 1$. We refer to the boundaries in $\omega > 1$ as "shear layer" instabilities. Indeed, an extreme case is μ near 1, $\omega >> 1$, in which case we are essentially dealing with a curved vortex sheet. The instabilities we obtained when ω is negative and below the boundaries in $\mu_m < \mu$ will be termed "splitting" instabilities, because these can be realized in the standard case. We show the standard case in Figure 3. It is a typical case in that a finite m-window of unstable modes exists (here $m = 2$ through 5), but it is interesting that the corresponding values of K, given in Table 1, reveal a balanced distribution of positive growth rates over all of these unstable modes. Thus there is available, on the basis of linear theory, two-dimensional modes capable of considerable distortion of coated vortices close to the standard case.

m	2	3	4	5	6
K	.828	1.389	1.368	.616	-.960

Table 1. Values of K (cf. (2.7)) for the standard case.

3. 3-D Instability for large m

There have apparently been few studies of non-axisymmetric 3-D instabilities of columnar vortices with general cores. Howard and Gupta [11] derived general stability criteria, including a sufficient condition for stability to non-axisymmetric disturbances. Recently Leibovich and Stewartson [12] gave a general sufficient condition for instability, and also carried out an asymptotic analysis for large m of the stability of a one-parameter model of the core of a trailing vortex. This method is remarkably successful down to $m = 3$, and we shall use it here to estimate the relative growth rates of 2- and 3-dimensional modes for the coated vortex.

Following the analysis in [12], the linear stability problem may be reduced to the following problem involving the radial velocity component $u(r) \exp i(st + m\theta + kz)$:

(3.1a) $$D^2\phi = F(r;m,\beta,s)\phi ,$$

$$(3.1b) \qquad \phi = \left(\frac{r^3}{1+\beta^2 r^2}\right)^{\frac{1}{2}} u \,, \qquad \beta = k/m$$

$$(3.1c) \qquad F = m^2 \frac{(1+\beta^2 r^2)}{r^2} \left\{1 - \frac{1+10\beta^2 r^2 - 3\beta^4 r^4}{4(1+\beta^2 r^2)^3 m^2} + \frac{r}{m\gamma} D\left(\frac{D_* V}{1+\beta^2 r^2}\right) + \frac{2\beta^2 V r}{\gamma^2 (1+\beta^2 r^2)}\right\} ,$$

$$(3.1d) \qquad \gamma = s + \frac{mV}{r} \,, \qquad D = \frac{d}{dr} \,, \qquad D_* = \frac{1}{r} Dr \,,$$

$$(3.1e) \qquad \phi(0) = 0 \,, \quad m \geq 2 \,; \qquad u(\infty) = 0 \,.$$

We shall study this problem for both m and kr_2 large, with βr_2 of order unity. For the study of unstable modes we may then replace the factor within braces on the right of (3.1c) by unity. Making a WKBJ substitution $\phi = e^{mf(r)}$ into the resulting equation, an asymptotic solution is obtained in the form

$$(3.2) \qquad \phi = \begin{cases} \Phi \equiv e^{mR} \left(\dfrac{R-1}{R+1}\right)^{\frac{m}{2}} , & r < r_1 \,, \quad R = \sqrt{1+\beta^2 r^2} \,, \\ = A\Phi + B/\Phi \,, & r_1 < r < r_2 \,, \\ \dfrac{C}{\Phi} \,, & r > r_2 \,. \end{cases}$$

From the continuity condition on ϕ and the jump conditions on $D\phi$ we may solve for the constants A,B,C, and thereby derive conditions for instability. At this point it is useful to make a further approximation, which is suggested from the curves in Figure 3. For large m in the 2-D problem, the splitting boundaries collapse onto the line

$\mu = 1$, $\omega < 0$. To resolve the boundaries for small βr_2 we are thus led to simultaneously take μ near to 1. The relevant parameter of order unity which replaces μ is then seen to be

$$\Delta = m(1 - \mu) . \tag{3.3}$$

We can immediately test the accuracy of the corresponding approximation to the 2-D theory from the results of the preceding section. For the splitting instabilities we obtain the asymptotic form of the neutral boundary:

$$\omega = \frac{2e^{-\Delta} - (2-\Delta)(1-\Delta) - 2e^{-\Delta/2}\sqrt{e^{-\Delta} + \Delta - 1}}{4e^{-\Delta} - (2-\Delta)^2} \tag{3.4}$$

In Figure 4 we compare (3.4) with the exact boundaries for $m = 2,3,4$. Note that for $\Delta < 1$ the curves are practically indistinguishable. Agreement for $\Delta > 1$ is not as good but there is nevertheless rapid convergence with increasing m.

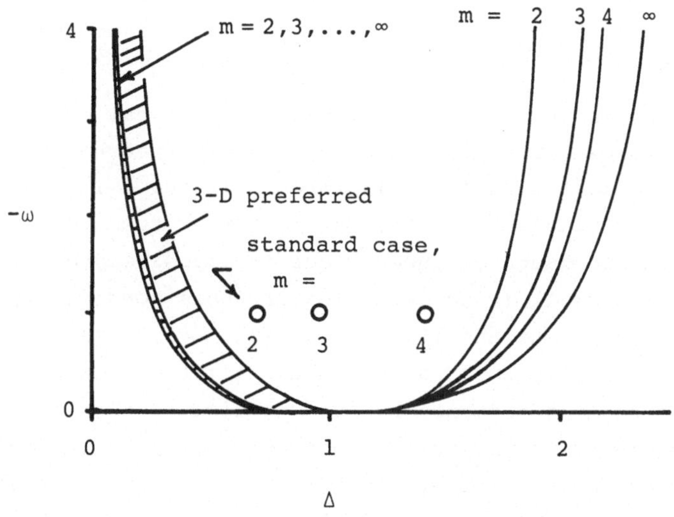

Figure 4. Instability boundaries in the 3-D, large m theory. Only the splitting modes are shown.

4. Axisymmetric modes, $m = 1$ modes, and stretching

We outline here a few partial results concerning the $m = 0$ and $m = 1$ modes, with emphasis on the effect which vortex stretching may have on the splitting process.

Since the $m = 1$ modes are neutrally stable in two dimensions, the results of the last section suggest that these modes might provide the dominant three-dimensional components to the splitting process. We have examined this problem only in the special case of the coated vortex where $\Omega \equiv \omega + \mu(1-\omega)$ is near zero. It is then possible to obtain an asymptotic dispersion relation incorporating three-dimensionality, by assuming that

$$\bar{s} \equiv \frac{2s}{\omega_1} = O(\Omega) , \qquad k^2 r_2^2 \equiv \delta = O(\Omega^2) . \tag{4.1}$$

The resulting expression for $\bar{s}$ is then

$$\bar{s}^2 + \Omega\bar{s} - \frac{\delta\omega}{4} (\mu + \omega \ln \mu) = 0 . \tag{4.2}$$

We omit details of this calculation. Note that (4.2) implies instability at finite δ whenever $\omega(\mu + \omega \ln \mu) < 0$. This establishes a case in which splitting can have a three-dimensional component, but unfortunately (4.2) does not yield an estimate of the axial wavelength of maximum growth rate. (It is likely that this occurs when kr_2 is of order unity.)

We note, in this connection, that the helical model shown in Figure 1 invokes three-dimensional modes to stretch active tubes and create new helical winding during the splitting. In the process the inviscid constraints on knottedness of vortex tubes are satisfied in the large (since daughter tubes occur in oppositely oriented pairs), but not in the small (since knotting of tubes does occur). If a three-dimensional instability such as an $m = 1$ mode cannot break these local helicity constraints in the limit of vanishing viscosity, on the time scale determined by tube vorticity, then it seems unlikely that any self-similar structure such as that of Figure 1 can be realistic. We therefore conjecture that if this fast reconnection of vortex tube cannot occur in the inviscid limit, then exactly self-similar γ-models are impossible. In the context of a β-model, this would imply a stochastic component in the cascade, which would in turn alter the intermittency corrections.

We have as yet not considered the $m = 0$ or axisymmetric modes, although the theory is far more developed in that case. Coated vortices are able to develop $m = 0$ instabilities with growth rates comparable to the $m \geq 2$,2-D instabilities, but the growth of these two kinds of modes responds quite differently to axial stretching of the vortex. If a columnar vortex is stretched (by a uniform, symmetric strain field) at a rate $a(t)$, the simultaneous evolution of 2-D modes in the transverse plane produces a modified transverse flow which can be mapped into n Eulerian 2-D flow by a simple change of variables [7]. With

$$(4.3a) \qquad A(t) = \exp\left(\int_0^t a(\tau)d\tau \right) ,$$

we set

$$(4.3b) \qquad r^* = A^{\frac{1}{2}}(t)r , \quad t^* = \int_0^t A(\tau)d\tau , \quad \theta^* = \theta ,$$

$$\psi(r,\theta,t) = \psi^*(r^*,\theta^*,t^*) .$$

In the starred variables the streamfunction satisfies the 2-D Euler equations. If t_0 is the termination time of the cascade, the simplest choice of active tubes with a γ-model is $a(t) = (t_0-t)^{-1}$. We then see from (4.3) that exponential growth of ψ^* relative to t^* leads to growth of ψ like a positive power of $t_0/(t_0-t)$.

For the $m = 0$ modes, however, no analogous mapping to classical dynamics occurs. With $z^* = z/A$, we have the unperturbed flow field

$$(4.4) \qquad \vec{u}_0 = (u_z, u_r, u\ ) = (a(t)z, -\frac{1}{2}\, a(t)r ,\ A^{\frac{1}{2}}V(A^{\frac{1}{2}}r)) .$$

We may then transform the equation for the perturbation radial velocity amplitude, $u(r,t)$, to starred variables. If the stars are dropped from the results expression we have

$$(4.5) \qquad \frac{\partial^2}{\partial t^2}\,[A^3 DD_* u] - k^2\,\frac{\partial^2 u}{\partial t^2} - k^2 \Phi u = 0 ,$$

where

$$(4.5b) \qquad \Phi = \frac{2V}{r} D_* V , \qquad D_* = \frac{1}{r} Dr .$$

We again focus on the choice $A(t) = t_\theta/(t_0 - t) \equiv T^{-1}$, and write (4.5) in the form

$$(4.6) \qquad \frac{\partial^2}{\partial T^2} [T^{-3} Lu - u] - \Phi u = 0 , \qquad L = \frac{1}{k^2} DD_* u .$$

Since the equation $Lu = 0$ has no nontrivial solutions which vanish at $r = 0 , \infty$, the series representing analytic functions of T just prior to the termination $T = 0$ have forms determined by the indicial equation in T. We thus obtain series representations

$$(4.7) \qquad u = u_j \equiv T^{2+j} \sum_{n=0}^{\infty} T^n u_j^{(n)} , \qquad j = 1,2 ,$$

for the asymptotic behavior of axisymmetric disturbances near the termination of the cascade.

Although these limited results are not conclusive on the point, it does appear that $m = 0$ modes are strongly suppressed by a factor no larger than $0(T^3)$, relative to the 2-D splitting modes. We have had to invoke stretching by as yet unspecified three-dimensional dynamics, but it seems difficult to go further than this without a more definite three-dimensional tube geometry.

5. The inertial range as "2+1"-dimensional

To the extent that the splitting of stretched vortices remains predominantly two-dimensional, it can be roughly approximated as a two-dimensional cascade, upon which is superimposed a one-dimensional elongation. While such a description cannot be exact, it is a natural hypothesis to consider when vortex tubes are the basic structures used to describe the three-dimensional inertial range.

If a β-model is invoked in three dimensions, the preceding viewpoint then induces a corresponding two-dimensional β-model. It follows that the three-dimensional intermittency correction is associated with a corresponding correction to the k^{-3} energy spectrum in the enstrophy cascade.

To compute this relationship between the two cascades we utilize the notation of the γ-models [9]. If M is the number of active daughter tubes and s is the stretching factor (both are 2 in the case shown in Figure 1), then $N^* = M$ is the number of active eddies produced at each step of the two-dimensional cascade. These decrease in size by a scale factor $\lambda^* = \left(\frac{\beta}{M}\right)^{\frac{1}{2}}$, where β is the volume reduction of active components in the β-model. In the enstrophy cascade the energy per unit area scales by the factor

$$(5.1) \qquad u_E = N^*\lambda^{*4} = \frac{\beta^2}{M} = (\lambda^*)^{e_2-1} = \left(\frac{\beta}{M}\right)^{\frac{e_2-1}{2}},$$

where the two-dimensional energy spectrum has the form $E_2(k) \propto k^{-e_2}$. Using the relation $\beta = \sqrt{M}/s$ from the γ-model we obtain

$$(5.2) \qquad e_2 = \frac{10 \ln s + \ln M}{2 \ln s + \ln M}.$$

The three-dimensional energy spectrum exponent $e_3 = \frac{5}{3} + \frac{(3-D)}{3}$ is determined in the γ-model by the dimension

$$(5.3) \qquad D = \frac{8 \ln s + 5 \ln M}{4 \ln s + \ln M},$$

yielding

$$(5.4) \qquad e_3 = \frac{8 \ln s + \ln M}{4 \ln s + \ln M}.$$

It is seen from (5.2) and (5.4) that e_2 is a function of e_3 given by

$$(5.5) \qquad e_2 = \frac{3e_3 - 1}{3 - e_3}.$$

A few of these values are shown in Table 2.

e_2	3	3.67	4	5
e_3	5/3	1.8	1.86	2
D	3	2.6	2.43	2

Table 2. The correspondence of spectra in the "2+1" model

Our result is, therefore, that the three-dimensional intermittency which occurs in the γ-models from the presence of residual inactive tubes can be consistent with near two-dimensional splitting only if the latter exhibits some dynamical intermittency. We emphasize that it is not at all clear at present how this hypothesis should be tested. Perhaps two-dimensional turbulence is the right model, but it would then appear to be inappropriate to model the cascade by active vortex tubes. It could also be that the "2+1" viewpoint should use an ensemble of splittings of essentially isolated vortex tubes (which have been stretched by various amounts during the course of the cascade) which implies a different simulation from that usually associated with two-dimensional turbulence. It is hoped that further work along these lines will suggest useful comparisons with the two-dimensional simulations discussed elsewhere in this volume.

6. Simulation of nonlinear evolution

As a general rule linearized stability theory cannot be relied upon to say much about evolution of unstable systems in the nonlinear range. It is therefore essential to simulate the splitting process numerically. We have made some preliminary and tentative calculations in two dimensions using the vortex method, and we describe these in the present section.

We used 197 vortices initially arranged on a square lattice, 129 carrying circulation +1/61 forming a central core, together with a coat of 68 vortices of strength -1/61. The lattice density is 64 sites/unit area. Thus the core is a rough representation of the standard case for a core of unit radius carrying a circulation +1. Each vortex is smoothed slightly by adopting the velocity field $(u,v) = (-y,x)(r^2+d)^{-1}$; in the calculations described below we took $d = .0005$.

This initial arrangement was subjected to a small initial perturbation. The positions of vortices are changed by an amount $\varepsilon(\psi_y,-\psi_x)$ where

$$\psi = \psi_1 + c\psi_2 \tag{6.1a}$$

$$\psi_1 = \begin{cases} y\cos\theta - x\sin\theta\,, & 0 \le r < .8\,, \\ 1.828(r^{-2}-1)(y\cos\theta - x\sin\theta)\,, & .8 \le r \le 1\,. \end{cases} \tag{6.1b}$$

$$(6.1c) \qquad \psi_2 = \begin{cases} x^2 - y^2 , & 0 \le r < .8 , \\ -.46(x^2-y^2) + 1.1\, xy + .61(x^2-y^2)r^{-4} - .46\, xy\, r^{-4} , & \\ & .8 \le r \le 1 . \end{cases}$$

This displacement, which depends on the parameters ε, c, and θ, consists of a linear combination of one $m = 1$ mode and the unstable $m = 2$ mode.

With the perturbed core as initial condition, the positions of the vortices are then advanced by the vortex method using an elementary second-order algorithm. For the standard case one computes an area-averaged eddy rotation time of 25 for the unperturbed core. In Figure 5 we show the vortex configuration at $t = 10$ for several choices of parameters in the perturbation. It is clear that significant breakup of the core has already occurred, even though it is too early to see well-defined isolated structures. We indicate in each case two oriented contours carrying oppositely signed circulation with numbers of vortices in roughly the correct proportion for forming two coated daughter vortices. In the standard case these daughters would contain 70 vortices, while the average number within the indicated contours is 35. This suggests that evolution to one or two eddy-turnover times would be needed to produce mature daughter vortices, should these in fact ever occur.

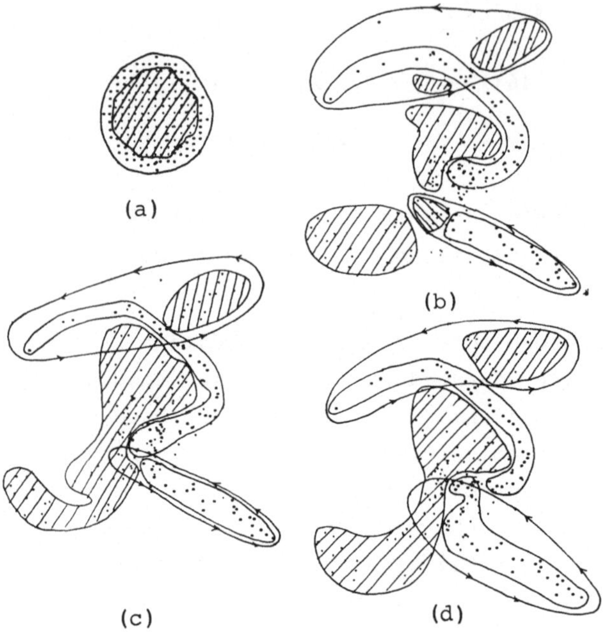

Figure 5. The splitting of a coated vortex.
(a) $t=0$, perturbed vortex with $\varepsilon = .02$, $c=1$, $\theta = 45°$. In (b), (c) and (d) the core is shown at $t=10$. Shaded areas consist of mainly + vortices, clear areas of mainly - vortices. The two oriented contours contain the numbers of vortices n_+^u, n_-^u for the upper contour, n_+^ℓ, n_-^ℓ for the lower one.

(b) $\varepsilon = .02$, $c = 1.$, $\theta = 45°$, $n_+^u = 21$, $n_-^u = 12$, $n_+^\ell = 14$, $n_-^\ell = 24$.

(c) $\varepsilon = .02$, $c = .5$, $\theta = 45°$, $n_+^u = 22$, $n_-^u = 11$, $n_+^\ell = 12$, $n_-^\ell = 24$

(d) $\varepsilon = .02$, $c = 1.$, $\theta = 0$, $n_+^u = 15$, $n_-^u = 8$, $n_+^\ell = 16$, $n_-^\ell = 33$.

7. Toward simulation of cascades

Although three-dimensional cascades must eventually include the stretching of vortex tubes, a preliminary step toward direct simulation involves only two-dimensional dynamics and a sequence of calculations of the kind outlined in the last section. The question then arises as to how a rescaling procedure should be carried out to reduce all steps of the cascade to an initial-value problem of universal form. One method would be to consider a configuration space consisting of a coated vortex perturbed by a linear combination of a finite number of modes. For example, in Section 6 we dealt with the four-parameter family determined (up to a change of scale) by μ, ε, c, and θ. As such a vortex evolves, subsidiary structures might eventually be formed which are essentially isolated from the other remnants of the parent structure. From such daughter structures one can deduce new values of the parameters, which can then be used to establish the initial configuration for the next step. From the sequence of isolated core problems we may set up a mapping $p_n \to p_{n+1}$ for the vector p_n for the configuration at the nth step. Scale factors for the core size are also then defined and can be used in the computation of intermittency. Some measure of activity or inactivity would have to be devised, presumably based upon p_n, and generation sizes M_n deduced. We cannot guess what universal configurations should actually be used in the place of the hypothetical perturbed coated core, but likely candidates might be suggested by watching the evolution of cores over several eddy-turnover times.

To gain experience with such renormalization techniques it would be useful to have an analogous dynamics in one space dimension. One can imagine the construction of various fractal curves in the plane as a sequential process terminating in a finite time, but it is another matter to devise a dynamical model which yields a cascading geometry within solutions of an initial-value problem. One possible approach is to move an oriented curve $C(t)$ in the plane so that a given material point moves according to

$$\frac{d\vec{x}}{dt} = -V\vec{n} \tag{7.1}$$

where $\vec{n}$ is a unit normal. A dynamical model is obtained by specifying V in terms of the geometry of the curve. Constructions of this kind occur, for example, in studies of interfacial pattern formation [14].

We have studied the example

$$V = \kappa^3 + b\,\frac{\partial^2\kappa}{\partial s^2} \tag{7.2}$$

where κ is curvature, s is arc length, and b is a positive constant. With (7.2) the dynamical model has an invariant scaling

$s \to A^{-1}s$, $\kappa \to A\kappa$, which would appear to be a necessary (although not sufficient) condition for a geometric cascade. One interesting property of (7.2) is that one elementary solution, namely an expanding circle, is linearly unstable to a finite m-window of modes depending upon b; there is thus a close analogy to the stability properties of the coated vortex. Other exact solutions correspond to $V = 0$ and have a periodic structure with amplitude-dependent wavelength. Numerical studies will be needed to determine if the curve simply locks onto structures determined by initial conditions, or if cascading to arbitrarily small scales actually occurs.

8. Concluding Summary

The inertial-range model studied in the present paper rests upon the analysis of stability and breakup of a class of vortex cores. Although the coated vortices present an oversimplified organization of eddy dynamics, numerical simulation using two-dimensional point vortices may be helpful in selecting more realistic structures. The most difficult question concerns the stretching of vortex tubes, which must be carried out in conjunction with the evolving core structure. It is suggested that the inertial range might be fruitfully viewed as a superposition of two-dimensional core dynamics and longitudinal stretching of tubes. Numerical simulations of these two processes might allow a kind of "calibration" of the inertial range, in which rescaling transforms a cascade into a discrete map of low dimension. It would be useful to apply such a procedure to a cascading dynamical model in one space dimension.

The author would like to thank P. Garabedian for stimulating discussions. This research was supported by the National Science Foundation under contract DMS-831 2229 at New York University.

REFERENCES

1. Burgers, J.M., Adv. in Appl. Mech. Vol. 1, 171-99, 1948.

2. Corrsin, S., Phys. Fluids 5, 1301-2, 1962.

3. Tennekes, H., Phys. Fluids 11, 669-71, 1968.

4. Saffman, P.G., Lecture on Homogeneous Turbulence, in Topics in Nonlinear Physics, N. J. Zabusky, ed., Springer-Verlag, 485-614, 1968.

References, continued

5. Synge, J.L. and Lin, C.C., Trans. Roy. Soc. Canada Sec. 3, 45-79, 1943.

6. Townsend, A.A., Proc. Roy. Soc. Lond. A, 208, 534-540, 1951.

7. Lundgren, T.S., Phys. Fluids 25, 2193-2203, 1982.

8. Frisch, U., Sulem, P.-L., and Nelkin, M., J. Fluid Mech. 87, 719-736, 1978.

9. Childress, S., Geophys. Astrophys. Fluid Dynamics 29, 29-64, 1984.

10. Kelvin, Lord, Phil. Mag. 5, 155-68, 1880.

11. Howard, L.N. and Gupta, A.S., J. Fluid Mech. 14, 463-476, 1962.

12. Leibovich, S. and Stewartson, K., J. Fluid Mech. 126, 335-356, 1983.

13. Drazin, P.G. and Reid, W.H., Hydrodynamic Stability, Cambridge University Press, 1981.

14. Ben-Jacob, E., Nigel Goldenfeld, J.S. Langer, and Gerd Schön, Phys. Rev. Lett. 51, 1930-32, 1983.

A STOCHASTIC SUBGRID MODEL FOR SHEARED TURBULENCE

J.P. Bertoglio
Laboratoire de Mécanique des Fluides
Ecole Centrale de Lyon
36 Avenue Guy de Collongue
69131 Ecully Cedex - France.

Abstract

A new subgrid model for homogeneous turbulence is proposed. The model is used in a method of Large Eddy Simulation coupled with an E.D.Q.N.M. prediction of the statistical properties of the small scales. The model is stochastic in order to allow a "desaveraging" of the informations provided by the E.D.Q.N.M. closure. It is basedon stochastic amplitude equations for two-point closures. It allows back-flow of energy from the small scales, introduces stochasticity into L.E.S., and is well adapted to non isotropic fields. A few results are presented here.

1. Introduction

Direct Numerical Simulation is certainly today one of the major tools for the study and prediction of turbulence. However, it is known that, for flows at large Reynolds numbers it is not feasible to make a Full Simulation over the whole range of scales of the turbulent spectrum. Only the large scales can be simulated, and the small eddies, or subgrid scales, have to be modelled (see for exemple Ferziger, 1982). From a spectral point of view, it is possible to explicity take into account the turbulent motion up to a wavenumber cutoff : K_c and the terms representing the exchanges across K_c have to be parameterized.

Two-point closures provide an helpful analytical framework in which to investigate and developp subgrid models. They are believed to take correctly into account the exchange of energy between eddies of various sizes, accordingly, they are considered as valuable tools for the evaluation of energy flows across the wave number cutoff. Kraichnan (1976), Leslie et al. (1979) and Chollet et al. (1981) have particularly used two-point theories to test and improve existing models in the caseof isotropic turbulence.

In Kraichnan (1976), Leslie (1979) and Chollet (1981) the existence of a universal behaviour for the small scale spectrum was assumed, which implies that K_c was supposed to be situated in the inertial range. It was then possible to derive subgrid models in which the parameterization was only made in terms of large scale quantities. Such models can therefore be used in L.E.S. without requiring an explicit computation of the small scale spectrum.

More recently, Chollet (1983) and Aupoix et al. (1983) have proposed method coupling L.E.S. and E.D.Q.N.M. This type of methods simultaneously involves a direct simulation of the large eddies and a two-point closure computation of the small scales. At each time step, an information concerning the whole range of the spectrum is then available. Such methods are well adapted to predict either situations in which the small scales are not in equilibrium or situations in which the wave number cutoff is not in the inertial range. They can also take into account finite Reynolds number effects.

However an important problem is encountered when applying two-point closures to subgrid models : the informations given by the closures are only statistically averaged informations, whereas one needs to account for the effect of small eddies on the particular realization of the field which is being simulated. The problem is known as the desaveraging problem (Basdevant et al. (1978)).

A simple way to solve the desaveraging problem is to introduce the concept of eddy viscosity. One of the advantages of subgrid models based on this concept is in particular that they ensure a drain of energy from the large scales to the small scales. In the case of isotropic three-dimensional turbulence eddy viscosity formulations were found to be consistent with the classical two-point closures (Kraichnan, 1976 ; Leslie, 1979 ; Chollet, 1981), and to lead to good predictions of the energy decay. The limitations on the use of eddy viscosity are however known. They have been pointed out by Kraichnan (1976) in the case of two-dimensional isotropic turbulence. For three-dimensional non isotropic turbulence we have shown, in an earlier study (Bertoglio and Mathieu, 1983), that the representation by an eddy viscosity was only justified for one part of the transfer.

Two deficiencies of eddy viscosity have in particular been pointed out : first, it does not provide for the possibility of back-flow of energy from the small scales, secondly it is not stochastic and does not account for the random forcing of small scales on large scales (Rose,1977). In the case of a non isotropic turbulence subjected to a uniform mean shear flow, the importance of the first deficiency was found to be particularly large (see Bertoglio & Mathieu, 1983). As for the lack of stochasticity, its consequences appear when the problem of predictability is considered. Let us for example consider two turbulent fields, which are initially supposed to have identical large scales and to differ only in their small scales. If the initial "error" is situated in the subgrid range of the spectrum, the eddy viscosity will never allow the "error" to contaminate the large eddies : the large scales of the two fields will always remain identical. This is known not to be a correct prediction.

In this paper, we present a new subgrid model in which the "desaveraging" operation is made by introducing a stochastic term. The model allows back-flow of energy from the small scales. It is based on the stochastic models for the analytical theories of turbulence (Kraichnan, 1961, 1971 ; Leith, 1971 ; Frisch et al., 1974). The cases of stochastic terms with and without memory are investigated.

The model is used in a "coupled" method in which the small scale spectrum is taken into account by the Eddy Damped Quasi Normal closure (Orszag, 1970), and the large scales are simulated by using a spectral algorithm. Results are first given in the case of isotropic turbulence. Extensions to the study of predictability and non isotropic sheared turbulence are shortly mentioned and a few results are presented.

2. Stochastic models for two-point closures

First introduced by Kraichnan (1961), stochastic models are equations for the turbulent fluctuation which are different from the original Navier-Stokes equations in a sense that the original non linear term has been altered in a random fashion. They are helpful tools for turbulence theory since they share with the original equations many interesting features. A remarkable property of stochastic models is that they lead to a closed set of equations for averaged quantities, such as double velocity correlations, without having to introduce further assumptions.

In particular, a stochastic model leading to the equations of the

D.I.A. was proposed by Kraichnan (1970), and Leith (1971) presented a stochastic equation corresponding to the E.D.Q.N.M. closure. We recall here both models. In the case of D.I.A. the amplitude equation for an isotropic turbulent field in which we include a solenoidal stirring force f_i in order to permit stastical stationarity, is :

$$\left\{\frac{\partial}{\partial t} + \nu K^2\right\} u_i(\vec{K},t) + \int_0^t \sigma(K,t,s)\; u_i(\vec{K},s)\, ds \;=\; q_i(\vec{K},t) + f_i(\vec{K},t) \quad (1)$$

where

$$\sigma(K,t,s) = \pi K \iint_\Delta b_{KPQ}\; G(P,t,s)\, U(Q,t,s)\, P Q\, dP\, dQ \quad (2)$$

is a quantity characterizing a damping and $q_i(\vec{K},t)$ is a stochastic force given by :

$$q_i(\vec{K},t) = -i\, P_{ijm}(\vec{K}) \sum_{\vec{P}+\vec{Q}=\vec{K}} \xi_j(\vec{P},t)\, \xi'_m(\vec{Q},t) \quad (3)$$

where ξ and ξ' are stochastic variables statistically independent of each other and such that :

$$\langle \xi_i(\vec{K},t)\, \xi_j(\vec{R},t')\rangle = \langle \xi'_i(\vec{K},t)\, \xi'_j(\vec{R},t')\rangle = \langle u_i(\vec{K},t)\, u_j(\vec{R},t')\rangle\,, \quad (4)$$

$G(P,t,s)$ is the average infinitesimal response fonction, $U(K,t,t')$ is the modal time covariance :

$$\left(\delta_{ij} - \frac{K_i K_j}{K^2}\right)\frac{U(K,t,s)}{2} = \lim_{L\to\infty}\left(\frac{L}{2\pi}\right)^3 \langle u_i(\vec{K},t)\, u_j(-\vec{K},t)\rangle\,,$$

the integration Δ is over all P and Q such that K, P, Q form a triangle, $P_{ijm}(\vec{K})$ satisfies :

$$P_{ijm}(\vec{K}) = \frac{1}{2}\left\{ K_\ell\left(\delta_{ij} - \frac{K_i K_j}{K^2}\right) + K_j\left(\delta_{i\ell} - \frac{K_i K_\ell}{K^2}\right)\right\}$$

and b_{KPQ} is a coefficient depending on the geometry of the triad :

$$b_{KPQ} \quad (P/K)(xy + z^3)$$

x, y, z are the interior-angle cosines opposite K, P, Q, respectively. L is the side of a cyclic box.

The model proposed by Leith for E.D.Q.N.M. can be considered as a degenerated form of (1), in which all the quantities are evaluated at the same time, and q_i is a white-noise process :

$$\left\{\frac{\partial}{\partial t}+\nu K^2+\sigma(K,t)\right\} u_i(\vec{K},t) = q_i(\vec{K},t) + f_i(\vec{K},t) \tag{5}$$

where

$$\sigma(K,t) = \pi K \iint_\Delta b_{KPQ}\ \theta_{KPQ}(t)\, U(Q,t)\, PQ\, dP\, dQ \tag{6}$$

and

$$q_i(\vec{K},t) = -\, i\, P_{ijm}(\vec{K}) \sum_{\vec{P}+\vec{Q}\ \vec{K}} w(t)\left[\theta_{KPQ}(t)\right]^{1/2} \xi_j(\vec{P},t)\ \xi'_m(\vec{Q},t) \tag{7}$$

in which w is a white noise process such that :

$$\langle w(t)\ w(t')\rangle = 2\,\delta(t-t')$$

It is worth noting that in the E.D.Q.N.M. formulation, the memory effects originally appearing in the D.I.A. equations have been artificially replaced by introducing a characteristic time $\theta_{KPQ}(t)$. We shall use here the expression (Pouquet et al. 1975) :

$$\theta_{KPQ}(t) = \frac{1 - e^{-\left[\eta(K,t)+\eta(P,t)+\eta(Q,t)\right]t}}{\eta(K,t)+\eta(P,t)+\eta(Q,t)} \tag{8}$$

where

$$\eta(K,t) = .355 \left\{\int_0^K P^2 E(P,t)\, dP\right\}^{1/2} + \nu K^2 \tag{9}$$

3. Application to subgrid models

As pointed by Kraichnan (1970), these stochastic model equations are a logical starting-point for using two-point closures for deriving subgrid scale models. A straighforward way to derive such a model would be to simply take into account the non linear interactions involving at least one wavenumber larger than K_c by a stochastic amplitude equation, meanwhile interactions involving only wavenumbers smaller than K_c would be explicitely computed in the original Navier Stokes formulation. In the case of E.D.Q.N.M., this would lead to :

$$\left\{\frac{\partial}{\partial t}+\nu K^2+\sigma^>(K,t)\right\} u_i^<(\vec{K},t) = -i P_{ijm}(\vec{K}) \sum_{\substack{\vec{P}+\vec{Q}=\vec{K} \\ P<K_c \text{ and } Q<K_c}} u_j^<(\vec{P},t)\, u_m^<(\vec{Q},t) + q_i^>(\vec{K},t) + f_i(\vec{K},t) \qquad (10)$$

for the explicit scales $u_i^<$ $(K<K_c)$, and :

$$\left\{\frac{\partial}{\partial t}+\nu K^2+\sigma(K,t)\right\} u_i^>(\vec{K},t) = q_i(\vec{K},t) + f_i(\vec{K},t) \qquad (11)$$

for the subgrid modes $u_i^>$ $(K>K_c)$, where σ and q_i are respectively given by (6) and (7), where :

$$q_i^>(\vec{K},t) = -i P_{ijm}(\vec{K}) \sum_{\substack{\vec{P}+\vec{Q}\ \vec{K} \\ P>K_c \text{ and/or } Q>K_c}} w(t)\left[\theta_{KPQ}(t)\right]^{1/2} \xi_j(\vec{P},t)\, \xi'_m(\vec{Q},t) \qquad (12)$$

and where :

$$\sigma^>(K,t) = \pi K \iint_{\Delta'} b_{KPQ}\, \theta_{KPQ}(t)\, U(Q,t)\, PQ\, dP\, dQ \qquad (13)$$

the integration being all over the part of the P, Q plane where $P>K_c$ and/or $Q>K_c$ and where K, P, Q can form a triangle. $U(Q,t)$ is defined by :

$$U(Q,t) = U(Q,t,t)$$

We can at this point remark that the only informations concerning the subgrid scales in the supergrid equation (10) are statistically averaged, therefore equation (11) can be replaced by the master equation for the kinetic energy spectrum $E(K,t)$:

$$\left\{\frac{\partial}{\partial t}+2\nu K^2\right\} E(K,t) = tr(K,t) \qquad (14)$$

in which tr is the usual E.D.Q.N.M. transfer term .

Unfortunately the complete evaluation of $q_i^>$ would require to much computational effort. Summing over all the subgrid-modes would result in a task comparable with the one required for a Full Direct Simulation. In the case of a D.I.A. type model, storage problems would furthermore be introduced by the explicit presence of memory effects.

Our prescription is to replace $q_i^>(\vec{K},t)$, which is a sum of products of stochastic processes, by a single stochastic process wich would satisfy relation (12) only in a statistically averaged way. If we name $T_i^+(\vec{K},t)$ this new stochastic process, the equation for the explicit scales (10) then becomes :

$$\left\{\frac{\partial}{\partial t}+\nu K^2+\sigma^>(K,t)\right\} u_i^<(\vec{K},t) = -i\,P_{ijm}(\vec{K}) \sum_{\substack{\vec{P}+\vec{Q}=\vec{K}\\ P<K_c \text{ and } Q<K_c}} u_j^<(\vec{P},t)\, u_m^<(\vec{Q},t) + T_i^+(\vec{K},t) + f_i(\vec{K},t) \qquad (K<K_c) \tag{15}$$

and the condition on T_i^+ can be written as :

$$\langle T_i^+(\vec{K},t)\, T_j^+(\vec{R},t')\rangle = \langle q_i^>(\vec{K},t)\quad q_j^>(\vec{R},t')\rangle \tag{16}$$

Using the definition of $q_i^>$, we can evaluate the two time correlations :

$$\langle q_i^>(\vec{K},t)\, q_j^>(\vec{R},t')+q_j^>(\vec{R},t)\, q_i^>(\vec{K},t')\rangle = \lim_{L\to\infty}\left(\frac{2\pi}{L}\right)^3 \delta(t-t')\left(\delta_{ij}-\frac{K_iK_j}{K^2}\right)\frac{t_r^{+>}(K,t)}{4\pi K^2} \quad \text{for } \vec{K}=-\vec{R}$$

$$= 0 \quad \text{for } \vec{K}\neq-\vec{R}$$

where the quantity $t_r^{+>}$, often called backscatter or input term, is given by :

$$t_r^{+>}(K,t) = \iint_{\Delta'} \theta_{KPQ}(t)\; b_{KPQ}\,\frac{K^3}{PQ}\; E(P,t)\, E(Q,t)\, dP\, dQ \tag{17}$$

Replacing in equation (16), we obtain:

$$\langle T_i^+(\vec{K},t)\, T_j^+(\vec{R},t')+T_j^+(\vec{R},t)\, T_i^+(\vec{K},t')\rangle = \left(\frac{2\pi}{L}\right)^3 \delta(t-t')\left(\delta_{ij}-\frac{K_iK_j}{K^2}\right)\frac{t_r^{+>}(K,t)}{4\pi K^2} \quad \text{for } \vec{K}=\vec{R} \tag{18}$$

$$= 0 \quad \text{for } \vec{K}=-\vec{R}$$

in which L is assumed to be large.

We have to specify now the statistical distribution of T_i^+. $q_i^>$ being a sum of a large number of stochastic terms, it seems reasonable to assume T_i^+ to be gaussian. We shall discuss how to generate T_i^+ in appendix 1.

The final set of equations is then (13), (14), (15), (17) and (18), together with the relations characterizing the stochastic process (A1.1) and (A1.2). Equation (15) is the equation governing the evolution of the large scales, (14) is the rate equation for the small scale spectrum, and relations (13), (17) and (18) are characterizing the coupling terms : they take into account interactions across the wavenumber cutoff.

The operator $\langle \; \rangle$ appearing in (18), as well as in the definition of U for $K<K_c$:

$$\left(\delta_{ij} - \frac{K_i K_j}{K^2}\right) \frac{U(K,t)}{2} = \left(\frac{L}{2\pi}\right)^3 \left\langle u_i^<(\vec{K},t)\, u_j^<(-\vec{K},t) \right\rangle ,$$

denotes ensemble average. In the case of L.E.S. we would have to consider N realizations, corresponding to N different initial conditions and N different sets of random numbers in the generation of T_i^+, and average over the results. On the basis of practical considerations, it was set N equal to 1.

An advantage of the model immediately appears : it will introduce stochasticity in L.E.S. as we shall see in section 5 when the model will be applied to the study of predictability. A second advantage appears when anisotropic turbulence is considered. The equivalent of equation (18) can then be written as :

$$\left\langle T_i^+(\vec{K},t)\, T_j^+(\vec{R},t') + T_j^+(\vec{R},t)\, T_i^+(\vec{K},t') \right\rangle = \left(\frac{2\pi}{L}\right)^3 \delta(t-t')\, T_{ij}^{+>}(\vec{K},t) \quad \text{for } \vec{K}=-\vec{R} \qquad (19)$$
$$= 0 \quad \text{for } \vec{K}\neq-\vec{R}$$

in which $T_{ij}^{+>}$ is the anisotropic backscatter (for a complete expression of $T_{ij}^{+>}$ see Bertoglio & Mathieu, 1983). Since it is possible to find T_i^+ satisfying (19), it then becomes possible to preserve completely the anisotropic characteristics of $T_{ij}^{+>}$ in the subgrid model. This was impossible when using the classical concept of eddy viscosity as we pointed out in an earlies paper (Bertoglio & Mathieu, 1983).
We shall shortly discuss the application of the model to non isotropic sheared turbulence in section 5.

4. Introduction of a stochastic term with memory

The presence of a white-noise process in T_i^+ can be judged to be rather unphysical, and one could try to improve the subgrid model presented in section 3 by introducing a memory effect in the stochastic process. Since memory is present in the model amplitude equation for D.I.A., relation (3) constitutes a logical starting point to make such an improvement.

It is however not our purpose here to work completely in the framework of D.I.A., computation of the small scales spectrum would be to cumbersome to permit extensions to anisotropic turbulence. Besides D.I.A. is not statistically Galilean Invariant.

Let us make crude assumptions and propose the following approach. We assume that the memory in T_i^+ is an exponentially decreasing function of time :

$$\langle T_i^+(\vec{K},t)\, T_j^+(-\vec{K},t')\rangle = \langle T_i^+(\vec{K},t)\, T_j^+(-\vec{K},t)\rangle\, e^{-\frac{t-t'}{t_{NL}(K)}} \quad (t \geq t') \tag{20}$$

in which $t_{NL}(K)$ is a characteristic time.

It is then possible to determine $t_{NL}(K)$ by using the D.I.A. stochastic model equation. This can be done in the case of a stationary turbulence, if we assume that the two-time velocity correlations are exponentially decreasing functions of time :

$$\langle u_i(\vec{K},t)\; u_j(-\vec{K},t')\rangle = \langle u_i(\vec{K},t)\, u_j(-\vec{K},t)\rangle\; e^{-\eta(K)(t-t')} \quad (t \geq t') \tag{21}$$

The calculation is presented in appendix 2.

The determination of the one time correlation $\langle T_i^+(\vec{K},t)\, T_j^+(-\vec{K},t)\rangle$ is some what more complex that in the case of the white-noise process, since E.D.Q.N.M. cannot now directly provide this quantity. Relation (18) was valid only for a white-noise. We have here to satisfy :

$$\langle T_i^+(\vec{K},t)\; u_j^<(-\vec{K},t) + T_j^+(-\vec{K},t)\; u_i^<(\vec{K},t)\rangle = \left(\frac{2\pi}{L}\right)^3 \left(\delta_{ij} - \frac{K_i K_j}{K^2}\right) \frac{t_r^{+>}(K,t)}{4\pi K^2} \quad (K < K_c) \tag{22}$$

where $t_r^{+>}$ is still given by (17). Equation (22) means that the amount of energy injected in the super-grid modes by the stochastic process must be statistically equal to its E.D.Q.N.M. determination.

Since the velocity fluctuation appears in it, equation (22) cannot be used in a straightforward way. We have to write the left hand side in the form

$$\langle T_i^+(\vec{K},t)\, u_j^<(-\vec{K},t) + T_j^+(-\vec{K},t)\, u_i^<(\vec{K},t) \rangle = \int_0^t G(\vec{K},t,t') \times \langle T_i^+(\vec{K},t)\, T_j^+(-\vec{K},t') + T_j^+(-\vec{K},t)\, T_i^+(\vec{K},t') \rangle\, dt' \quad (K<K_c) \qquad (23)$$

in which the infinitesimal response function was assumed to be statistically independent of T_i^+.

Assuming furthermore that :

$$G(\vec{K},t,t') = e^{-\frac{t-t'}{\theta_G(K)}} \qquad (K \leq K_c\,,\; t \geq t') \qquad (24)$$

and using (20) and (23) give for a stationary turbulence :

$$2\langle T_i^+(\vec{K},t)\, T_j^+(-\vec{K},t) \rangle = \left(\frac{2\pi}{L}\right)^3 \left[\frac{1}{\theta_G(K)} + \frac{1}{t_{NL}(K)}\right] \left(\delta_{ij} - \frac{K_i K_j}{K^2}\right) \frac{t_r^{+>}(K,t)}{4\pi K^2} \qquad (25)$$

$\theta_G(K)$ is a characteristic time of the response of the large scale field, it can therefore be estimated at each time step from the result of L.E.S.

$t_{NL}(K)$ is the correlation time of the stochastic process. It can be determined by using (A2.3). It could also be arbitrarily fixed and used as the only adjustable parameter in the model. In the limit $t_{NL} \longrightarrow 0$ equations (25) and (20) degenerate into equation (18), which means that the memory model degenerated into the white-noise model.

The damping term appearing in (15)

$$\sigma^{>}(K,t)\ u_i^{<}(\vec{K},t)$$

is not modified here. We do not use a term involving a time integration such as the one encountered in (1) on the basis of practical considerations.

The final set of equations for the model with memory is then : (13), (14), (15), (17), (20), (25), (A1.1) and (A1.2), together with (A2.3) which specifies the correlation time of the subgrid term.

5. Results

The model has first been used to compute the decay of an isotropic turbulence. A numerical code has been written for the simulation of the large eddies, following the method proposed by Orszag & Patterson (1972). We used a 16^3 grid. A few runs were performed with a 32^3 grid. The initial data are generated as suggested by Rogallo (1981). Since the triple velocity correlations are initially equal to zero, we used relation (8) to specify the E.D.Q.N.M. characteristic time. This form ensures initial compatibility between L.E.S. and E.D.Q.N.M. since it also corresponds to $\langle uuu \rangle = 0$ at $t = 0$ (Aupoix, personnal communication).

The first results showed a discontinuity in the slope of the energy spectrum at the wavenumber cutoff. Our interpretation was that the effect of the damping term $\sigma^{>}$ on the characteristic correlation times of the supergrid modes was to strong to be compatible with E.D.Q.N.M.. Transfers between supergrid modes were then underestimated in comparison with E.D.Q.N.M., and accordingly they were unable to balance the energy drain across the cutoff in a correct way.

This undesirable behaviour was cured by a modification of the E.D.Q.N.M. characteristic time for the supergrid modes. Our proposition is to identify $\eta(K,t)$ with the inverse of the characteristic time of the average response fonction, a quantity which is directly deduced from the computed scales :

$$\eta(K,t) = \frac{1}{\theta_G(K,t)} \qquad \text{for } K \leq K_c$$

In order to ensure continuity of η at $K = K_c$ we furthermore write :

$$\eta(K,t) = \left[\frac{1}{\Theta_G^2(K_c,t)} + [.355]^2 \int_{K_c}^{K} P^2 E(P,t)\,dP\right]^{1/2} + \nu K^2$$

$$\text{for } K > K_c$$

With this modification, the behaviour of the energy spectrum $E(K,t)$ appears to be correct. In figure 1 we have plotted a spectrum obtained in the case of a freely evolving isotropic turbulence. The 16^3 grid has been used and the stochastic term includes memory effects (t_{NL} arbitrarily fixed).

On figure 2 the backscatter predicted by E.D.Q.N.M. $t_r^{+>}(K,t)$ is compared with the effective value actually injected in L.E.S. by the stochastic term, called $t_{r\,eff}^{+>}$. The agreement is reasonably good. In this case the 32^3 grid has been used. The stochastic process is without memory ($t_{NL} = \Delta t$). In the case of a stochastic process including memory effects, the agreement would be less satisfactory since, for a freely evolving turbulence, equation (25) is not exactly statisfied. On figure 2, the term characterizing the drain of energy from the large scales to the small scales :

$$t_r^{->}(K,t) = 2\,\sigma^{>}(K,t)\;E(K,t)$$

has also been plotted.

Results concerning the turbulent kinetic energy decay are presented on figure 3. Evolutions of the turbulent kinetic energy are plotted. We compare the kinetic energy of the supergrid field to the kinetic energy of the full field. The stochastic process used here includes memory effects, the characteristic time t_{NL} is given by (A2.3).

The advantage of including memory in T_i^+ does not appear clearly when comparing the results. Quantities like double velocity correlations, energy spectra, are only very slightly modified. Only the skewness of the large scale field seems to be affected.

It is beyond the scope of the present paper to give a full presentation of the extensions to the study of predictability and to the

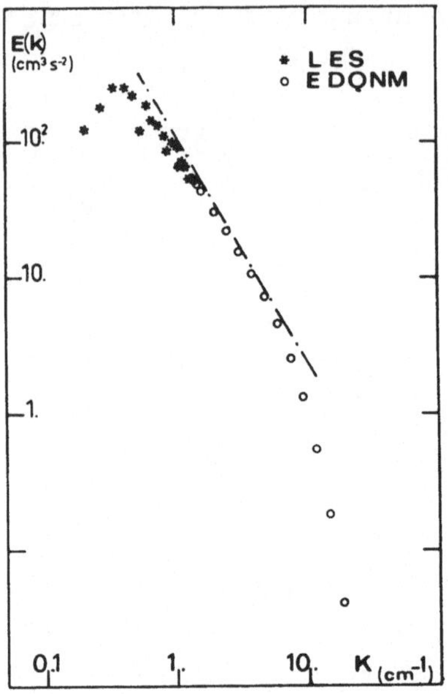

Figure 1 - Kinetic energy spectrum. Results of L.E.S. : *, and E.D.Q.N.M. o ; t = 0.42s, K_c = 1.21 cm^{-1}, 16^3 grid, t_{NL} = 0.15 s.

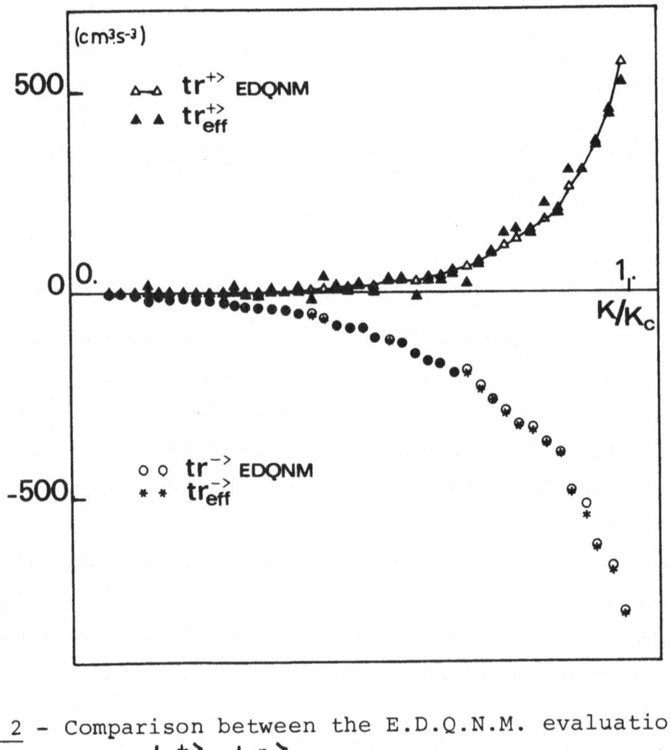

Figure 2 - Comparison between the E.D.Q.N.M. evaluation of the subgrid terms $t_r^{+>}$, $t_r^{->}$, and the values effectively injected in the simulation : $t_{r_{eff}}^{+>}$, $t_{r_{eff}}^{->}$; t = 0.5 s, 32^3 grid, $t_{NL} = \Delta t$.

prediction of non-isotropic homogeneous turbulence. Nevertheless some results are presented here as examples.

In the case of predictability two realizations of the large eddies have to be simulated, simulatneously with a closure computation of the subgrid energy spectrum and of the "error" spectrum. The correlation between the stochastic subgrid terms acting on the two realizations is fixed by the closure. On figure 4, the growth of the "error" appears in the spectra. On the first stage, only the subgrid eddies are contaminated. Latter the error affects the supergrid field. It is worth pointing out that the error spectrum $E_\Delta(K,t)$ keeps a K^4 slope on both sides of the cutoff.

When non-isotropic turbulence is investigated, a coordinate transformation have to be introduced in the spectral simulation (Rogallo, 1981). In the E.D.Q.N.M. computation, simplifications are introduced in order to reduce the computational cost (Bertoglio, 1981). The model is applied to an initially isotropic turbulence subjected to a uniform mean shear flow. Spectral results are plotted in figure 5, they are compared with results obtained with an eddy viscosity formulation, also coupled with E.D.Q.N.M.. The improvement due to the stochastic modelling of the backscatter appears on the component normal to the velocity in the plane of the shear $\varphi_{33}(K,t)$. Another interesting result is that the slope of the Reynolds stress spectrum tends to a $K^{-7/3}$ behaviour on both sides of K_c (figure 6).

6. Conclusion

A new subgrid model has been presented. It has been shown to give satisfactory results in the case of isotropic turbulence. Since the model is stochastic it permits to take into account effects neglected by eddy viscosity formulations, for example in the case of predictability studies. When non isotropic turbulence is considered, the model seems to do better than eddy viscosity.

The model have been used in a method involving both L.E.S. and E.D.Q.N.M. computations. It is therefore quite uneasy to apply to industrial flows. It can however be used as a guidance to developp simpler model in which the stochastic modelling of energy back-flows could be retained. It can also be used to test "defiltering" methods (Bardina et al., 1983).

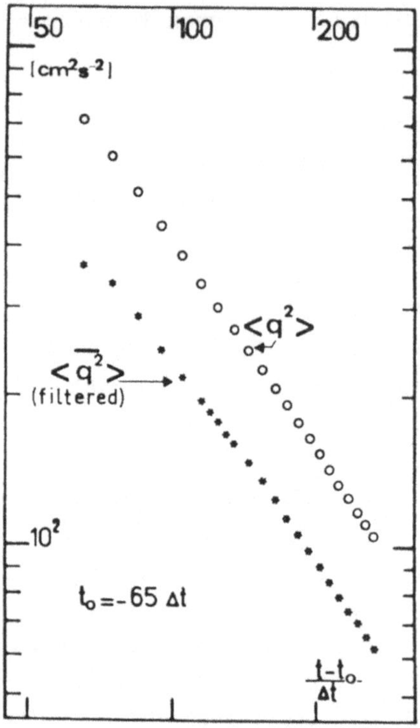

Figure 3 - Evolution with time of the turbulent kinetic energy. Contribution of large eddies only : ✱ , and full field value : ○ ; t_{NL} given by (A2.3).

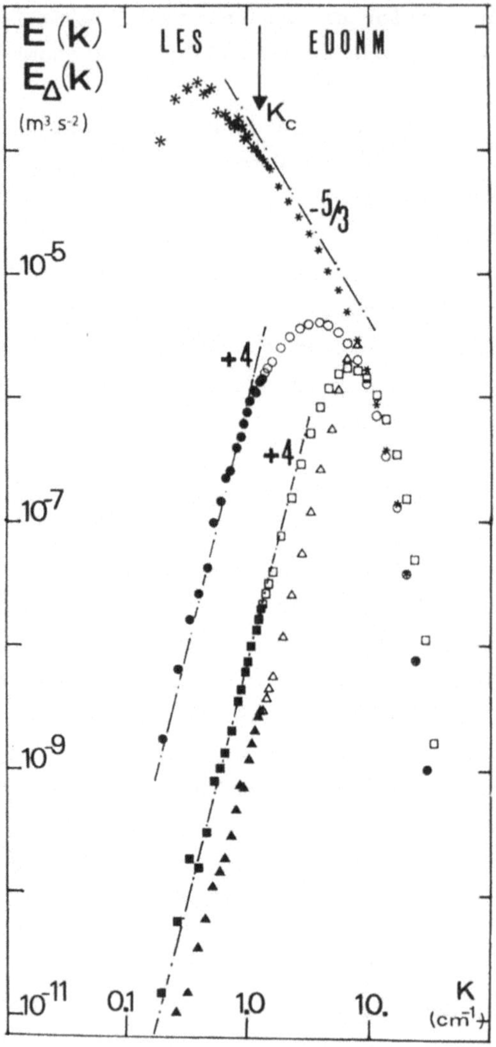

Figure 4 - Energy spectrum $E(K,t)$ for t = 0.1 s, L.E.S. prediction : ✱ , and E.D.Q.N.M. : * ; and error spectrum $E_\Delta(K,t)$ for t = 0.025 s, L.E.S.: ▲ and E.D.Q.N.M. : △ ; for t = 0.05 s, L.E.S. : ■ and E.D.Q.N.M. : □ ; and for t = 0.1 s, L.E.S. ● and E.D.Q.N.M.: ○ .

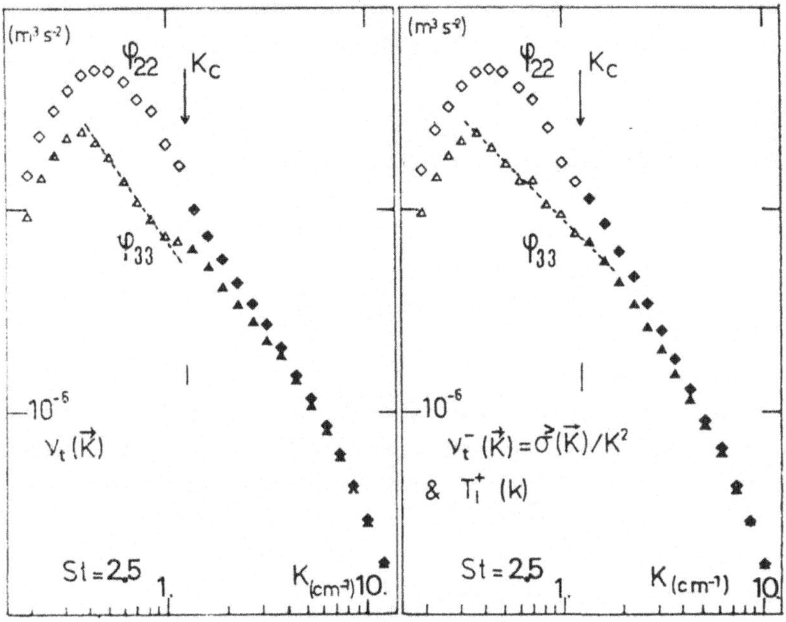

Figure 5 - Comparison between eddy viscosity and the present model. φ_{22} and φ_{33} are two components of the double correlation spectrum integrated over a spherical shell of radius K. The value of the shear is $S = 12.5\,s^{-1}$.

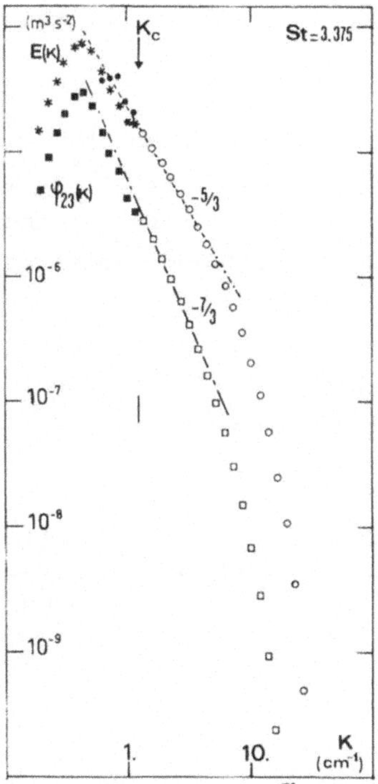

Figure 6 - Energy spectrum : $E(K,t)$, L.E.S. : * and ⁎, and E.D.Q.N.M. : O, and Reynolds stress spectrum : $\varphi_{23}(K,t)$ L.E.S. : ■ , and E.D.Q.N.M. □ .
$St = 3.375$, in which S is the value of the shear : $S = 12.5\ s^{-1}$.

Extensions to two-dimensional turbulence could be of interest since in this case back-flows of energy are important.

Appendix 1 - Generation of a stochastic process

We propose here a method to generate a stochastic terme such as the averaged quantity :

$$\langle T_i^+(\vec{K},t)\, T_j^+(-\vec{K},t') \rangle$$

satisfies relation (18), or, more generally, relations (20) and (25). In other words, the stochastic process must have a given characteristic time and its single time correlation must be equal to a given value. We call this value $f_{ij}(\vec{K},t)$.

Since incompressibility implies :

$$K_i\, f_{ij}(\vec{K},t) = 0$$

it is convenient to introduce a new frame, an axis of which is parallel to $\vec{K}$, for example K_3 . In this new frame all components corresponding $i=3$ vanish. We can then consider that i and j take only the values 1 and 2.

In this new frame f_{ij} is an isotropic tensor if turbulence is isotropic. However, since the model is supposed to be valid for non isotropic fields, we consider the general case in which f_{ij} is not isotropic. f_{ij} is real and symmetrical.

Details concerning the generation of are given in Bertoglio and Mathieu (1984). The final result is written in a discrete form and for a fixed value of $\vec{K}$. It is, at the time step number (n) :

$$T_1^+{}_{(n+1)} = \left(1 - \frac{\Delta t}{t_{NL}}\right) T^+_{1(n)} + \beta_{11(n+1)} \left(f_{11(n)} \frac{\Delta t}{t_{NL}}\right)^{1/2} e^{i2\pi\varphi_{1(n+1)}}$$

$$+ \beta_{12(n+1)} \left(f_{22(n)} \frac{\Delta t}{t_{NL}}\right)^{1/2} e^{i2\pi\varphi'_{(n+1)}}$$

(A1_1a)

$$T^{+}_{2\,(n+1)} = \left(1 - \frac{\Delta t}{t_{NL}}\right) T^{+}_{2\,(n)} + \beta_{22(n+1)} \left(f_{22(n)} \frac{\Delta t}{t_{NL}}\right)^{1/2} e^{i2\pi \varphi_{2(n+1)}}$$

$$+ \beta_{12(n+1)} \left(f_{11(n)} \frac{\Delta t}{t_{NL}}\right) e^{i2\pi \varphi'_{(n+1)}}$$

(A 1.1b)

where Δt is the time step, where the β_{ij} are solutions of :

$$\begin{cases} \beta^2_{11(n+1)} = \dfrac{\left(f_{11(n+1)} - f_{11(n)}\right) t_{NL}/\Delta t - f_{22(n)} \beta^2_{12(n+1)}}{f_{11(n)}} + \left(2 - \Delta t/t_{NL}\right) \\ \beta^2_{22(n+1)} = \dfrac{\left(f_{22(n+1)} - f_{22(n)}\right) t_{NL}/\Delta t - f_{11(n)} \beta^2_{21(n+1)}}{f_{22(n)}} + \left(2 - \Delta t/t_{NL}\right) \\ \beta_{12(n+1)}\, \beta_{21(n+1)} = \dfrac{\left(f_{12(n+1)} - f_{12(n)}\right) t_{NL}/\Delta t}{\left(f_{11(n)}\, f_{22(n)}\right)^{1/2}} + \dfrac{f_{12(n)}}{\left(f_{11(n)}\, f_{22(n)}\right)^{1/2}} \left(2 - \Delta t/t_{NL}\right) \\ \beta^2_{12(n+1)} = \beta^2_{21(n+1)} \end{cases}$$

(A 1.2)

$\varphi_{1(n)}$, $\varphi_{2(n)}$ and $\varphi'_{(n)}$ constitute a set of random numbers independent of each other, with uniform distributions between 0 and 1.

Such a process is not gaussian, but it has been found to be convenient, at least for $t_{NL} \gg \Delta t$ (Bertoglio & Mathieu, 1984).

Appendix 2 - Determination of characteristic time $t_{NL}(K)$

Starting from equation (20), and replacing $q_i^{>}$ by its D.I.A. expression

$$q_i^{>}(\vec{K},t) = -i\, P_{ijm}(\vec{K}) \sum_{\substack{\vec{P}+\vec{Q}=\vec{K} \\ P>K_c \text{ and/or } Q>K_c}} \xi_j(\vec{P},t)\, \xi'_m(\vec{Q},t) \;,$$

give :

$$\langle T_i^+(\vec{K},t)\, T_j^+(-\vec{K},t')\rangle = P_{ilm}(\vec{K})\, P_{jnr}(\vec{K}) \sum_{\substack{\vec{P}+\vec{Q}=\vec{K} \\ P>K_c \text{ and/or } Q>K_c}} \tag{A2-1}$$

$$\times \Big(\langle u_l(\vec{P},t)\, u_n(-\vec{P},t')\rangle \langle u_m(\vec{Q},t)\, u_r(-\vec{Q},t')\rangle + \langle u_l(\vec{P},t)\, u_r(-\vec{P},t')\rangle \langle u_m(\vec{Q},t)\, u_n(-\vec{Q},t')\rangle \Big)$$

For a stationary turbulence, replacing the two time correlations by the exponential forms (20) and (21), then integrating between $t'=0$ and $t'=t$ give:

$$t_{NL}(K) \langle T_i^+(\vec{K},t)\, T_j^+(-\vec{K},t)\rangle = P_{ilm}(\vec{K})\, P_{jnr}(\vec{K}) \tag{A2-2}$$

$$\sum_{\substack{\vec{P}+\vec{Q}=\vec{K} \\ P>K_c \text{ and/or } Q>K_c}} \frac{1}{\eta(P,t)+\eta(Q,t)} \Big\{ \langle u_l(\vec{P},t)\, u_n(-\vec{P},t)\rangle \langle u_m(\vec{Q},t)\, u_r(-\vec{Q},t)\rangle + \langle u_l(\vec{P},t)\, u_r(-\vec{P},t)\rangle \langle u_m(\vec{Q},t)\, u_n(-\vec{Q},t)\rangle \Big\}$$

for large values of t .

We can now use (A2.1) for $t=t'$ to express $\langle T_i^+(\vec{K},t)\, T_j^+(-\vec{K},t)\rangle$ in (A2.2). We finally obtain that t_{NL} is equal to the R.H.S. of (A2.2) divided by the R.H.S. of (A2.1). In the case of an isotropic turbulence it comes :

$$t_{NL}(K) = \frac{\iint_{\Delta'} \theta_{PQ}(t)\, b_{KPQ} \frac{K^3}{PQ} E(P,t)\, E(Q,t)\, dP\, dQ}{\iint_{\Delta'} b_{KPQ} \frac{K^3}{PQ} E(P,t)\, E(Q,t)\, dP\, dQ} \tag{A2-3}$$

where :

$$\theta_{PQ}(t) = \frac{1}{\eta(P,t)+\eta(Q,t)}$$

References

Aupoix B., Cousteix J. and Liandrat J., 1983, Effects of rotation on isotropic turbulence. Fourth Int. Symp. Turb. Shear Flows, Karlsruhe.

Bardina J., Ferziger J.H. and Reynolds W.C., 1983, Improved turbulence models based on large eddy simulation of homogeneous, incompressible, turbulent flows. Stanf. Univ. Report NOTE-19, May 1983.

Basdevant D., Lesieur M. and Sadourny R., 1978, Subgrid-scale modeling of enstrophy transfer in two-dimensional turbulence. Journal of Atm. Sci., Vol. 35, pp. 1028-1042.

Bertoglio J.P., 1981, A model of three-dimensional transfer in Non-isotropic homogeneous turbulence. Third Int. Symp. Turb. Shear Flows, Davis, Sept. 81, Springer-Verlag, 1982.

Bertoglio J.P. and Mathieu J., 1983, Study of subgrid models for sheared turbulence. Fourth Symp. on Turb. Shear Flows, Karlsruhe, Sept. 83.

Bertoglio J.P. and Mathieu J., 1984, Modélisation stochastique des petites échelles de la turbulence : génération d'un processus stochastique. C.R.Acad. Sci., to be published.

Chollet J.P. et Lesieur M., 1981, Parameterization of small scales of three-dimensional isotropic turbulence utilizing spectral closures. J. Atm. Sci., Vol. 38, pp. 2747-2757.

Chollet J.P., 1983, Two-point closure as a subgrid scale modeling for Large Eddy Simulations. Fourth Symp. on Turb. Shear Flows, Karlsruhe.

Frish U., Lesieur M. and Brissaud A., 1974, A Markovian random coupling model for turbulence. J. Fluid Mech., Vol. 65, part 1, pp. 145-152.

Ferziger J.H., 1982, State of the art in subgrid scale modeling.Num.and Phys. Asp. Aerod. Flows, T. Cebeci, pp. 53-68. New-York : Springer 636.

Kraichnan R.H., 1961, Dynamics of nonlinear stochastic systems. Journ. Math. Phys. Vol. 2, n° 1, pp. 124-148.

Kraichnan R.H., 1970, Convergents to turbulence functions. J. Fluid Mech., Vol. 41, part 1, pp. 189-217.

Kraichnan R.H., 1971, An almost-Markovian Galilean-invariant turbulence mode. J. Fluid Mech., Vol. 47, part 3, pp. 513-524.

Kraichnan R.H., 1976, Eddy viscosity in two and three dimensions. Journ. Atm. Sci., Vol. 33, pp. 1521-1536.

Leslie D.C. and Quarini G.L., 1979, The application of turbulence theory to the formulation of subgrid modelling procedures. J. Fluid Mech., Vol. 91, part 1, pp. 65-91.

Leith C.E., 1971, Atmospheric predictability and two-dimensional turbulence. Journ. Atm. Sci., Vol. 28, n° 2, pp. 145-161.

Orszag S.A., 1970, Analystical theories of turbulence. J. Fluid Mech., Vol. 41, part 2, pp. 363-386.

Orszag S.A. and Patterson G.S., 1972, Numerical simulation of three-dimensional homogeneous isotropic turbulence. Phys. Rev. Letter.

Pouquet A., Lesieur M., André J.C. and Basdevant C., 1975, Evolution of high Reynolds number two-dimensional turbulence. J. Fluid Mech., Vol. 72, part 2, pp. 305-319.

Rogallo R.S., 1981, Numerical experiments in homogeneous turbulence. NASA Techn. Mem. n° 81315, sept. 81.

Rose H.A., 1977, Eddy diffusivity, Eddy noise and subgrid-scale modeling. Journal of Fluid Mech., Vol. 81, pp. 719-734.

SOME CHALLENGES FOR MODELLING OF TURBULENCE AND INTERNAL WAVES IN STABLY STRATIFIED FLUIDS

C. W. Van Atta
Scripps Institution of Oceanography and
Department of Applied Mechanics and Engineering Sciences
University of California, San Diego, La Jolla, CA 92093/USA

1. Introduction

The gross effects of stratification on turbulent shear flow have been modelled with varying degrees of success for some time. However, most of the models have been crude, designed mainly to predict the effects of stratification on the mean properties, and not concerned with a prediction of the detailed behavior of the fluctuating velocity and density fields. In a sufficiently strongly stratified turbulent flow the density is an active scalar. Experiments are difficult to carry out, and the many experimental and theoretical results available for passive scalar fields do not apply.

The present review is intended to briefly present some of the physical ideas which need to be considered by prospective modelers of turbulence in stratified flows, to suggest some simple possible test cases for which substantial comprehensive laboratory data is available for detailed comparison with modelling computations, to describe the qualitative features of the data and discuss modelling attempts to date for the same flows, and finally to discuss some recent geophysical data which provide further challenges for modelling stratification effects on higher Reynolds number turbulent flows.

2. Governing equations and physical mechanisms for stratified turbulence and internal waves

To restrict the discussion to stratification effects, the effects of rotation will, for the most part, be ignored here, effectively limiting the discussion to those scales that are small compared with the Rossby deformation radius (see e.g. Gill (1982)). This restriction thus avoids a discussion of the coupling of lower frequency rotational and stratification effects which are encountered in many geophysical situations but which are absent in the laboratory situations to be discussed here.

Stably stratified turbulent flows differ fundamentally from those in homogeneous fluids because buoyancy forces provide an additional physical mechanism for production, transfer, and radiation of kinetic and potential energy. In addition to the normal "chaotic" fluctuations of turbulent motions, in a stably stratified fluid internal wave motions coexist with the turbulence so that measured fluctuating velocity and scalar fields contain contributions from both kinds of disturbances. It thus is necessary to simultaneously consider the individual behavior and interactions of two fluctuating fields having their own characteristic length and time scales. Where the ranges of these scales overlap, it is difficult to distinguish wave contributions from turbulence in both geophysical and laboratory flows.

Compared with the large body of laboratory turbulence measurements in homogeneous fluids, detailed definitive studies of stratification effects on turbulence in both the laboratory and field are rare. Most laboratory experiments on the structure of stratified turbulence have been carried out either in stationary towing tanks or in flow systems employing at most only two of three discrete layers of differing densities. Experimental facilities designed to produce well controlled continuously stratified flows in which detailed measurements of turbulence and internal waves can be made are extremely rare. This has retarded progress in recognizing and understanding the physical processes at work in many stratified turbulent flows of engineering and geophysical interest. Another effective deterrent to understanding has been the tendency of turbulence modelers to treat stable stratification as just another parasitic effect which might be modelled in a conventional way without recognizing the fundamentally different physical processes at work.

One of the main gross physical effects of stratification on turbulent motions is the tendency of stratification to inhibit vertical motions, thereby decreasing the intensity of vertical turbulent velocity fluctuations and the vertical range of fluid particle excursions. There is an associated leveling out or decrease in characteristic vertical length scale as measured by a turbulent integral scale or by an overturning scale related to the extent of fluid particle vertical migrations. The profound effects on turbulent diffusion are familiar from pictures of thin, slowly diffusing horizontal smoke layers originating from smoke stacks under stable atmospheric conditions, and in the sudden decrease in vertical extent or "wake collapse" of turbulent wakes behind bodies moving in a stratified fluid.

The velocity field in incompressible stratified turbulent flows may be decribed by the Navier-Stokes equations with the Boussinesq approximation, i.e.

$$\frac{\partial \tilde{u}_i}{\partial t} + \tilde{u}_j \frac{\partial \tilde{u}_i}{\partial x_j} = -\frac{1}{\rho_o}\frac{\partial p}{\partial x_i} + g\frac{\rho}{\rho_o} + \nu \frac{\partial^2 \tilde{u}_i}{\partial x_j^2} \qquad (1)$$

the diffusion equation for the density perturbation

$$\frac{\partial \rho}{\partial t} + \tilde{u}_i \frac{\partial \rho}{\partial x_i} + w \frac{\partial \bar{\rho}}{\partial z} = \kappa \nabla^2 \rho \tag{2}$$

and the incompressibility condition

$$\frac{\partial \tilde{u}_i}{\partial x_i} = 0 \tag{3}$$

where $\tilde{u}_i$ are the velocity components, p pressure, ρ density, g gravity, ν kinematic viscosity, ρ the density perturbation in the total density field $\bar{\rho} + \rho$. The linearized versions of Eqs. (1) and (2), neglecting molecular diffusion, are satisfied by linear internal waves of the form

$$w = w_o \cos(kx + \ell y + mz - \omega t) \tag{4}$$

where w_o is the amplitude of vertical velocity fluctuations, $\tilde{k} = (k, \ell, m)$ is the wavenumber of the disturbance, and ω is the frequency. The disperson relation is:

$$\omega = N \cos \phi \tag{5}$$

$$N = \left(- \frac{g}{\bar{\rho}} \frac{\partial \bar{\rho}}{\partial z}\right)^{1/2} \tag{6}$$

where N is the Brunt-Väisälä frequency and ϕ is the angle between $\tilde{k}$ and the horizontal. Note that ω is independent of the magnitude of $\tilde{k}$ and depends only on the angle ϕ.

Several criteria are available for distinguishing turbulent motions from those caused by linear internal waves. The maximum internal wave frequency is N. Although both turbulence and linear internal wave fields have three dimensional velocity fluctuations, the vertical vorticity $\partial v/\partial x - \partial u/\partial y$ is identically zero for the internal wave field. The average vertical buoyancy flux $\overline{\rho w}$ is zero for linear internal waves, whereas it is often robustly non-zero (usually positive) for a turbulent field. The dispersion relation can also be used to distinguish turbulence from internal waves of multipoint measurements are available. Finally the linearized continuity equation gives a simple relation between the fluctuating density and vertical velocity (w) fields for an internal wave, which in terms of the time spectra E(f) of each variable is

$$E_\rho(f) = \frac{\bar{\rho}^2}{g^2} N^4 f^{-2} E_w(f) \tag{7}$$

The total velocity and density fluctuation fields may be considered to be the sum of a "turbulent" part and a wave part: $u_i = u_{ti} + u_{wi}$ and $\rho = \rho_t + \rho_w$. The total kinetic energy equation is:

$$\left(\frac{\partial}{\partial t} + U_j \frac{\partial}{\partial x_j}\right)\frac{1}{2}\overline{u_i^2} + \frac{\partial}{\partial x_j}\left\{\overline{u_j\left(p/\rho_o + \frac{1}{2}u_i^2\right)}\right\} = -\overline{u_i u_j}\frac{\partial U_i}{\partial x_j} + \frac{g}{\bar{\rho}}\overline{\rho w} - \nu\overline{\left(\frac{\partial u_i}{\partial x_j}\right)^2} \qquad (8)$$

and that for the total potential energy fluctuation, which is an active rather than passive scalar, is:

$$\left(\frac{\partial}{\partial t} + U_j \frac{\partial}{\partial x_j}\right)\overline{\rho^2} = -2\overline{\rho w}\frac{\partial\bar{\rho}}{\partial z} - 2\kappa\overline{\left(\frac{\partial \rho}{\partial x_j}\right)^2} + \frac{\partial}{\partial x_i}[\overline{u_i \rho^2}] \qquad (9)$$

For a linear internal wave field, the kinetic energy $2\rho\overline{u_i^2}$ and fluctuation potential energy $(1/2)(g^2/\bar{\rho})(\overline{\rho^2}/N^2)$ are equal.

Although a few studies have addressed the problem of the breakdown of one or more internal waves into turbulence, and the generation of internal waves by an isolated patch of turbulence such as a turbulent wake behind a moving body, there appear to be no studies which have comprehensively investigated the continuous interaction of coexisting internal wave and turbulence fields, each containing a sizeable number of Fourier components which could interact.

Despite the complexity of stratified turbulent flows, some simple basic ideas about the range of scales which might be influenced by buoyancy are available. Experimental tests of these ideas, and attempts to make their predictions quantitative, are quite recent. Stillinger, Helland, and Van Atta (1983) experimentally produced some simple quantitative criteria for estimating when buoyancy effects and internal waves would become important for decaying homogeneous turbulence. These criteria explained the differences noted in earlier towed grid experiments and produced a logical framework for interpreting ocean turbulence measurements.

The basic idea is that buoyancy effects act to place a limiting value of the largest turbulent scales which can be produced or maintained. Simple physical arguments suggest that the largest overturning eddy motions which turbulence can produce in the presence of stable stratification have a scale L_b equal to

$$L_b = w'/N \qquad (10)$$

where w' is the rms vertical velocity fluctuation. A closely related scale, which can be employed when the turbulent dissipation rate ε is known, is the Ozmidov scale

$$L_O = (\varepsilon/N^3)^{1/2} \qquad (11)$$

Buoyancy will strongly affect those scales with wavenumbers smaller than the buoyancy wavenumber $k_b = 2\pi/L_O$. When shear is also present, the Richardson number R_i is also important, where

$$Ri = N^2 \Big/ \left(\frac{dU}{dz}\right)^2 \tag{12}$$

The characteristic length scale of the turbulence is the overturning scale

$$L_T = 2(\overline{\rho^2})^{1/2} \Big/ \frac{\partial\bar{\rho}}{\partial z}$$

If the integral or overturning scale of the turbulence is much less than the buoyancy scale, then buoyancy effects will be unimportant for the turbulence. This imples that the development of a stratified turbulent flow will depend sensitively on the initial conditions, especially on the initial ratio of L_T/L_O and on the initial relative energies of the turbulence and internal wave fields. For turbulence in which L_T/L_O is initially small, the onset of appreciable buoyancy effects might be expected whenever L_T/L_O locally reaches some critical value, which may differ for different flows. In general, buoyancy forces will be important for turbulent scales which are larger than some constant factor times the buoyancy scale.

3. Some recent laboratory studies and related numerical simulations

Experiments in progress over the last few years at UCSD have been designed to study the behavior of the simplest possible stably stratified turbulent flows, beginning with the cases of unsheared linearly stratified grid turbulence and then proceeding to the case of a constant velocity gradient (linear velocity profile) linearly stratified turbulent shear flow. All the experiments were done in a continuous flow ten layer salt stratified water channel described by Stillinger, Head, Helland, and Van Atta (1983). Some of the results for the evolution of unsheared grid generated turbulence have been reported in Stillinger, Helland, and Van Atta (SHV, 1983), Van Atta, Helland, and Itsweire (1984), and by Itsweire (1983). Initial results for the constant shear cases will appear in Rohr, Helland, Itsweire, and Van Atta (1985). In these experiments simultaneous single-point measurements of the horizontal and vertical velocity and density fluctuations were obtained, so that direct measurements of the buoyancy flux, dissipation rate of turbulent kinetic energy, and Reynolds stress were also obtained. The evolution of the buoyancy and overturning scales was studied for different degrees of stratification.

For decaying turbulence, when the buoyancy length scale was initially substantially larger than the largest turbulent overturning scales, the initial behavior of the velocity and density fluctuations was similar to that in the non-stratified (passive scalar) case. With further downstream development, the buoyancy length scale decreased while the turbulence scale increased, deviations from neutral behavior first becoming noticeable when these two length scales became of the same order. As the decay proceeded further, the overturning scale locked in on and remained proportional to the Ozmidov scale, a feature also found to be characteristic of ocean turbulence in the vertically sampled ocean microstructure measurements of Dillon (1982). Spectral measurements showed that buoyancy forces produced anisotropy in the largest scales first, preventing them from overturning, while smaller scale isotropic turbulent motions remained embedded within the larger scale internal waves. These small scale motions exhibited classical turbulent behavior and scaled universally with Kolmogorov length and velocity scales. Eventually even the smallest scales of the turbulence were affected by buoyancy and Kolmogorov scaling failed. The buoyancy flux $\overline{\rho w}$ decreased to zero, indicating the initially turbulent field had been completely converted to a random internal wave field. This transition from a fully turbulent state to one of internal waves occurred rapidly (a fraction of a Väisälä period). After buoyancy forces became important, the transition time was also less than the characteristic time of the largest scales in the turbulence at transition. The experiments determined quantitative limits on the range of active turbulent scales in homogeneous stratified turbulence, in terms of an upper limit near the buoyancy length scale and a lower limit determined by viscosity in the usual way. This simple length scale description has since been used to interpret results of a number of ocean and laboratory experiments. In particular, the strong effect, both qualitative and quantitative, of varying the initial ratio of buoyancy length scale to overturning length scale on the evolution of the velocity fluctuations have been reported in Van Atta et al. (1984). This sensitive dependence on initial conditions may be traceable to the different relative amounts of initial internal wave energy and turbulent kinetic energy.

SHV compared their data with results of the direct numerical simulations of Riley, Metcalfe, and Weissman (RMW, 1981), which were done with a 32 cubed spectral code in which unstratified isotopic turbulence was allowed to evolve naturally up to a certain time at which the density gradient was suddenly "turned on." Although the simulations and laboratory experiments employed quite different initial conditions, many similarities between them are evident. In both cases, the density fluctuation and buoyancy flux increased smoothly from initial to maximum values, with the peak value of the buoyancy flux occurring somewhat earlier than that for the mean square density fluctuation. In both cases, the "collapse time" measured from the initial onset of buoyancy effects to the time when the buoyancy flux reached a zero value was about 0.18 Brunt-Väisälä periods. An interesting difference in interpretation of results arose between the experiments and simulations. SHV interpreted the final stage of decay as a purely random internal wave field in which turbulent motions had disappeared. On the other hand, although RMW observed wave-like behavior, they found that the dissipation and nonlinearity remain fairly strong during the decay, and that no sharp transition from a fully turbulent state to an internal wave field occurred. RMW also presented a theoretical analysis which interpreted the field in the final stage of the decay as a superposition of an internal wave field and a quasi two-dimensional turbulent field. Such a field would have a vanishing buoyancy flux but a non-vanishing vertical vorticity. The first criterion is satisfied by the experiments of SHV, but the value of the vertical vorticity, which would be very difficult to measure, is not known. For the decaying stratified grid turbulence data, Itsweire, Helland, and Van Atta (1985) have proposed two operational techniques for determining the individual contributions of the internal wave and turbulence components to the turbulent kinetic energy. These two techniques give mutually consistent results but do not furnish any information on vertical vorticity. Perhaps in future laboratory measurements, vorticity criteria, as well as multipoint phase information utilizing the dispersion relation (eq. (6)), spectra or other moments, may be used to experimentally resolve the question of which interpretation is the most appropriate. A case in which multiple probes and cospectral techniques have been used to decompose oceanic velocity and density fluctuation fields into internal wave and "vortical mode" components will be discussed in the next section.

A serious shortcoming of the RMW simulations was the small range of scales dictated by the 32 cubed restriction as compared with the larger range of the experiments. Current efforts are being made to remedy this shortcoming using various approaches. Metais (1985) has carried out direct numerical simulations of the type used by RMW, but he has managed to increase the range of scales and to therefore achieve higher Reynolds and Peclet numbers by using a subgrid scale modelling developed by Chollet and Lesieur (1981, 1982) for isotropic turbulence, and which (according to Metais) can be shown to be valid under conditions of stable stratification. When initial conditions for the dissipation rate, mean square density fluctuation, and velocity fluctuation intensity were chosen equal to those for one of the laboratory runs, good agreement was found between the simulations and experimental data for the evolution of the dissipation rate and velocity fluctuations. However, the mean square density fluctuation for the simulation increased more rapidly and peaked at a larger value than the experimental data, while the simulated normalized density flux, which was not matched to the experiment initially, started at a value about twice as large as the experimental value and exhibited a large oscillation at the B-V frequency not present in the experimental data. Since, from eq. (9), the buoyancy flux is a source term for the mean square density fluctuation, it seems likely that matching of the initial condition on the buoyancy flux, if this proves possible for the simulation, would lead to closer agreement for the mean square density fluctuation evolution. Regarding the question of the nature of the later stages of decay, by comparing spectral data for stratified and unstratified runs, Metais finds that they are very similar, and concludes that there does not seem to be any tendency towards a two-dimensional state, since no energy transfer towards large scales is observed. These initial results of Metais suggest that such simulations may be on the verge of producing a substantial advance in closing the gap between observations and direct numerical modelling of stratification effects on one of the simpler turbulent flows, as was qualitatively suggested by the earlier results of Riley et al. It will be interesting to see if this promise is realized by further comparisons using more completely matched initial conditions and larger (64 cubed or greater) computational schemes.

For any numerical simulation capable of successfully describing the simplest unsheared case discussed above, a next logical step would seem to extend the simulation to compare with experimental data for a stratified shear flow with uniform (constant, independent of z) gradients in both mean density and mean velocity. Suitable data, in which N and R_i have been systematically varied, have been recently obtained at UCSD by Rohr et al. (1985), and will soon be available for comparison with modelling results.

For these uniform shear experiments immediately downstream of the turbulence generator, for all values of N and R_i the turbulent kinetic energy decayed for a short distance in x. However, at larger values of x the shear and buoyancy forces became important and for sufficiently small values of R_i (less than about 0.25), the turbulent kinetic energy grew with increasing x, with a growth rate depending on N. The range of actively turbulent scales was found to be consistent with the criteria established from the decaying grid turbulence measurements. When shear is present and $R_i > 0.25$, for large x the ratio L_t/L_O approaches the same ratio observed for grid turbulence. When $R_i < 0.25$ L_O grows and L_T is locked into the growing L_O, with the ratio L_T/L_O increasing with increasing x and N and reaching an asymptotic value for large x. As for the decaying grid turbulence data, the proportionality of L_T and L_O for the sheared data is also similar to that of Dillon's ocean microstructure data. The additional dependence of the ratio L_T/L_O of the value of N for the shear flow data might prove useful in interpreting the ocean data, for which there is a very large observed range for the L_T/L_O ratio, which has up till now been attributed to experimental scatter. The laboratory results suggest that the "scatter" in the ocean data might be a real effect related to different values of N, and it will be interesting to see if this is indeed the case. The results suggest that the laboratory shear flow data will provide flow fields of direct relevance to the interpretation of oceanic data obtained under a great variety of conditions.

4. Some recent measurements of turbulence and internal waves in the ocean and in tidal inlets

The mechanisms principally responsible for vertical mixing in the ocean have not been positively identified. In fact, a controversy persists on the relative importance of small scale vertical mixing over large areas in the ocean interior and larger scale advective mixing of fluid stirred up only near the continental boundaries (Armi, 1978). This basic issue has literally launched a thousand ships to deploy instruments of many kinds to sample fluctuations of velocity, temperature, and salinity over a wide range of scales. The oceanic internal wave field has been studied much more fully than ocean turbulence. In contrast to the ubiquitous internal wave field, the turbulent fluctuations are highly variable in both space, time (i.e. intermittent) and in amplitude. It has been only in the past decade that suitable instrumentation has been developed to measure the smaller amplitudes and small length and time scales associated with ocean turbulence. As these measurements reached smaller scales the associated variations found were first called "finestructure" (say scales down to about 1 meter) and finally microstructure (scales down to 1 millimeter). While the physical processes controlling ocean mixing in this range can be duplicated and studied in laboratory experiments, understanding of the effects of the prevalent stratification of the ocean on its turbulent mixing is still in a primitive state, largely because of a lack of such basic laboratory experiments.

To infer mixing rates from vertical profiles requires a number of assumptions about the dynamical state of the turbulence. With the assumption of vertical homogeneity, the equation for the variance of the temperature fluctuation T' analogous to eq. (9) becomes simply

$$\frac{d}{dt}\,\frac{1}{2}\,(\overline{T'^2}) = -\overline{wT'}\,\frac{\partial \overline{T}}{\partial z} - \chi/2$$

Since the rate of change of temperature variance is not obtained in a single dropsonde, it is often assumed to be zero and the heat flux per unit mass is calculated from the remaining terms as

$$\overline{wT'} = \overline{(\nabla T')^2} \Big/ \frac{\partial \overline{T}}{\partial z}$$

This is quite a far reaching assumption, as one would get widely divergent estimates if a decaying or growing turbulent field was alternatively assumed. Oceanographers presently appear to favor the steady state assumption. Gibson (1982) has reinterpreted a number of oceanic temperature microstructure measurements assuming the observed temperature fluctuations to have been produced by decaying turbulence originally generated by energetic mixing events. This interpretation is thus very sensitive to the critical values of L_T/L_O assumed for the onset of buoyancy effects. Alternate views have been advanced by Caldwell (1983) and Dillon (1983), who argue for a more continuous steady state generation of microstructure in which turbulent production is nearly equal to dissipation.

Small scale ocean turbulence is embedded in a ubiquitous background of internal wave motions. It is a remarkable and curious fact that the deep ocean internal wave energy level varies by less than an order of magnitude over the entire deep world ocean. The energy in the wave field decreases monotonically with increasing vertical wavenumber, and any ocean turbulence present shows up as a high wavenumber bump for scales smaller than about 1 meter (Gargett et al., 1982). As noted earlier, the mechanisms and degree of interaction between the wave field and turbulence is an area of active speculation, referred to as a "no man's land" by Gargett et al. since so little is known or understood. To avoid contamination of the measured turbulent structure by internal waves, rapid profilers which ascend or descend in a time less than the Brunt-Väisälä period are employed. Caldwell and Dillon (1980) and others have found that for large enough Cox numbers a classical Batchelor spectrum is observed in rapid profiling measurements of the vertical temperature gradient. They also noted that the degree of stratification produces changes from classical passive scalar form in their spectra, which they were unable to correlate with N or other parameters with existing data. A major problem is that in most field experiments not enough parameters can be simultaneously measured to allow assessment of the dynamical state of the flow. One does not usually know whether or not the turbulence is decaying with time, is in near steady state equilibrium in which production might be nearly equal to dissipation, or in a growing mode in which the production terms are larger than the combined buoyancy flux and dissipation sink terms in eq. (8). Recently, Gargett, Osborn, and Nasmyth (1984) found that the turbulent flow fields generated by tidally forced flow over the sill in Knight Inlet generate decaying stratified turbulence with many aspects reminiscent of laboratory flows used to study the decay of grid generated turbulence in the presence of stable stratification. A major advantage of the Knight Inlet flow is the large Reynolds number of the flow (equivalent mesh Reynolds number on the order of 4×10^4, so that when the stratification is small inertial subrange behavior is observed in the velocity spectra, as found earlier in the relatively unstratified tidal channel experiments of

Grant, Stewart, and Moilliet (1962). The Gargett et al. measurements were made using the small two-person research submersible Pisces instrumented for measuring fine scale turbulent velocity and temperature fluctuations, as well as some parameters of the larger scale motions. The longitudinal turbulent velocity component was measured using heated film sensors, the cross stream components with two single axis airfoil probes, and temperature fluctuations with thermistors. It was found that the energy spectra of the velocity could be classified according to observed departures from the homogeneous case and the departures could be parameterized by I, the ratio of the Kolmogorov wavenumber k_s to the buoyancy wavenumber k_b, which is also the reciprocal of the ratio of the Ozmidov length to the Kolmogorov scale. The data were collected in a breaking internal wave train adjacent to the sill. For large k_s/k_b ($I \simeq 0(3000)$, called class A) the spectra exhibit classical inertial subrange -5/3 behavior, but for decreasing k_s/k_b ($I \simeq 50$-100, class B) the vertical velocity spectral levels at the lower wave numbers are systematically lowered. As I decreases, the vertical velocity breaks away from the universal passive spectrum at an increasing value of wavenumber roughly equal to k_b. The effect on the horizontal spectra is much smaller and systematic effects are not apparent. This behavior is similar to that found previously for laboratory grid turbulence by Stillinger, Helland and Van Atta (1983), but is perhaps much more convincing because of the large range of scales achieved.

Similar effects on oceanic velocity gradient spectra have been found by Osborn and Lueck (1984) using the research submarine USS Dolphin 555, a 55 meter long diesel electric submarine specifically designed for research work. The horizontal variation of $\partial w/\partial x$ and $\partial v/\partial x$, where w is the vertical velocity and v the lateral horizontal velocity component, was measured over scales ranging from 5 m to 1 cm. The data show that stratification suppresses the low wavenumber portion of the spectrum of the vertical velocity shear relative to the horizontal velocity component, which exhibits inertial subrange behavior. The slope of the horizontal velocity shear spectrum is +1/3 when the slope of the vertical velocity shear spectrum has increased to +1, consistent with the Gargett et al. data. The buoyancy wavenumber k_b is found to be the appropriate scaling parameter for buoyancy effects as in the Gargett et al. and Stillinger et al. data. The temperature spectra peak at a lower wavenumber than the shear spectra. Osborn and Lueck suggest that a possible explanation for the difference of the temperature spectra from the Batchelor prediction is that the temperature gradient variance is dominated by anisotropic features associated with the fine structure, the edges of the turbulent patches, and the large scale structures that generate the patches.

More evidence for the possible non-universal nature of the small scale temperature field in ocean turbulence comes from the data for the temperature field (Gargett, 1984) associated with the previously discussed velocity data of Gargett et al. (1984). These temperature data are most interesting, but quite perplexing. The observations suggest that Corrsin-Obukhov theory is essentially incorrect as a description of the spectrum of temperature fluctuations in water. The Class A scalar spectra which have the largest value of I and hence should be the least buoyancy dominated, have neither $k^{-5/3}$ nor k^{-1} subranges and a Batchelor spectrum fit to the high wavenumber roll-off region yields a very large value for the "universal" constant. However, the buoyancy-affected Class B spectra exhibit a clear -5/3 subrange, an approach to a k^{-1} subrange, and a value of the constant which is in rough agreement with previous estimates. These results are very surprising, especially since the earlier Discovery Passage towed body measurements of Grant et al. (1968) have for some time been interpreted as evidence for both the -5/3 and -1 ranges in the passive scalar spectrum for large Reynolds numbers. Perhaps these earlier data were in fact buoyancy influenced. Gargett points out that eddies of scale k_b^{-1} and larger can lose energy either by supporting an energy cascade to smaller scales or by generating internal waves, whereas in the classical model of the k^{-1} subrange it was assumed that the energy was exclusively transferred to turbulent motions at higher wavenumbers. As the velocity field becomes influenced by buoyancy, increasing the anisotropy of the forcing near wavenumber k_b, the forced internal wave field approaches the limit of horizontal plane motion. This suggests two-dimensional turbulence considerations may be appropriate and Gargett further argues that such behavior would remove the inconsistency found in the k^{-1} range and help to explain why temperature spectra are considerably less universal in character, even at the smallest scales, than associated velocity spectra. The decomposition of observed finescale oceanic velocity and density fluctuations into internal wave and "vortical mode" components requires measurements of the potential vorticity. Muller et al. (1978) attempted to perform the decomposition from an analysis of horizontal currents and vertical displacements from the three legged IWEX mooring. Using consistency tests and a least squares fit to a total of 1444 cross spectra, they determined the amount of energy in the internal wave field and in the current and temperature finestructure, interpreted as being due to the vertical advection of passive finestructure past the sensors by mainly high-frequency internal waves. The advected finestructure is a two-dimensional field and has vertical coherence scales smaller than 10 m and horizontal coherence scales larger than a few hundred meters, an anisotropy one might expect in buoyancy dominated stratified turbulence. Muller (1984) interprets the current finestructure as a manifestation of the vortical mode, which at the small scales is equivalent to stratified two-dimensional turbulence. We note that wave-turbulence decompositions of this type (and those used in laboratory measurements by Itsweire et al. (1985)) which yield information only for Fourier amplitudes and not for phases can produce no direct insight into the nature or degree of interaction of the turbulent or vortical modes with the internal wave gravity modes. As this nonlinear interaction is of central importance for the overall energy and enstrophy cascades and hence for mixing in the ocean, it is a problem on which future research efforts should be focused.

Clearly these results raise fascinating and difficult challenges for modelling of the behavior of turbulence and internal waves in stably stratified fluids.

Acknowledgments. This review was prepared in connection with research on turbulent flows funded by the National Science Foundation under Grants OCE82-05946 and MEA81-00431.

References

Armi, L. (1978) Some evidence for boundary mixing in the deep ocean, J. Geophys. Res. 83, 1971-79.

Caldwell, D. R. (1983) Oceanic turbulence: big bangs or continuous creation? J. Geophys. Res. 88, 7543-7550.

Chollet, J. P. and M. Lesieur (1981) Parameterization of small scales of three-dimensional isotropic turbulence utilizing spectral closures, J. Atmos. Sci. 38, 2747.

Chollet, J. P. and M. Lesieur (1982) Modelisation sous-maille des flux de quantite de mouvement et de chaleur en turbulence tridimensionelle isotrope, La Meterologie, VI serie, No. 29 et 30, p. 138.

Dillon, T. M. and D. R. Caldwell (1980) The Batchelor spectrum and dissipation in the upper ocean, J. Geophys. Res. 85, 1910-1916.

Dillon, T. M. (1982) Vertical overturns: a comparison of Thorpe and Ozmidov length scales, J. Geophys. Res. 87, 9601.

Dillon, T. M. (1983) The energetics of overturning structures: implications for the theory of fossil turbulence, J. Phys. Ocean. 14, 541-549.

Gargett, A. E. (1984) Evolution of scalar spectra with the decay of turbulence in a stratified fluid, preprint.

Gargett, A. E., T. R. Osborn and P. W. Nasmyth (1984) Local isotropy and the decay of turbulence in a stratified fluid, J. Fluid Mech. 144, 231, 280.

Gargett, A. E., P. J. Hendricks, T. B. Sanford, T. R. Osborn and A. J. Williams III (1984) A composite spectrum of vertical shear in the upper ocean, J. Phys. Ocean. 11, 1258-1271.

Gibson, C. H. (1980) Fossil temperature, salinity, and vorticity in the ocean, in Marine Turbulence (ed. J. Nihoul), Elsevier, 221-257.

Gill, A. E. (1982) Atmosphere-Ocean Dynamics, Academic Press.

Grant, H. L., R. W. Stewart and A. Moilliet (1962) Turbulence spectra from a tidal channel, J. Fluid Mech. 12, 241-263.

Grant, H. L., B. A. Hughes, W. M. Vogel and A. Moilliet (1968) The spectrum of temperature fluctuations in turbulent flows, J. Fluid Mech. 34, 423-442.

Itsweire, E. C. (1983) Measurements of vertical overturns in a stably stratified turbulent flow, Phys. Fluids 27, 764-766.

Itsweire, E. C., K. N. Helland and C. W. Van Atta (1985) Evolution of a grid-generated turbulence in a stably stratified fluid, submitted to the Journal of Fluid Mechanics.

Metais, O. (1985) Influence of stable stratification on three-dimensional isotropic turbulence, Abstract submitted to Fifth Symposium on Turbulent Shear Flows, Cornell University, August 1985.

Muller, P. (1984) Small-scale vortical motions, in Internal Gravity Waves and Small-Scale Turbulence, Proceedings of the 'Aha Huliko'a' Winter Workshop, Jan. 1984 (eds. P. Muller and R. Pujalet).

Osborn, T. R. and R. G. Lueck (1984) Oceanic shear spectra from a submarine, in "Internal Gravity Waves and Small-Scale Turbulence," Proceedings of Hawaiian Winter Workshop, Jan. 1984 (eds. P. Muller and R. Pujalet).

Riley, J. J., R. W. Metcalf and M. A. Weissman (1981) Direct numerical simulations of homogeneous turbulence in density-stratified fluids, in Nonlinear Properties of Internal Waves (ed. B. J. West), AIP Conference Proceedings No. 76.

Rohr, J. J., K. N. Helland, E. C. Itsweire, and C. W. Van Atta (1985) Turbulence in a stably stratified shear flow: a progress report, Abstract submitted to Fifth Symposium on Turbulent Shear Flows, Cornell University, August 1985.

Stillinger, D. C., M. R. Head, K. N. Helland and C. W. Van Atta (1983) A closed-loop gravity-driven water channel for density-stratified shear flows, J. Fluid Mech. 131, 73-89.

Stillinger, D. C., K. N. Helland and C. W. Van Atta (1983) Experiments on the transition of homogeneous turbulence to internal waves in a stratified fluid, J. Fluid Mech. 131, 91-122.

Van Atta, C. W., K. N. Helland and E. C. Itsweire (1983) The influence of stable stratification on spatially decaying vertically homogeneous turbulence, in Turbulence and Chaotic Phenomena in Fluids (ed. T. Tatsumi), North-Holland, 519-528.

NUMERICAL SIMULATION OF HOMOGENEOUS TURBULENCE

K. DANG and Ph. ROY
Office National d'Etudes et de Recherches Aérospatiales
BP 72 - 92322 Châtillon Cedex, FRANCE

Abstract

This paper describes the main results of direct simulation and large eddy simulation of homogeneous turbulence submitted to two kinds of constant mean velocity gradients. The Taylor microscale Reynolds number is in the range 20-70. The two strains considered are plane strain and solid body rotation. For the plane strain case, the two described simulations show clearly the reorganizing processes of the turbulent field after each abrupt change of the imposed strain. For the rotation case, three different effects of the rotation are enhanced by a judicious choice of appropriate initial turbulent (isotropic and anisotropic) or random initial conditions.

1. Introduction

The analysis of homogeneous turbulence submitted to constant mean velocity gradient has received a large development in recent years. We can quote the works made at NASA-Ames Research Center [1, 17, 18] by means of Full Turbulence Simulation (F.T.S.) and Large Eddy Simulation (L.E.S.), at Ecole Centrale de Lyon [3, 4] using rapid distorsion theory, spectral closures and L.E.S.. At ONERA, this subject has been studied since 1981 [7, 8, 9, 13, 19, 20]. In this short text, we will sum up the work done and point out the main results obtained for incompressible flows. For complete details, the reader is invited to look at the previous references. A thesis written by Ph. ROY will be available by mid 1985 and a review paper of our work, written by the two authors will supplement this thesis.

In the following, the three first paragraphs are devoted to a rapid description of the numerical method. Paragraphs 5 to 9 give the main results obtained for the numerical simulation of the following turbulence experiments :

i) isotropic turbulence submitted to two successive plane strains
ii) the return to isotropy of an initial anisotropic turbulence
iii) isotropic turbulence submitted to rotation
iv) random quasi 2-D field submitted to rotation
v) initially strained turbulence submitted to rotation.

2. Numerical model

The model uses the change of variables first introduced by R.S. ROGALLO [17] for the numerical simulation of homogeneous turbulence. The particularity of our method compared to ROGALLO's one is that the rapid distorsion theory is not included in our equations and that we carry out also a change of functions which furnishes equations very similar to the Navier-Stokes equations of a rotating fluid [19]. Furthermore, our method is well adapted to treat general mean flows including superimposition of strain and rotation. This formulation allows also the use of arbitrary predistorted domains of computation, which enables a better treatment of highly strained flows [7, 20].

The numerical scheme is based on a pseudo-spectral method in space combined with a second order finite difference time scheme.

3. Codes and software

Two codes are available. The first one is implemented on a CRAY 1 computer. Its modularity and the efficiency of the CFT compiler make very easy the incorporation of different and elaborate subgrid-scale models. The evolution of a passive scalar can be followed too. The outputs are very complete including a lot of second and third order momenta and spectra, and 3-D graphics. The number of points is limited to 40^3. The CPU per time step is then .15 s with simple subgrid-scale model.

The other code is implemented on a MIMD computer [10], built with four AP120B array processors managed by a GOULD SEL 32/77 mini computer. The four AP120B are connected to a sharable memory. Due to the programming complexity, only simple subgrid-scale models are implemented in this code, and no passive scalar.

The outputs are here limited to second order momenta and correlation spectra, and 3-D graphics. The maximum discretisation is 64×128^2 and will be upgrated to 128^3 in July 1985. In this case, the executing time per time step will be of 2 mm. At present, 100 hours of computations per week are available on this system for our application. Typical runs are 200-1000 time steps long.

The graphics furnish 512 x 512 pixels images with 8 color bits. The written software allows the following visualisations.

- isovorticity or isodensity surfaces,
- velocity or vorticity vectors,

- 2-D isovorticity contours,
- repartition of energy in spectral space,
- repartition of velocity phases in spectral space.

4. Subgrid-scale models

Only eddy-viscosity type subgrid-scale models have been used at present, with a sharp cut-off filter in spectral space at cut-off wave-number k_c.

From general results obtained with two-points closure in spectral space [14] for isotropic turbulence, the eddy viscosity (as a function of the wave-number k) is known to exhibit - a time dependent constant part ν_e and - an important cusp near k_c responsible for the local drain of energy through the filter, towards the subgrid-scales.

The effective viscosity ν_e, responsible for non local interactions between wave numbers far from k_c, must be related to mean values representative of the actual state of the turbulence. Several expressions have been tried for isotropic turbulence giving equivalent results [7]. One simple relation proposed by CHOLLET and LESIEUR [6] relates ν_e to the energy $E(k_c)$ at the cut-off.

The cusp localised near k_c has been approximated by a power law $\lambda(k/k_c)^\mu$, extending the superdissipativity model of Basdevant and Sadourny [2] to three-dimensional turbulence.
The model can be written in the form :

$$\nu_T = C1 \sqrt{E(k_c)/kc}\ (1 + \lambda\ (k/k_c)^\mu)$$

In the isotropic case and independently of us, CHOLLET [5] has made some numerical simulations with the same analytical formulation of the cusp. The constants of the model are obtained by matching the power law to the results obtained with the EDQNM closure. As pointed in [5], the constants of the EDQNM closure have been derived for a $k^{-5/3}$ spectrum so that, the constants of the eddy-viscosity model are well suited to a cut-off in the inertial range. As k_c is situated very near the production range in our 16^3 and 32^3 simulations and a little farther in the beginning of the inertial range in our 64×128^2 simulations, it is not surprising that the constants we used are somewhat different from those resulting from the EDQNM closure, and that they must be varied according to the discretization. We intend to address to this problem in more detail later on, when F.T.S with 128^3 points will be available to test our lower discretization L.E.S..

Nevertheless, the subgrid-scale model, as described above, has been extended to the anisotropic case with distorted computing domains. Results obtained with 16^3, 32^3, 64^3 and 64×128^2 discretizations are coherent and coherent with the F.T.S. results.

Subgrid-scale modelling has thus allowed to double approximately the Reynolds numbers of the simulations. The Reynolds numbers R_λ reached (based on the Taylor microscale λ and the effective ν_e) vary from 40 to 72 for the 64×128^2 discretization ($20 < R_\lambda < 30$ for the 64^3 F.T.S).

5. Experiment 1 : isotropic turbulence submitted to two successive plane strains

A turbulent flow passing through two distorting ducts is submitted to two successive plane strains of same rate of strain [11]. The two ducts are connected by a circular section (figure 1) ; this allows to modify the direction of the second strain by rotating the second duct. The main interest of this experiment is to understand which tensor is responsible for the reorientation process of the Reynolds stress tensor observed during the second strain. The experiment cannot furnish this result because the searched tensor is related to the pressure. We found that, as suggested by Lumley and Newman [15], the "return to isotropy tensor" ϕ_{ij} defined by a combination of the slow pressure strain correlation tensor and the dissipation tensor :

$$\phi_{ij} = \langle - 2\, p^{(1)} \; s_{ij} \rangle / \varepsilon + (\varepsilon_{ij} - 2/3\, \varepsilon\, \delta_{ij}) / \varepsilon$$

is linearly related to the anisotropy tensor b_{ij} :

$\phi_{ij} = \beta\ b_{ij}$.

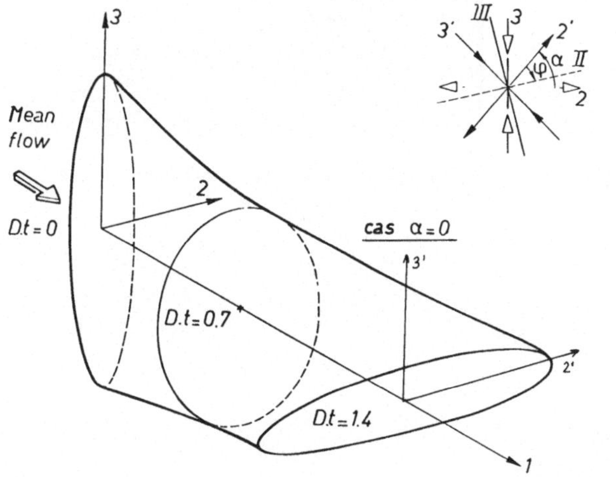

Fig. 1 – Gence's experiment wind tunnel.

This experiment has been performed with 16^3, 32^3 and 64×128^2 L.E.S. and 64^3 F.T.S. giving coherent results. We show on figure 2 the isovorticity surfaces for the initial condition, after the first plane strain, after the second plane strain, for three values of the angle α : $\alpha = 0, \Pi/4, \Pi/2$.

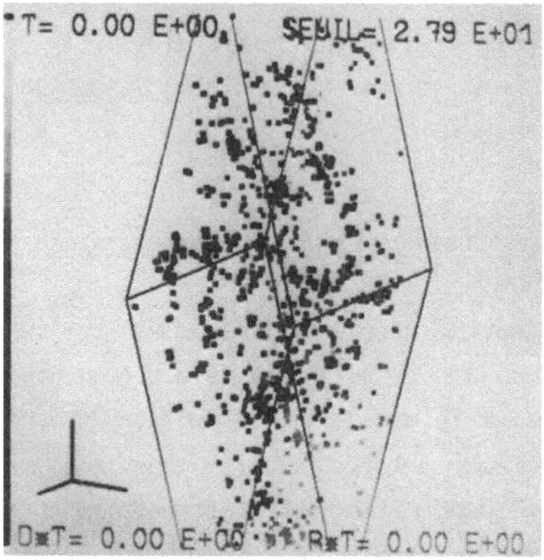

Fig. 2a – Isovorticity surfaces : initial condition (64^3 FTS).

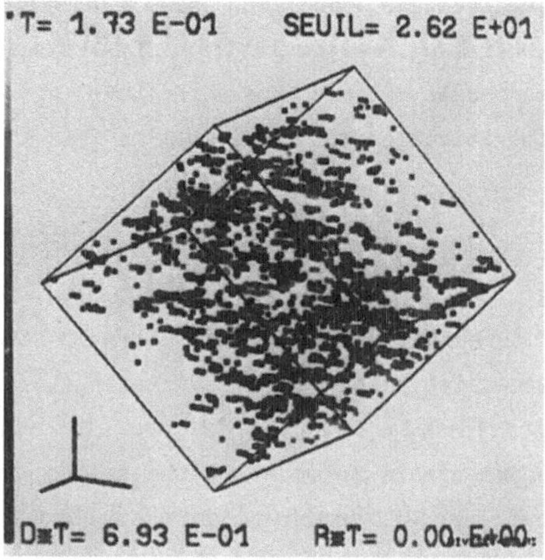

Fig. 2b – Isovorticity surfaces : after the first plane strain (64^3 FTS).

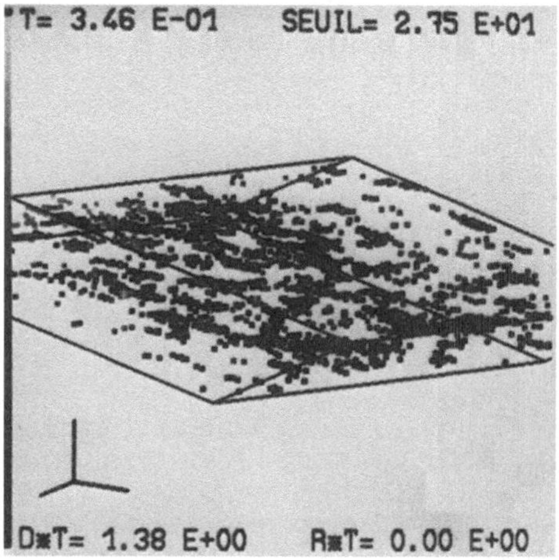

Fig. 2c – Isovorticity surfaces : after the second plane strain ($\alpha = 0$, 64^3 FTS).

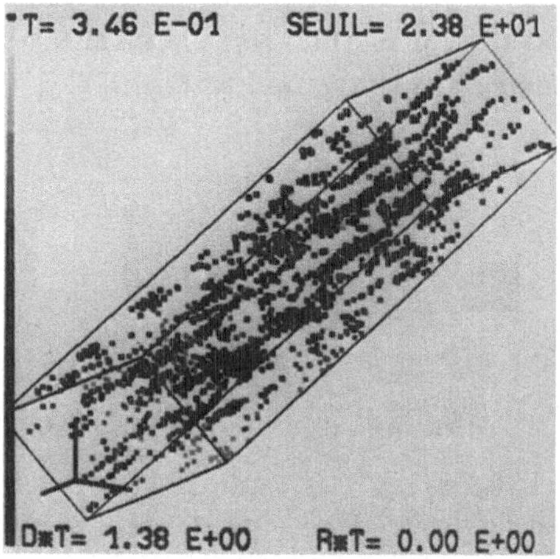

Fig. 2d – Isovorticity surfaces : after the second plane strain ($\alpha = \pi/4$, 64^3 FTS).

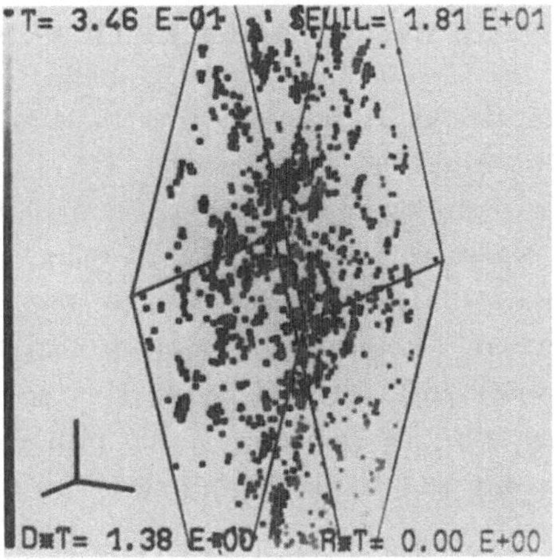

Fig. 2e – Isovorticity surfaces : after the second plane strain
$(\alpha = \pi/2, 64^3$ FTS).

6. Experiment 2 : the return to isotropy of an initial anisotropic turbulence

We have made some return to isotropy runs, starting from the final field of the previous experiment. In each case, after a transitional state (during which, small scales return first to isotropy), large scales return to isotropy following approximately a linear process. Comparisons with Rotta's model :

$\phi_{ij} = \beta b_{ij}$,

give a constant value :

2.6 < β < 2.8.

The computed rate of return to isotropy defined by the ratio τ_D/τ_R of time scales related to the decay of turbulence energy and the decay of the second invariant II of b_{ij} are coherent with the simulations of Rogallo for axisymmetric strain and with the experiment of Gence (figure 3).

$$\tau_D = -\frac{\overline{q^2}}{\frac{d\overline{q^2}}{dt}} \quad \tau_R = \frac{II}{\frac{dII}{dt}} \quad q = <u_i u_i> \\ II = -b_{ij} b_{ij}/2$$

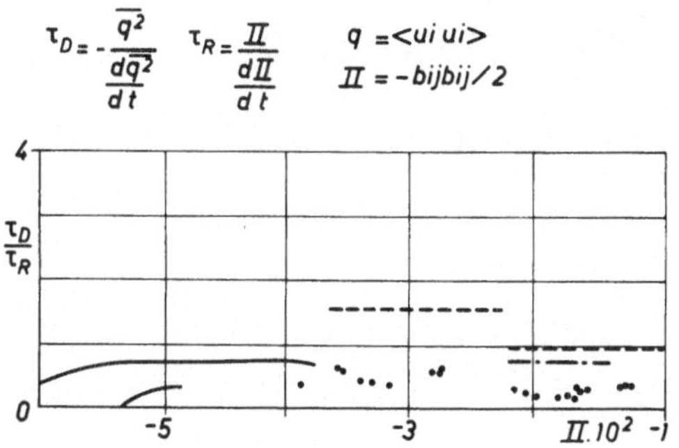

Fig. 3 – Rate of return to isotropy.

7. Experiment 3 : isotropic turbulence submitted to rotation

This simulation has been done to resolve some contradictions which appeared in WIGELAND and NAGIB's experiment [21], i.e. an increase or decrease of energy decay rate according to the rotation rate. Our Rossby number varies from 1. to 10^{-3}. As it decreases the decay rate of energy always decreases (figure 4). No anisotropy appears in the velocity spectra for the rotating turbulence case. Neither is apparent a tendency to two-dimensionality (PROUDMAN-TAYLOR theorem). CAMBON [4] suggests that rotation should be imposed during a very long time before this phenomenon can be observed. But in numerical simulation, the energy decay is very fast and quickly the flow is no longer turbulent.

On the other hand, we have observed differences on the shape of energy spectra (figure 5) and also of isovorticity surfaces (figure 6), the vortices being less stretched than in cases without rotation. A detailed analysis of the contribution of each spectral plane normal to the rotation axis to the total energy spectrum shows an anisotropic distribution which could perhaps be related to an evolution towards two-dimensionality [4]. The time evolution of transfer spectra shows the modulation of the transfer rate by the rotation (figure 7).

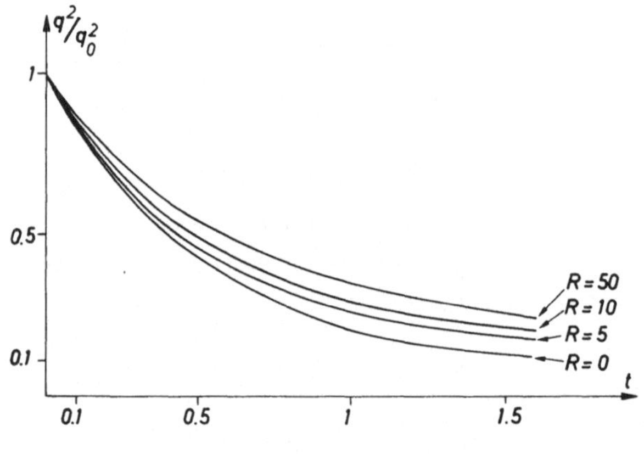

Fig. 4 – Time evolution of the energy (64^3 FTS) (R = dimensionless rotationrate).

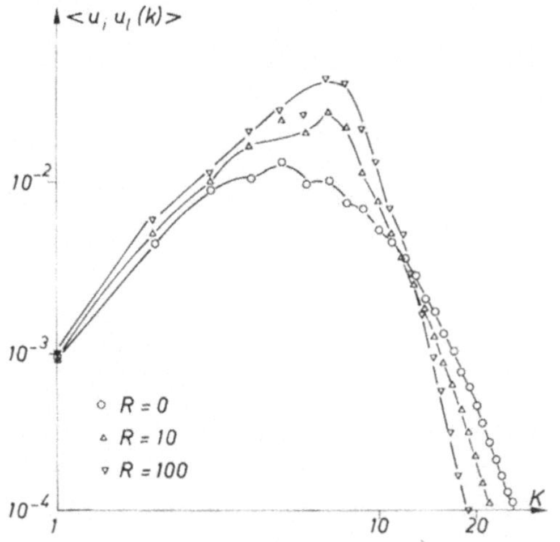

Fig. 5 – Energy spectra at T = 1.6 for R = 0, 10 and 100 (64^3 FTS).

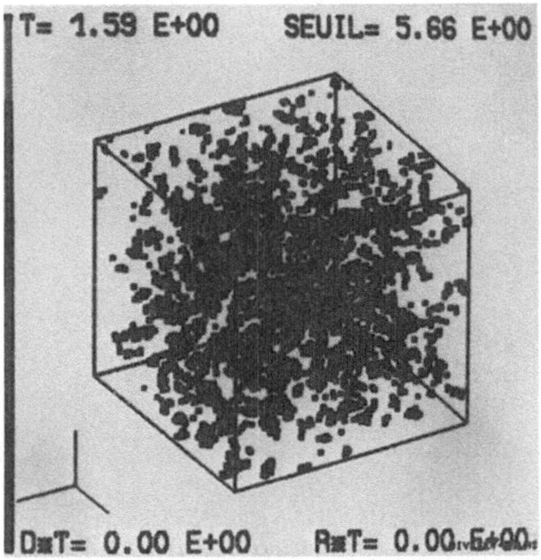

Fig. 6a – Isovorticity surfaces – The rotation axis is vertical : T = 1.6 – R = 0. (64^3 FTS).

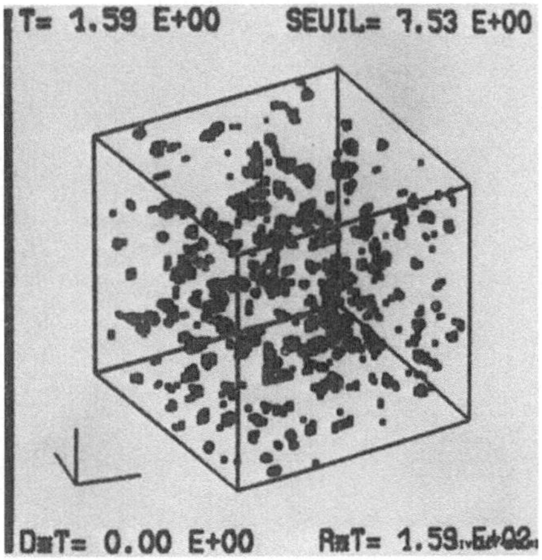

Fig. 6b – Isovorticity surfaces – The rotation axis is vertical : T = 1.6 – R = 100. (64^3 FTS).

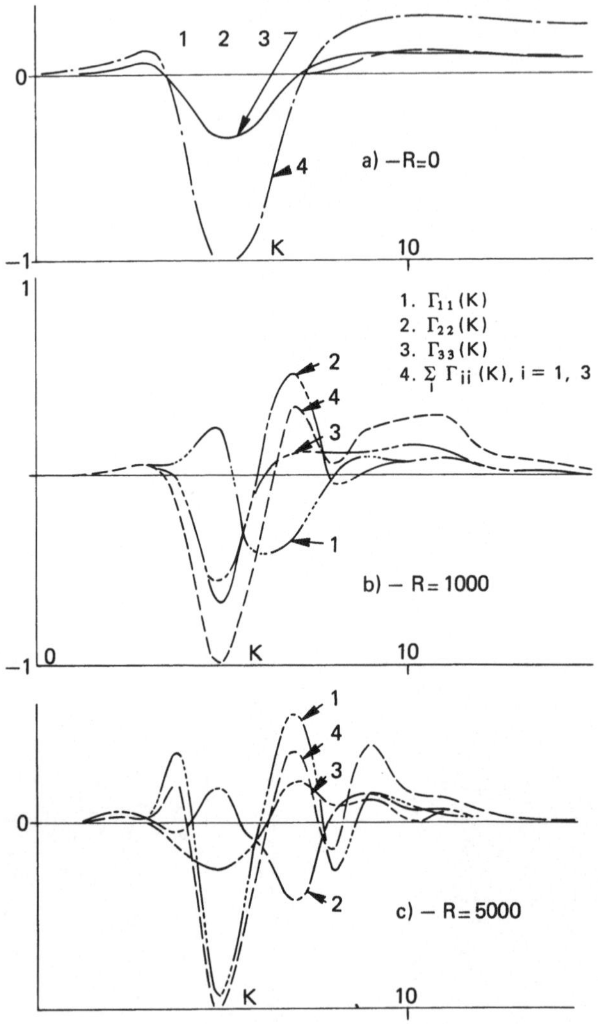

Fig. 7 – Inertial transfer spectra. t = 0.025. (32^3 LES).

8. Experiment 4 : random quasi 2-D field submitted to rotation.

The special initial field of this simulation has been introduced in order to enhance the transition phenomenon from 3-D to 2-D turbulence in presence of rotation. It can be seen as an attempt to simulate (using a homogeneous turbulence code) the experiment of HOPFINGER et al [12, 16]. The initial condition is a random field that is 2-D for large scales ($5 < k < 8$, 85 % of total energy) and 3-D for small scales ($15 < k < 17$, 15 % of total energy). The corresponding isovorticity surfaces are shown on figure 8a. In the absence of rotation, the flow evolves towards a 3-D turbulent flow (isovorticity surfaces on figure 8.b), with isotropic small scales. The large 2-D structures oscillate and break down. The 3-D perturbation propagates through all scales.

In the presence of rotation, this 3-D perturbation cannot propagate and is only dissipated. The flow evolves towards a 2-D configuration (figure 8c). This situation can also be observed in spectral space. We present on figure 9 the distribution of energy in the three planes $k_i = 0$ (the rotation axis is vertical) for two cases, with and without rotation.

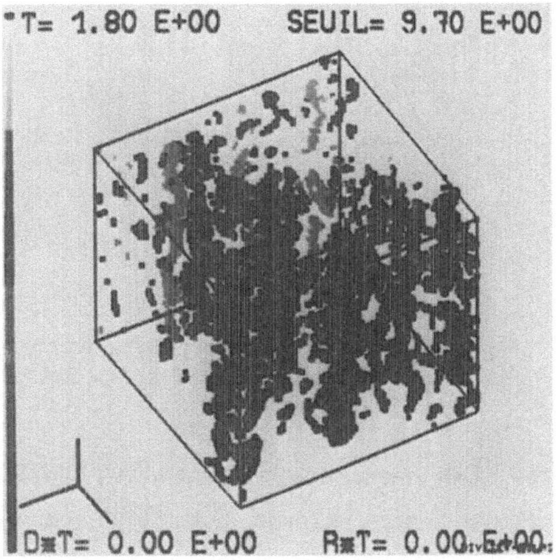

Fig. 8a – Isovorticity surfaces – The rotation axis is vertical : T = 0 – initial condition (64^3 FTS).

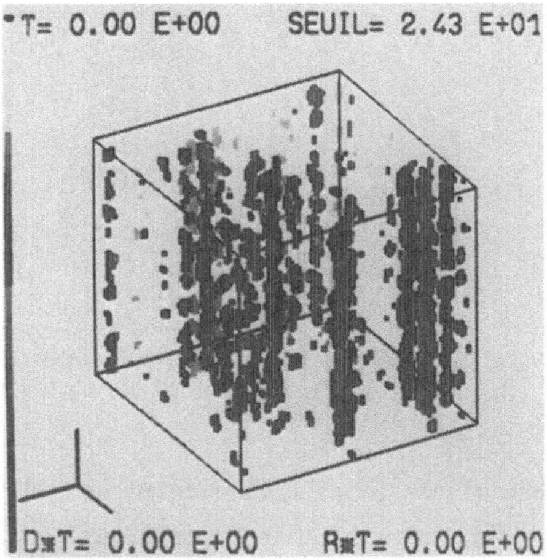

Fig. 8b – Isovorticity surfaces – The rotation axis is vertical : T = 1.8 – R = 0 (64^3 FTS).

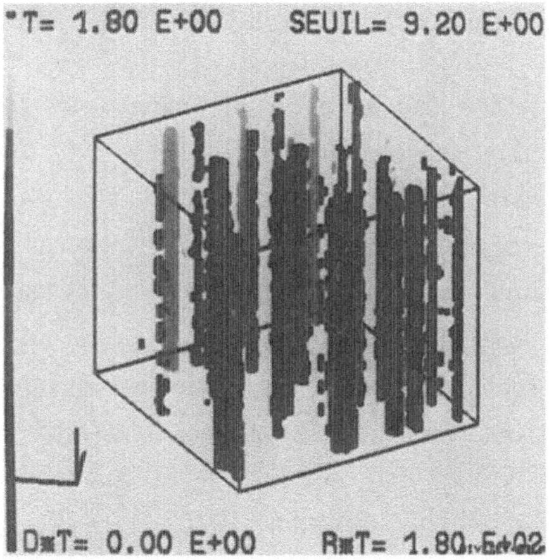

Fig. 8c – Isovorticity surfaces – The rotation axis is vertical : T = 1.8 – R = 100 (64^3 FTS).

Without rotation, the energy distribution is isotropic, while with rotation it is mainly located in the plane normal to the rotation axis. This is verified as well for the total energy ($u_i u_i$) as for the three velocity components. The 3-D part of the energy remaining in the rotation case surrounds a domain through which linear inviscid theory indicates an inhibition of energy transfer (this domain is composed of two spheres centered on the rotation axis and tangent to the plane normal to rotation axis).

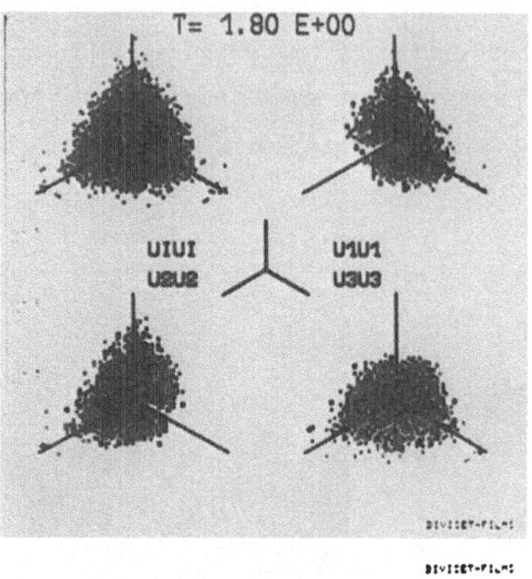

Fig. 9a – Energies in the planes $k_i = 0$ (k_3 vertical) – The rotation axis is vertical : R = 1.8 – R = 0 (64^3 FTS).

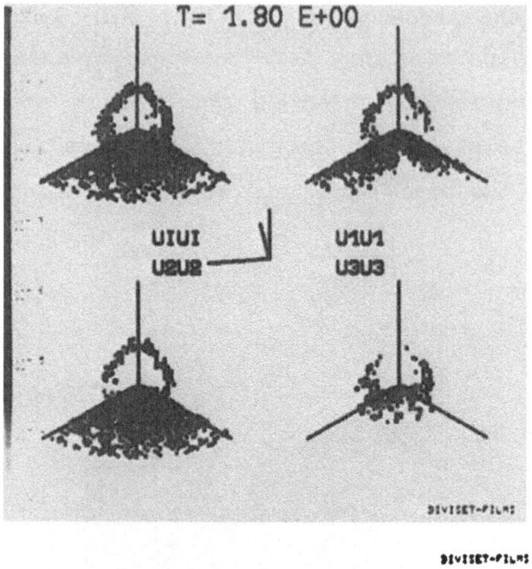

Fig. 9b – Energies in the planes $k_i = 0$ (k_3 vertical) – The rotation axis is vertical : T = 1.8 – R = 100 (64^3 FTS).

This simulation, first carried with 64^3 F.T.S., has been repeated with 64^3 L.E.S.. In this case, the 3-D perturbation is dissipated more slowly but still cannot propagate.

9. Experiment 5 : initially strained turbulence submitted to rotation

The initial condition is the field obtained after the first plane strain in the simulation of the plane strain experiment (§ 5), with 32^3 and 64×128^2 L.E.S. The rotation axis is normal to the plane of strain. All the mean values (Reynolds stress deviator and invariants) indicate a very fast and oscillatory return to some isotropy properties. In fact, it is not an evolution towards isotropic turbulence but only an equirepartition of energy between the three velocity components, the energy keeping roughly in spectral space the same anisotropic distribution of the initial field. This kind of result has also been observed by CAMBON [4] using spectral closures.

For comparison with the return to isotropy experiments described in § 6, we present on figure 10 the evolution of the ratio τ_D/τ_R. It also shows a very rapid and oscillatory return to isotropy.

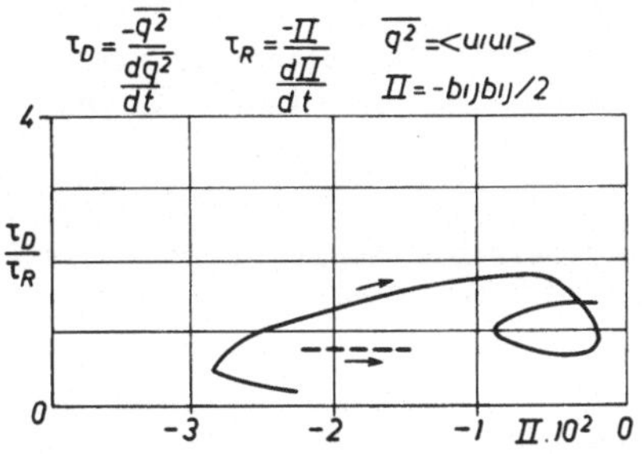

Fig. 10 – Rate of return to isotropy.

This experiment is a sample case where statistical quantities are inadequate to describe the real state of the flow. The simulation will be continued removing the rotation to observe how this pseudo isotropic field returns to isotropy.

10. Conclusion

The numerical simulations of a number of homogeneous turbulence experiments involving plane strains and rotation have demonstrated :

- the feasibility of the eddy-viscosity subgrid-scale model for the simulation of anisotropic turbulence (for the types of strains considered here).
- the interest of considering anisotropic initial turbulent or random fields.

Results obtained at low Reynolds numbers $R_\lambda < 70$ concern principally :

- the response of turbulence characteristic tensors when submitted to abrupt changes of the imposed strains.
- the inhibition of turbulent energy transfer by a solid-body rotation.
- the return to isotropy after an imposed strain with and without rotation.

With the new 128^3 F.T.S. and L.E.S. codes, we intend :

- to confirm the previous results for higher Reynolds number ($R_\lambda < 200$).
- to test more elaborate subgrid-scale models in the more difficult cases of highly sheared turbulence with or without extra rotation.

References

[1] BARDINA, J., FERZIGER, J.H., and REYNOLDS, W.C. :
"Improved turbulence models based on Large Eddy Simulation of homogeneous, incompressible, turbulent flows",
Report TF-19, Stanford University, Stanford, Calif., May 1983.

[2] BASDEVANT, C., and SADOURNY, R. :
"Parametrisation of virtual scale in numerical simulation of two dimensional turbulent flows",
J. Méc. Th. et Appl. n° spécial, 1983.

[3] BERTOGLIO, J.P. :
"A stochastic subgrid model for sheared turbulence". Proceedings of workshop on macroscopic modelling of turbulent flows and fluid mixtures - Springer-Verlag, 1985, Ed. O. PIRONNEAU.

[4] CAMBON, C. :
"Etude spectrale d'un champ turbulent incompressible soumis à des effets couplés de déformation et de rotation, imposés extérieurement". Thèse de doctorat d'Etat Univ. Cl. Bernard, Lyon, 1982.

[5] CHOLLET, J.P. :
"Statistical closure to derive a subgrid-scale modeling for large eddy simulations of three-dimensional turbulence", NCAR Technical Note, TN-206-STR, Boulder, Colorado, 1983.

[6] CHOLLET, J.P., LESIEUR, M. :
"Paramaterization of small scales of three-dimensional isotropic turbulence utilizing spectral closures", J. of Atmos. Sciences, vol. 38, 1981.

[7] DANG, K. :
"Evaluation of simple subgrid-scale models for the numerical simulation of anisotropic turbulence", AIAA paper 83-1692, to be published in the AIAA Journal, 1983.

[8] DANG, K. :
"Numerical simulation of homogeneous turbulence submitted to plane strains and rotation". Euromech 180 Colloquium on "Turbulence Modelling", Karlsruhe (RFA) 4-6 July 1984. TP ONERA 1984-54.

[9] DELORME, Ph. :
"Simulation numérique de turbulence homogène, isotrope, bidimensionnelle pour un fluide compressible". La Recherche Aérospatiale, 1984-1.

[10] ENSELME, M., FRABOUL, C., LECA, P. :
"A MIMD architecture system for PDE numerical simulation". 5th IMACS Symposium on Computer Methods for Partial Differential Equations Proceedings, Betlehem USA, 19-21 June 1984.

[11] GENCE, J.N. :
"Action de deux déformations planes sucessives sur une turbulence isotrope", Thèse de Doctorat d'Etat, Université Claude Bernard, LYON, 1979.

[12] HOPFINGER, E.J., BROWAND, F.K., GAGNE, Y. :
"Turbulence and waves in a rotating tank", J.F.M., vol. 125, 1982.

[13] LECA, P., ROY, Ph. :
"Simulation numérique de la turbulence sur des mini-systèmes à processeurs attachés (en configuration mono ou multi-processeur), La Recherche Aérospatiale n° 1983-4. French and English editions.

[14] LESLIE, D.C., QUARINI, G.L. :
"The application of turbulence theory to the formulation of subgrid modelling procedures", J.F.M. 91, 65-91 (1979).

[15] LUMLEY, J.L., NEWMAN, G.R. :
"The return to isotropy of homogeneous turbulence, J.F.M. 82, 161-178, 1977.

[16] MORY, M. and HOPFINGER, E.J. :
"Rotating turbulence evolving freely from an initial quasi 2-D state", Proceedings of Workshop on macroscopic modelling of turbulent flows and fluid mixtures. Springer-Verlag, 1985, Ed. O. PIRONNEAU.

[17] ROGALLO, R.S. :
"Numerical experiments in homogeneous turbulence", NASA TM 81315-1981.

[18] ROGALLO, R.S., MOIN, P. :
"Numerical Simulation of Turbulent Flows", Ann. Rev. Fluid Mech., Vol 16, 1984.

[19] ROY, Ph. :
"Numerical simulation of homogeneous anisotropic turbulence", 8th Int. Conference on Numerical Methods in Fluids Dynamics, Proceedings, Springer-Verlag, Lecture Notes in Physics vol 170, Ed. E. KRAUSE.

[20] ROY, Ph. :
"Numerical simulation of homogeneous turbulence submitted to two successive plane strains and to solid-body rotation", Proceedings, 4th Seminar on "MHD Flows and Turbulence", Beer Sheva (Israel), 1984.

[21] WIGELAND, R.A., NAGIB, H.M. :
"Grid generated turbulence with and without rotation about the streamwise direction". IIT Fluids and heat transfer report R78-1. Illinois institute of technology, 1978.

TIME-DEPENDENT RAYLEIGH-BENARD CONVECTION IN LOW PRANDTL NUMBER FLUIDS

M. Meneguzzi,(1) C. Sulem,(2) P.L. Sulem(3) and O. Thual(4).

(1) CNRS and Service d'Astrophysique, CEA, Saclay, France.
(2) Dept. of Mathematics, Ben Gurion University of the Negev, Beer Sheva, Israel and CNRS, Université de Nice, France.
(3) School of Mathematics, Tel Aviv University, Israel and CNRS, Observatoire de Nice, France.
(4) Centre National de Recherches Météorologiques, Toulouse-Mirail, France.

ABSTRACT

We present three-dimensional numerical simulations of time-dependent convection in low Prandtl number fluids confined between two infinite horizontal bounding surfaces maintained at constant temperatures. We consider the case of free slip boundary conditions for a fluid of Prandtl number $Pr = 0.2$ and that of nonslip boundary conditions for a fluid with $Pr = 0.025$. In the former situation, we observe stationary, periodic, bi-periodic and chaotic regimes as the Rayleigh number is increased. In the latter situation, the characteristic times have different orders of magnitude and the transients have a long persistence. The first bifurcations to oscillatory regimes are obtained in this case.

1. INTRODUCTION

In connection with the development of the theory of dynamical systems, time-dependent convection in low Prandtl number fluids has recently been the object of experimental studies. Libchaber and Maurer [1] investigated the convection in a small box of helium (Prandtl number Pr=0.1) and observed a transition to chaos through a sequence of period doubling bifurcations. More recently, experiments were performed with mercury (Pr=0.025) by Libchaber, Fauve and Laroche [2]. They investigate the effect of a magnetic field parallel to the rolls and observed different routes to chaos, depending on the intensity of the magnetic field and on the values of the aspect ratios.

The influence of lateral boundaries is generally not considered in theoretical or numerical investigations: periodicity is often imposed in the horizontal directions. On the upper and lower boundaries, the fluid is assumed to slip freely or to be at rest. In the former situation, assuming perfectly conducting boundaries, Schluter, Loltz and Busse [3] computed analytically the steady solution corresponding to two-dimensional rolls for Rayleigh number slightly in excess of the onset of convection. Its stability for low Prandtl number fluids was discussed in Busse [4]. In the case of no-slip boundary conditions, the steady two-dimensional solution was computed numerically by Clever and Busse [5],[6], who also analyzed its stability in different ranges of Prandtl number.

Three-dimensional simulations of the evolution problem have been performed with impermeable free-slip boundary conditions [7] and rigid boundary conditions [8],[9] at moderate or order one Prandtl number. Transitions to chaos by scenarios which support the theory of Ruelle, Takens and Newhouse [10] were observed. In this paper, we use similar numerical methods to investigate time-dependent convection at low Prandtl numbers. We use $Pr = 0.2$ in the case of free-slip conditions and $Pr = 0.025$ for rigid boundary conditions. One of the interests of low Prandtl number fluids is that the dynamics are time-dependent as soon as the two-dimensional convective rolls become unstable (oscillatory instability). On the other hand, the difference of magnitude of the time-scales makes numerical simulations more delicate.

2. DYNAMICAL EQUATIONS AND COMPUTATIONAL TECHNIQUES

We use the thickness d of the fluid layer as length unit and the vertical viscous time d^2/ν (where ν is the kinematic viscosity) as time unit. In addition the temperature deviation θ from the diffusive profile is measured in units $\Delta T\,Pr$ where ΔT is the temperature difference between the lower and upper boundary. The Boussinesq equations then read:

$$\begin{aligned} \frac{\partial v}{\partial t} + v\cdot\nabla v &= -\nabla p + \nabla^2 v + R\theta\,\hat{e}_3 \\ \frac{\partial \theta}{\partial t} + v\cdot\nabla\theta &= \frac{1}{Pr}(\nabla^2\theta + v\,\hat{e}_3) \\ \nabla\cdot v &= 0 \end{aligned} \tag{1}$$

In eq.(1), $\hat{e}_3$ denotes the upward vertical unit vector. The Rayleigh number and the Prandtl number are defined as

$$\begin{aligned} R &= \frac{\alpha g \Delta T d^3}{\nu\kappa\rho_0} \\ Pr &= \kappa/\nu \end{aligned} \tag{2}$$

where κ is the thermal diffusivity, g the gravity field, ρ_0 the mass of the unit volume and α the coefficient of thermal expansion. The conditions on the bounding planes $z = \pm 1/2$ and $\theta = v_3 = 0$ and $v_1 = v_2 = 0$ for no-slip conditions and $\partial v_1/\partial z = \partial v_2/\partial z = 0$ for free-slip conditions.

The numerical computations were performed using pseudo-spectral methods in space and second order finite differences in time. In the horizontal directions, the solutions are expanded in exponential Fourier series (24×24 modes for free-slip boundary conditions and 16×10 or 16×16 modes for rigid boundary conditions are retained). In the vertical direction, they are expanded in Chebyshev polynomials

(33 polynomials retained) in the case of rigid boundary conditions, and in cosine series for the horizontal velocity components or in sine series for the vertical velocity component and the temperature deviation in the case of free-slip boundary conditions (12 modes are then retained). Concerning the temporal scheme, for free-slip boundary conditions, the dissipative terms are treated exactly (through exponential factors in Fourier space); the advection terms are treated by a leap-frog scheme and the buoyancy force by a Crank-Nicolson scheme. In the case of no-slip boundary conditions, Adams-Bashforth is used for the advection and buoyancy terms, Crank-Nicolson for the dissipative terms. To solve the Poisson equation obeyed by the pressure, we impose that the divergence of the velocity field vanishes on the boundary of the domain [11]. This condition ensures that the incompressibility of the flow is preserved by the temporal scheme. Technical details on the implementation of the method can be found in [12], where preliminary results have been reported in the case of rigid boundary conditions.

3. TIME-DEPENDENT CONVECTION WITH FREE-SLIP BOUNDARY CONDITIONS

When the Rayleigh number reaches a critical value $R_{cr} = 27\pi^2/4 \approx 657.5$, disturbances of wavenumber $k_x = \pi/\sqrt{2}$ become unstable [13], independently of the value of the Prandtl number. There is then a range of Rayleigh numbers where the system evolves to a steady state where the convective patterns are two-dimensional rolls. At low Prandtl number, this range is limited by the oscillatory instability which occurs at $R_{osc} \approx R_{cr}(1+0.3\ Pr^2)$ [4]. New patterns are then generated by modes which were damped at lower Rayleigh numbers and become coupled with the basic two-dimensional rolls when the Rayleigh number is increased. The dominant effect is due to the inertial term $v \cdot \nabla v$ which couples vertical vorticity modes with basic two-dimensional rolls.

The numerical simulation was done at $Pr = 0.2$, assuming basic wavenumbers $k_x = k_y = \pi/\sqrt{2}$. At $R = 700$, we observe that the rolls are stable. When the Rayleigh number is increased to $R = 710$, we observe the development of the oscillatory instability and its saturation leading to a time-periodic solution (Fig.1). In the physical space, we observe the propagation of waves travelling in the direction of the rolls.

When the Rayleigh number is increased to $R = 800$, the solution is again asymptotically time-periodic but we observe a transient where the amplitude of the oscillation is slowly modulated (Fig.2). Fig.3 illustrates the dynamics of the convective patterns in the physical space. Note that in addition to horizontal oscillations of the rolls, vertical collective motions are visible.

At $R = 900$, we observe a bi-periodic regime as clearly seen by considering the time Fourier transform of the temperature of a velocity component at a given point of

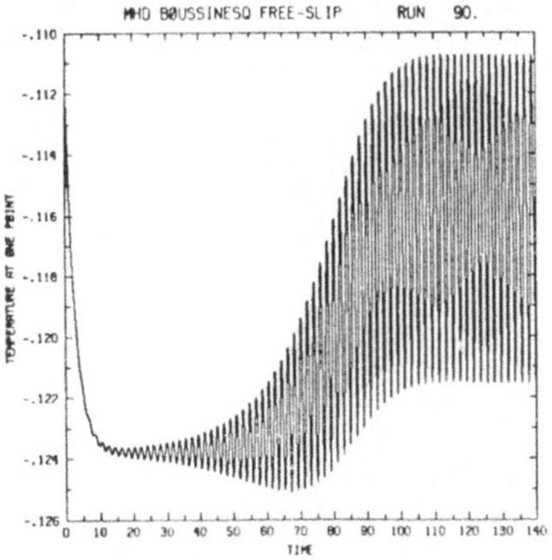

Fig.1: Temperature at a given point as a function of time (in thermal diffusive time units) at R=710. The initial condition is the steady state at R=700 (free slip B.C., Pr=.2).

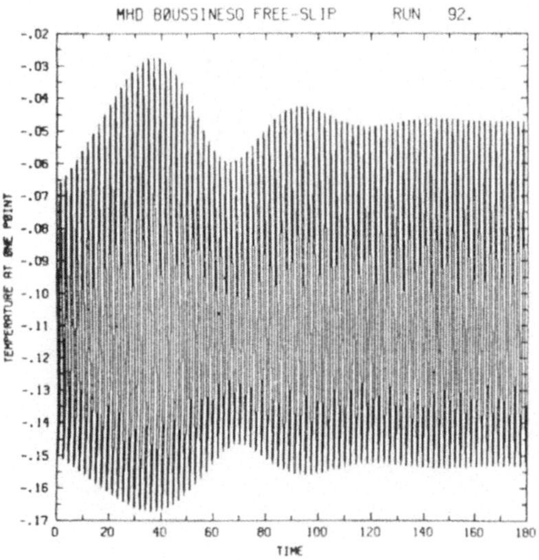

Fig.2: Temperature at a given point as a function of time (in thermal diddusive time units) at R=800 (free slip B.C., Pr=.2).

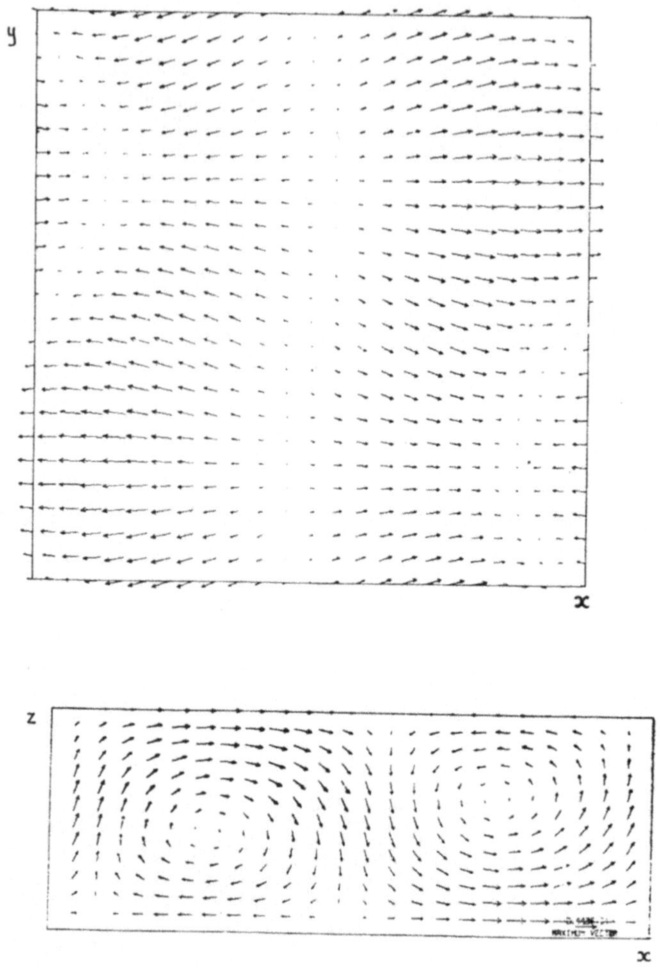

Fig. 3: Velocity fielf for R=800 (free slip B.C.; Pr=.2).

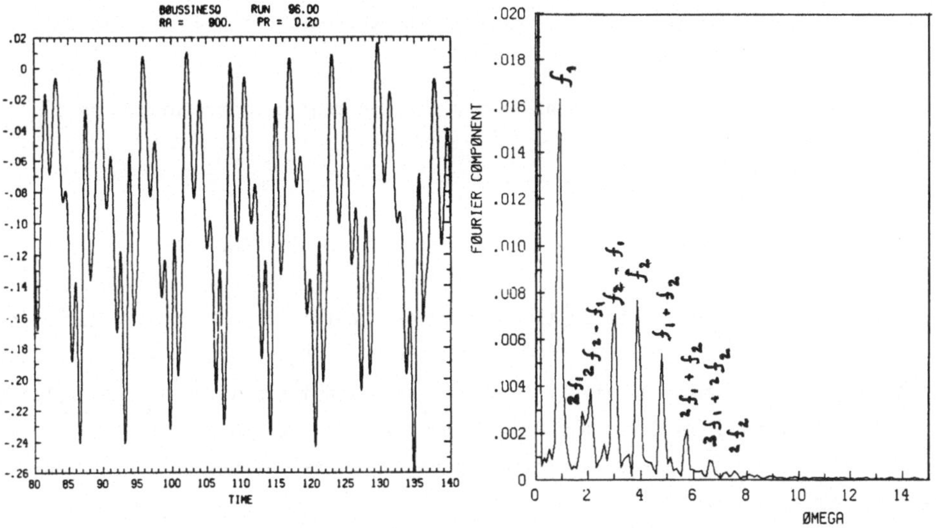

Fig. 4: Temperature at a given point as a function of time (in thermal diffusive time units) and its Fourier transform ar R=900 (free-slip B.C., Pr=.2).

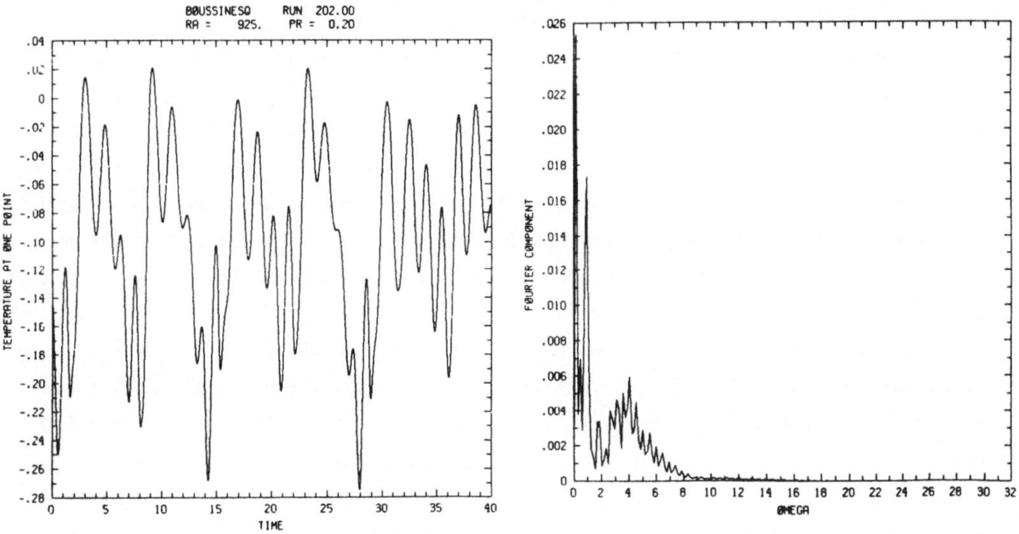

Fig. 5: Same as Fig.4 at R=925.

the flow (Fig.4). At this Rayleigh number, the two basic frequencies are in the ratio $f_1/f_2 = 0.19$.

Finally, for $R = 925$, we observe a chaotic regime characterized by a continuous spectrum (Fig.5).

4. TRANSITION TO OSCILLATORY REGIMES WITH RIGID BOUNDARY CONDITIONS

When the flow is at rest on the boundaries, convection occurs at a Rayleigh number $R_{cr} \approx 1708$ [13]. Disturbances of wavenumber $k_x = 3.117$ are then linearly unstable. In the numerical simulations, we use wavenumbers which are multiples of $k_x = 3.117$ and $k_y = 2.5$; the Prandtl number $Pr = 0.025$ is that of the mercury. Fig.6 shows the steady state corresponding to two-dimensional rolls which establishes at $R = 1800$.

In the conditions we have considered, the two-dimensional patterns become unstable for three-dimensional disturbances when the Rayleigh number reaches $R_{osc} \approx 1885$ (oscillatory instability) [14]. The development of this instability at $R = 2000$ is seen of Fig.7 which represents the temperature at a point of the flow as a function of time. At early time, we observe an exponential growth of the amplitude and a period of oscillation $T_{osc} \approx 0.065$ viscous times or equivalently 2.6 thermal diffusive time. This corresponds to a frequency $\sigma_{osc} = 2\pi/T_{osc} = 2.4$ thermal units, in agreement with Busse and Clever [5]. At later time, we see a non-linear saturation of the amplitude. The amplitude is, however, strongly modulated and the question arises whether this is a transient analogous to that was observed with free-slip boundary conditions at $Pr = 0.2$ when $R = 800$. Investigation of this point requires a very long integration time (several tens of hours on a Cray 1 computer). Such a long computation was done at $R = 1925$ where we observe a similar modulation (Fig.8). The computation was performed during 28 viscous times (corresponding to 1120 thermal diffusive times), starting from the three-dimensional solution at $R = 1950$ at a given instant of time. We observe that the amplitude of the modulation is damped at a very slow rate and that the solution becomes eventually time-periodic. The dynamics is thus similar to that observed at $R = 800$ and $Pr = 0.2$ with free-slip boundary conditions, but the transient is much longer. Fig.9 shows the oscillation in physical space. The main observation is that the rolls oscillate not only horizontally but also vertically.

When the Rayleigh number is reduced to $R = 1900$, we observe a weak and slow modulation. But when starting from the solution for $R = 1900$ at an instant of time, the Rayleigh number is reduced to $R = 1895$, the amplitude of the oscillation relaxes exponentially to a constant, leading to a time-periodic solution (Fig.10). Finally, when R is reduced to 1880, we observe that the oscillation is exponentially damped, in agreement with the linear stability analysis of the two-dimensional

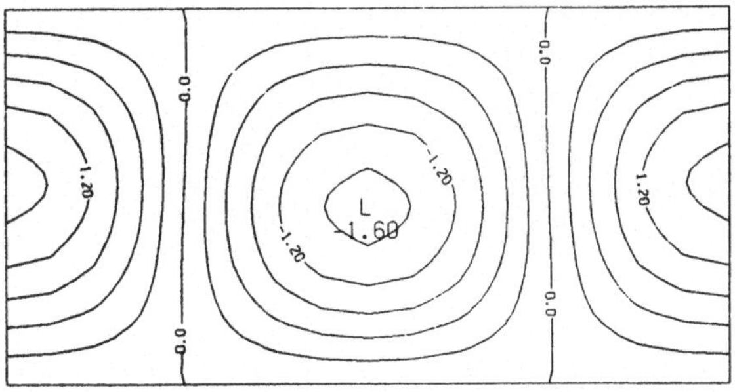

Fig. 6 a

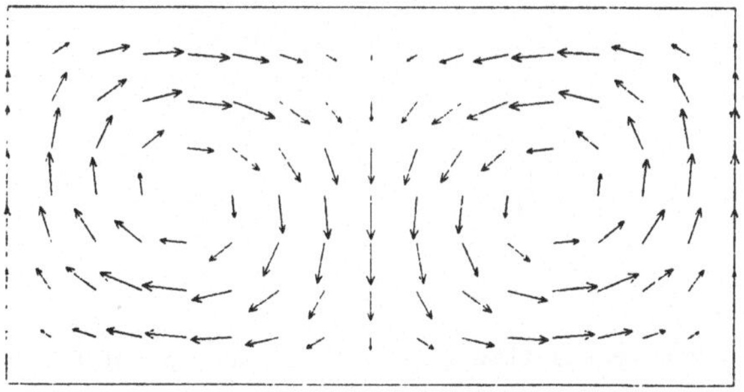

Fig. 6 b

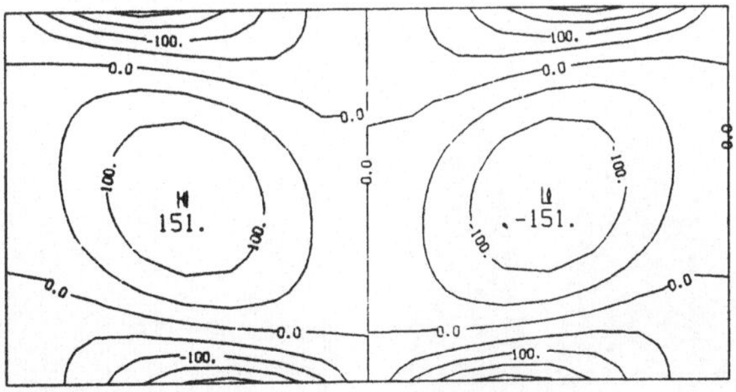

Fig. 6 c

Fig.6: Steady two-dimensional solution at R=1880 (no slip B.C., Pr=.025): (a) temperature deviation from the diffusive profile; (b) velocity; (c) vorticity.

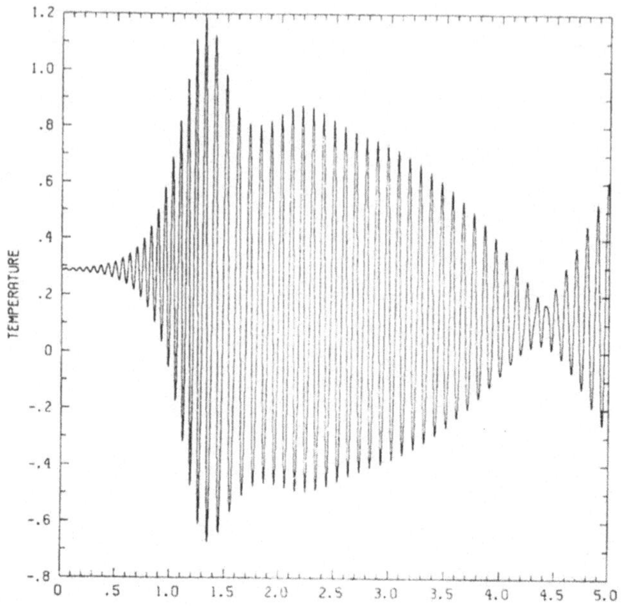

Fig.7: Temperature **versus** time (viscous time units) at R=2000 (no slip B.C.; Pr=.025). The initial condition corresponds to the two-dimensional **steady** solution with slight three-dimensional disturbances.

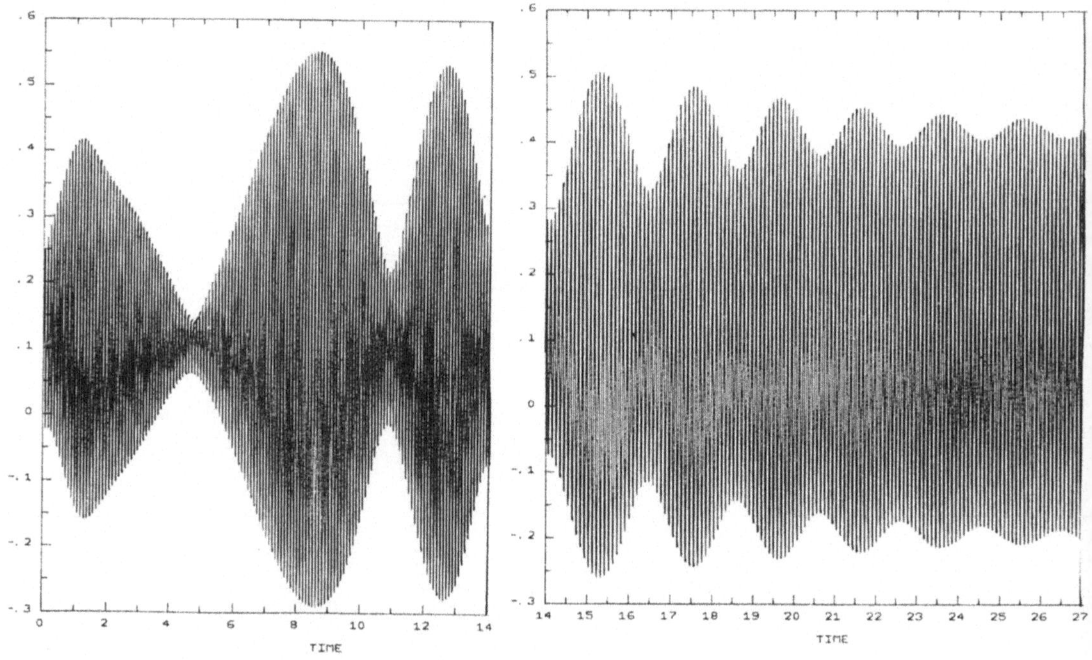

Fig.8: Temperature versus time (viscous time units) at R=1925 (no slip B.C.; Pr=.025).The initial condition corresponds to the state of the flow at a given time for R=1950.

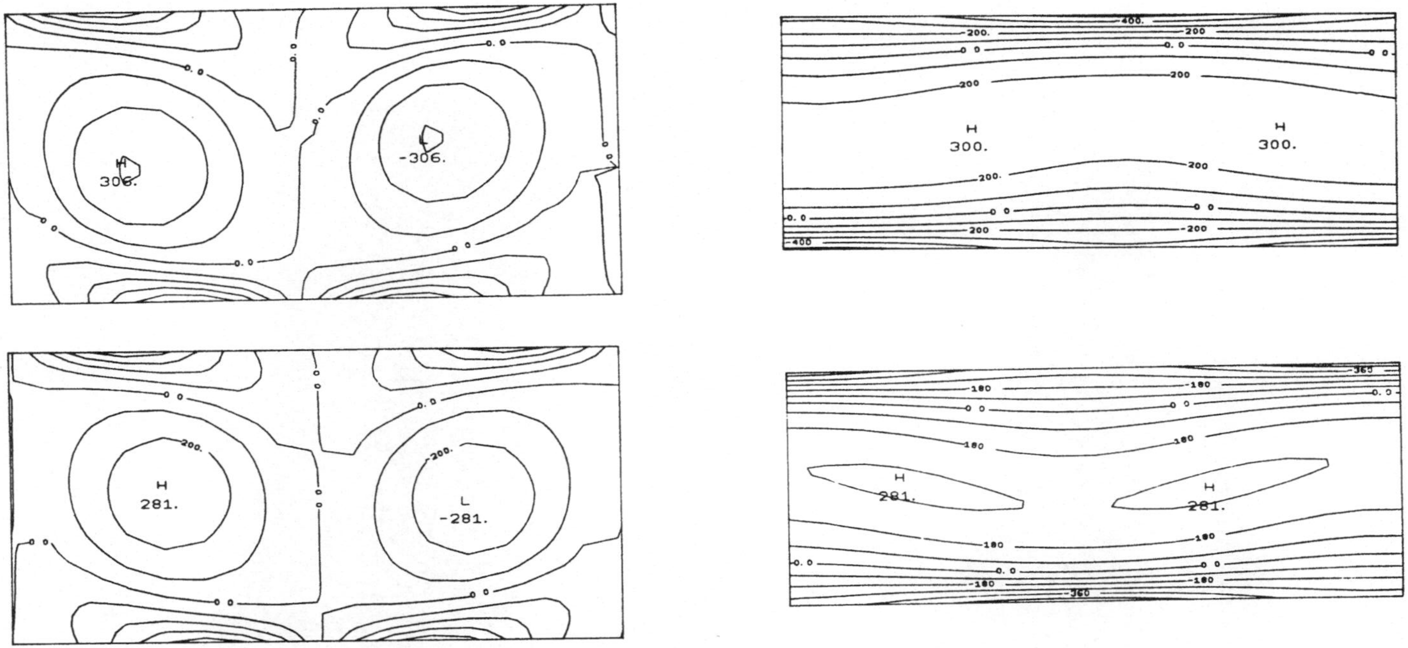

Fig. 9: **Contours** of the vorticity component in the roll direction at two different times (no slip B.C.; Pr=.025).

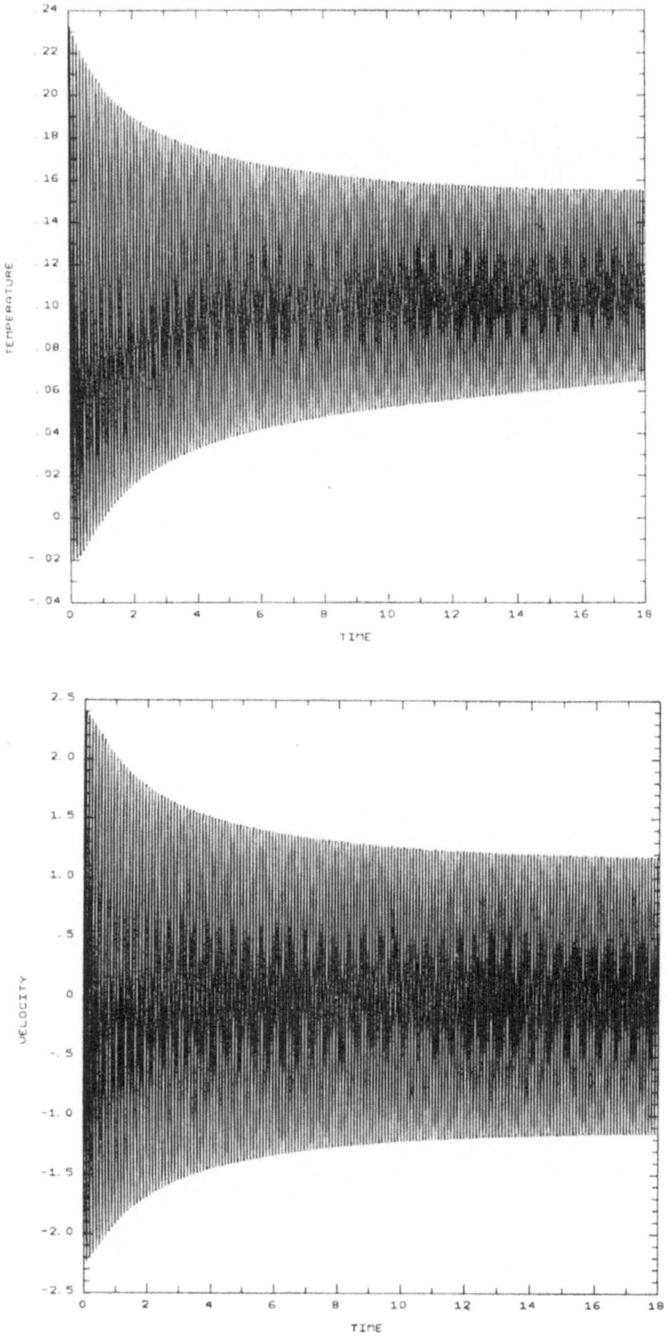

Fig. 10: Temperature and velocity in the roll-direction as functions of time (viscous time units) at R=1895 (no slip B.C., Pr=.025).

rolls [14].

5. CONCLUDING REMARKS

One of the main difficulties in the numerical simulation of convection at very low Prandtl number is the rapid rotation of the rolls. This imposes a drastic constraint on the time step to insure the stability of the numerical scheme when the advection terms are treated with the usual explicit scheme. For $Pr = 0.025$ and rigid boundary conditions, this makes calculations at high Rayleigh number prohibitively expensive in computer time: the spatial resolution has to be increased with the Rayleigh number and the transients persist for very long times. As a consequence, only the first bifurcations to time-dependent convection have been computed in this situation. We also investigated the influence of a magnetic field parallel to the convective rolls when the magnetic Prandtl number is very small (10^{-6} in mercury). The magnetic field then essentially produces an anisotropic dissipation which tends to stabilize the two-dimensional solutions. In the range of Rayleigh numbers we have considered ($R \lesssim 2000$), we observe that the effect of the magnetic field is to increase the threshold for the oscillatory instability (in agreement with the linear analysis [14]) without drasting modification of the dynamics. Some details were reported in [12].

At $Pr = 0.2$ and free-slip boundary conditions, we observe a transition to chaos, resulting from the destabilization of a bi-periodic regime. Similar observations were reported in [15]. Note that the period doubling cascade observed in the experiments [1],[2] was not obtained in the simulations, suggesting that an important role may be played by the boundary conditions.

Acknowledgments.

We acknowledge very useful discussions with A. Arneodo, M.E. Brachet, P. Coullet, S. Fauve, U. Frisch, S. Orszag and E. Spiegel. The computations were done on the Cray 1 of the CCVR (Palaiseau). We also benefit of the support of a DRET contract. Figures were done using subroutines from the NCAR library.

REFERENCES

[1] Libchaber A. and Maurer J., J. Physique, C3, 41 (1980), 51.

[2] Libchaber A., Fauve S. and Laroche C., Physica 70 (1983), 73.

[3] Schluter A., Loltz D. and Busse F.H., J. Fluid Mech. 23 (1965), 129.

[4] Busse F.H., J. Fluid Mech. 52 (1972), 97.

[5] Clever R.M. and Busse F.H., J. Fluid Mech. 65 (1974), 625.

[6] Clever R.M. and Busse F.H., J. Fluid Mech. 102 (1981), 61.

[7] Curry J.H., Herring J.R., Loncaric J. and Orszag S.A., J. Fluid Mech. 147 (1984), 1.

[8] Lipps F.B., J. Fluid Mech. 75 (1976), 113.

[9] McLaughlin J. and Orszag S.A., J. Fluid Mech. 122, (1982),123.

[10] Ruelle D., Takens F. and Newhouse S., Comm. Math. Phys. 64 (1978), 35.

[11] Kleiser L. and Schumann U., Notes on Num. Fluid Mech. Vol. 2, E.H. Hirschel ed. Proc. Num. Meth. in Fluid Mech. DFVLR, Cologne (1979).

[12] Sulem P.L., Sulem C. and Thual O., Proc. 4th Beer Sheva Seminar on MHD flows and Turbulence (1984), AIAA Progress in Astronautics and Aeoronautics, in press.

[13] Chandrasekhar, S. Hydrodynamic and Magnetohydrodynamic stability, Oxford University Press (1961).

[14] Busse F.H. and Clever R.M., J. Meca. Theor. et Appl. 2 (1983), 495.

[15] Palm E., Skogvang A. and Tveitereid M., IUTAM conference (August 1984).

SPECTRAL CLOSURES TO DERIVE A SUBGRID SCALE MODELING FOR LARGE EDDY SIMULATIONS

Jean-Pierre CHOLLET
Institut de Mécanique de Grenoble
B.P. 68, 38402 Saint-Martin d'Hères Cedex

Abstract.

Spectral two-point closures are used to derive a subgrid scale model for large eddy simulations of velocity and passive scalar fields of fully developed three-dimensional isotropic turbulence. This model is a "full coupling" of explicit calculations of large scales and statistical evolutions of small scales. It can be simplified into a formalism of eddy-viscosity and eddy diffusivity. The predictability of the large scales is impaired by the growth of errors which can appear in small scales ; this error-growth is studied by comparing two realizations of the velocity field.

1. Introduction.

We address almost every point mentioned in the goals of the workshop : the use of analytical and numerical methods, interactions among a wide range of scales, large scales transport of turbulence and even turbulence with a spectral gap.

Statistical theories can be used to calculate high Reynolds number turbulence. Such statistical closures are also liable to analytical formulations of some parts of the energy transfers. We use the eddy-damped quasi normal markovian (EDQNM) approximation (Leith, 1971, Herring et al., 1982) which, being markovian, does not take a full account of time-space correlations but is especially suitable to study the dynamics of inertial ranges, without requiring big amounts of computing power, at least for isotropic turbulence considered in the present paper.

Improvements in memory size and in the calculating power of computers make possible direct numerical integrations of the Navier-Stokes equations. Nevertheless, such full simulations of three-dimensional turbulence are beyond the capabilities of current computers as long as the Reynolds number is large : in this case we must choose the range of scales for the explicit computation of the flow field, the other scales being handled statistically. We are interested in a detailed calculation of the large scales : a large eddy simulation (LES). The subgrid scales (small scales) are not handled explicitely and have to be modeled statistically through a subgrid scale (SGS) modeling. Interactions among a wide range of scales have to be considered in order to derive a SGS modeling. Spectral space is then of special convenience for both statistical closures and large eddy simulations.

Statistical two point closure (here, the EDQNM) provides a tool to model the small scales. Assuming that the cutoff wavenumber k_c is in the middle of a long inertial range, eddy-viscosity and -diffusivity can be derived and then used in large eddy simulations. Less drastic assumptions are required by a procedure of full coupling between the large eddy simulation and the statistical (here, the EDQNM) evolution of the small scales. The formalism of statistical theories can be used to infer specific behaviours of eddy-quantities depending on the location of the cutoff wavenumber. We focus on free decaying turbulence whose initial energy is concentrated only in the large scales.

Let us notice that these methods of SGS modeling are used in this workshop by Bertoglio (1984) and Aupoix (1984) and we do not repeat some general features which are presented in their papers.

In order to interpret the results of numerical simulations, the predictability of the large scales has to be considered. The growth of an error initially introduced in the small scales is studied by comparing the evolution in time of two realizations of the velocity field.

Turbulent fields with spectral gaps cannot be computed with numerical simulations ; statistical closures can generate such spectra and give some insight into the dynamics which dominate this particuliar kind of spectra and is closely related to the dynamics of the interactions involved in subgrid scales modeling.

2. Spectral eddy-viscosity and eddy-diffusivity.

Two point statistical closures have the ability of handling interactions between various scales of turbulence and then yield a self consistent method to model the energy transfers from large to small (subgrid) scales.

Energy transfers at the wavenumber k are induced by triad interactions between k and two other wavenumbers p and q such that k, p and q form a triangle. It means that these transfers depend on the distribution of energy among the whole range of scales from integral to dissipative ones. Moreover some of these interactions involve small scales (such that p or (and) $q > k_c$) and can be analysed in some details to derive a subgrid scale modeling.

Let us first assume that a $k^{-5/3}$ energy spectrum represents a real inertial range provided it is long enough (high values of the Reynolds number). Thanks to the statistical closure, the transfers associated to small scales can be calculated and rewritten -at first rather arbitrarily- in terms of a generalized eddy viscosity (Kraichnan, 1976, Leslie and Quarini, 1979, Chollet and Lesieur, 1981). An eddy-diffusivity can be derived in a similar way (Chollet and Lesieur, 1982).

For $k^{-5/3}$ energy and scalar variance spectra these eddy quantities are conveniently written in a non dimensional form ν_t^+ and D_t^+ when considering the characteristic eddy viscosity $(E(k_c)/k_c)^{1/2}$ which is proportional to the turbulent velocity $(k_c\, E(k_c))^{1/2}$ times the mixing length $1/k_c$. Then an eddy-Prandtl number can be defined, as $P_{rt}(k/k_c) = \nu_t^+(k/k_c) \,/\, D_t^+(k/k_c)$. These quantities are plotted on fig. 1 versus k/k_c with their origin at their asymptotic values ($k/k_c \longrightarrow 0$). The numerical values are well defined for the particuliar EDQNM closure and a given set of closure constants. We have obtained :

$\nu_t^+(0) = 0.27$ $\quad D_t^+(0) = 0.42$ and $\quad P_{rt}(0) = 0.6$.

These asymptotic values ($k \ll k_c$) feature nonlocal interactions which are typically eddy-viscous since a large eddy (wavenumber k) is damped by much smaller eddies (wavenumbers p and q). The constant values of $\nu_t^+(k/k_c)$ and $D_t^+(k/k_c)$, for $k \ll k_c$, give the justification to the eddy-viscosity (diffusivity) formulation. Nevertheless it will be noticed, below, that this behaviour is significantly altered by the strong cusp in the neighbourhood of the cutoff ($k \longrightarrow k_c$), and depends on a $k^{-5/3}$ inertial range.

The generalized eddy viscosity (in fig. 1) can be split into two parts : the asymptotic value, for $k \ll k_c$, which is the "true" eddy viscosity and the contribution of the cusp

$$\nu_t^+(k/k_c) = \nu_t^+(0) + \nu_{t\,cusp}^+(k/k_c) \qquad (1)$$

The cusp cannot be reduced to a simple exponential or a power law as easily demonstrated by plotting $\nu_{t\,cusp}^+(k/k_c)$ in semi-log or log-log coordinates (Chollet, 1983 a). Nevertheless, in order to interpret the contribution of the cusp as a high order dissipativity, we attempt to adjust a power law by plotting in fig. 2 the value of the exponent of the tangent power law for each value of k/k_c. We observe, in fig. 2, that the value of this exponent is not far from 4. Actually, a log-log regression with a least-square adjustment gives (Chollet, 1983 a) :

$$\nu_{t\,cusp}(k/k_c) = 0.4724\ (k/k_c)^{3.742}, \quad \text{For: } 0.4 < k/k_c < 0.9 \qquad (2)$$

Consequently, if we do not consider the range which is very close to the cutoff ($0.9 < k/k_c < 1$), the contribution of the cusp can be interpreted as a high order dissipativity (tri-laplacian) to be added to a viscosity. The contribution of the cusp is quite significant ; when using the expression (2) and still for a $k^{-5/3}$ spectrum, the cusp contributes to 31,7 per cent -against 68,2 per cent for the asymptotic value $\nu_t^+(0)$- of the energy flux $\Pi(k_c)$ at the cutoff k_c.

The cusp of the eddy diffusivity (fig. 2) is close to a k^2 power law and can be interpreted as a scalar dissipativity (bi-laplacian) to be added to a diffusivity, nevertheless this interpretation does not held for $0.7 < k/k_c < 1$. Let us notice that the value of the asymptotic eddy-viscosity (diffusivity) and the shape of the cusp depend on arbitrary constants included in the formalism of the EDQNM closure. For the energy transfer, there is one constant λ and two constants (λ', λ'') for the transfer of passive scalar. In the present paper we use the values $\lambda = 0.361$,

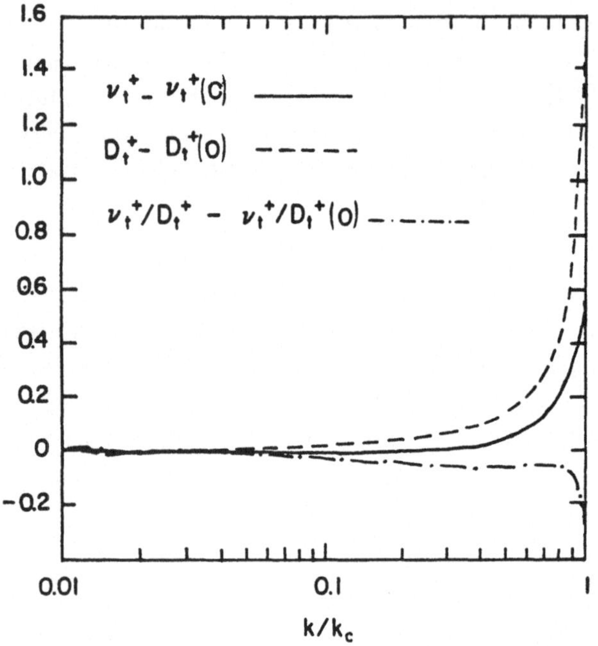

Fig. 1 : Eddy-viscosity, eddy-diffusivity and eddy-Prandtl number, from $k^{-5/3}$ energy and scalar variance spectra. Values are normalized by $(E(k_c)/k_c)^{1/2}$ with the origin at the asymptotic value ($k/k_c = 0$).

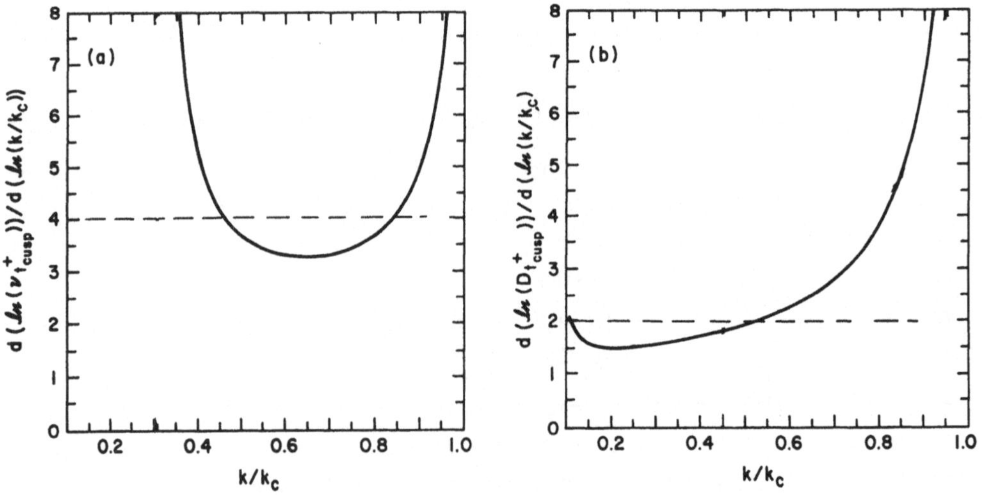

Fig. 2 : Exponents of a power law for the cusp of :
(a) the eddy-viscosity
(b) the eddy-diffusivity.

$\lambda' = 0$, $\lambda'' = 3.61\lambda$ as suggested by Herring et al. (1982) who emphasize that the proper choice is much more arbitrary for the scalar than for the energy. Using other sets of constants (λ', λ'') give noticeably different results for the eddy diffusivity as presented by Chollet (1983 a, 1984). There are obvious limitations to the universality of such eddy-viscosity and -diffusivity since their calculation was associated to long inertial ranges with $k^{-5/3}$ spectra. Analytical handling of expressions of the energy transfers in the frame of the statistical closure (Lesieur and Schertzer, 1978) lead to some remarks about departures of real values from those which were above-calculated.

First, when the cutoff scale is not far from the Kolmogorov dissipative scale, as it is usual at low Reynolds number, we can calculate the rate $\nu_t^{(s)}(0) / \nu_t(0)$; $\nu_t^{(s)}(0)$ is calculated when assuming no energy beyond the Kolmogorov wavenumber k_s ; $\nu_t(0)$, above calculated, corresponds to an infinite value of k_s, that is an infinite value of the Reynolds number. $\nu_t^{(s)} / \nu_t$ (k_c/k_s) is calculated analytically for an inertial $k^{-5/3}$ spectrum in Chollet (1983 a) and the result is plotted in fig. 3. The decrease of the eddy-viscosity, due to a lack of energy in the small scales, is quite significant over almost two decades of wavenumbers.

When using subgrid scale models in large eddy simulations, in section 3, there is a strong evidence of the interest of locating the cutoff wavenumber k_c as close as possible to k_I where k_I is the wavenumber where the energy is maximum. With the analytical formulations of transfer terms which are dominant in this range of scales, the eddy quantities ν_t^+ and D_t^+ are modified into (Chollet, 1983 a) :

$$\nu_{tg}^+ (k/k_c) = \nu_t^+ (k/k_c) - 0.149 \, \frac{k^2}{E(k)} \, \frac{E(k_c)}{k_c^2} \tag{3}$$

$$D_{tg}^+ (k/k_c) = D_t^+ (k/k_c) - 0.118 \, \frac{k^2}{E_\theta(k)} \, \frac{E_\theta(k_c)}{k_c^2} \tag{4}$$

with the same set of closure constants as above and $k^{-5/3}$ spectra for $k > k_c$. Consequently, the cusp is smoothed at $k/k_c = 1$ as sketched in fig. 4. Nevertheless these numerical values cannot be directly used in calculations since they do not take a full account of all the interactions. When $k \longrightarrow 0$, the behaviour of ν_{tg}^+ (or D_{tg}^+) depends on the shape of the spectrum. If this spectrum obeys a power law k^s, when $s < 2$, ν_{tg}^+ and D_{tg}^+ tend to the asymptotic values $\nu_t^+(0)$ and $D_t^+(0)$, which is illustrated by a numerical example in Aupoix (1984). On the contrary, if $s > 2$ $\nu_{tg}^+(0)$ and $D_{tg}^+(0)$ tend to negative values as illustrated numerically in Chollet and Lesieur (1981) with $s = 4$. These negative viscosities are associated with the inverse transfer of energy at very low k when $E(k) \sim k^s$ and $s \geqslant 4$, through "beating interactions". It has nothing to do with the negative viscosities which are derived in two-dimensional turbulence.

A universal formulation of eddy-viscosity (-diffusivity), like ν_t^+, D_t^+, is closely associated with the existence of a rather long inertial range. It cannot correctly handle energy distributions which depart significantly from this inertial behaviour.

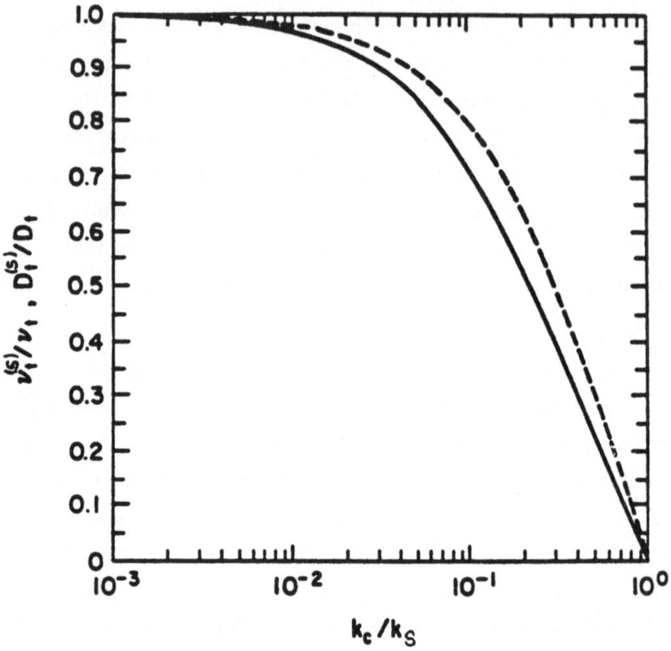

Fig. 3 : Rate of the asymptotic eddy viscosity $\nu_t^{(S)}(0)$ for a finite value of the Reynolds number to its value $\nu_t(0)$ for an infinite value of the Reynolds number

——— eddy viscosity rate : $\nu_t^{(S)}(0)/\nu_t(0)$

- - - eddy-diffusivity rate : $D_t^{(S)}(0)/D_t(0)$

(k_S is the Kolmogorov wave-number).

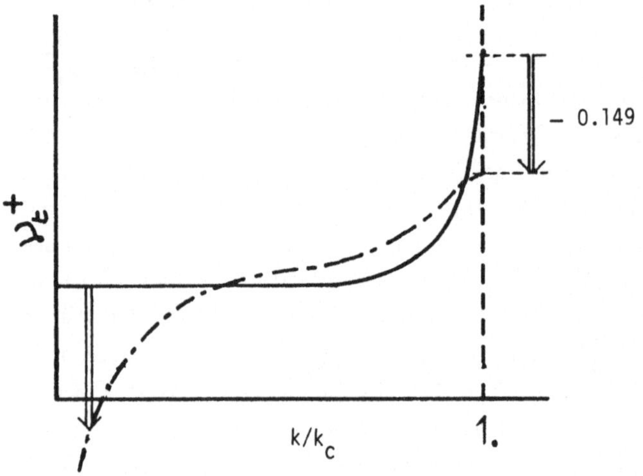

Fig. 4 : Schematic interpretation of the change of $\nu_t(k/k_c)$ when k_c goes to k_I (the energy is maximum at $k = k_I$).

In such a case, the statistical closure can be used to calculate a generalized eddy-viscosity from the complete actual energy spectrum as presented later in section 3.

3. Large eddy simulations.

The numerical simulations we consider here are limited to homogeneous isotropic turbulence in a cubical box with periodic boundary conditions. The resolution of the pseudo spectral code is 32^3.

The eddy viscosity :

$$\nu_t\,(k|k_c,t) = \nu_t^+(k/k_c).\,(E\,(k_c,t)\,/\,k_c)^{1/2} \tag{5}$$

mentioned in the previous section is suitable to model the small scales of a freely-decaying turbulence since it depends on the evolution of the small scale energy through $E(k_c,t)$.

We have focussed our attention on the decay of high Reynolds turbulence whose initial energy is located in the large scales only ($k < k_c$). No experiments are available to be compared with numerical results but we can expect good inertial and inertio-convective spectra. We have already checked these spectra in Chollet and Lesieur (1981), (1982) for both velocity and scalar fields.

The evolution of the energy with time is plotted in fig. 5 for a particular realization of the turbulent field. The energy of the large scale field is complemented by assuming a $k^{-5/3}$ energy spectrum in the small scales ($k > k_c$) ; this small scale energy $\frac{3}{2} k_c E(k_c)$ is of course overestimated at the early stage of the evolution when an inertial range is not yet established. We can consider that the energy-decay actually starts at the time t_* which is equal to about 4 initial large eddy turn-over times ,in the result plotted on fig. 5. This fig. 5 of the energy-decay illustrates the trouble due to the assumption of $k^{-5/3}$ inertial spectra to derive subgrid scale eddy-viscosities. The evolution of the distribution of energy among small scales can be taken into account by coupling a statistical (EDQNM) description of small scales to the explicit calculation of the large scales (Chollet, 1983 b, Aupoix, 1984). This coupling procedure can be used with any kind of spectra possibly departing from a $k^{-5/3}$ as it necessarily happens at the beginning of a decay experiment. Moreover, the cutoff k_c can be located close to the dissipative range, for computations at low or moderate Reynolds numbers in order to compare with experiments. For high Reynolds number turbulence, the cutoff k_c can be located at the very beginning of the inertial range.

Energy spectra from two numerical calculations are compared on fig. 6. They are averaged on two realizations starting from the same initial velocity field. The first calculation uses the eddy viscosity (ν_t^+), the other the "full coupling" procedure. There is a slight discrepancy between spectra at high values of k since the eddy

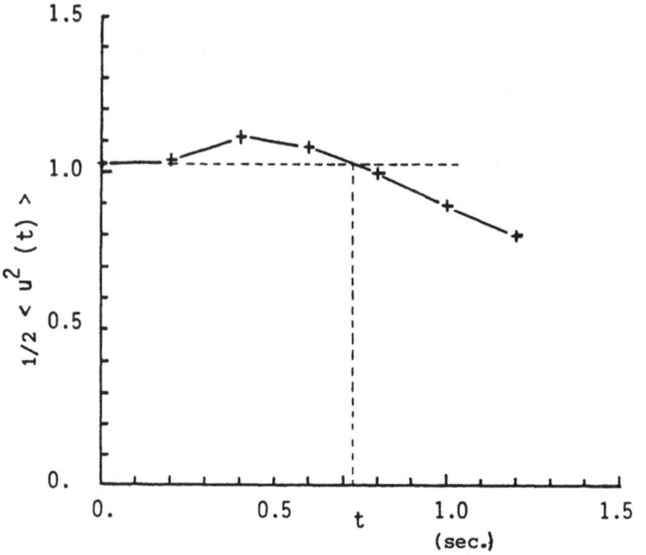

Fig. 5 : Decay of energy from a large eddy simulation (the initial eddy turn over time is about 0.18 sec).

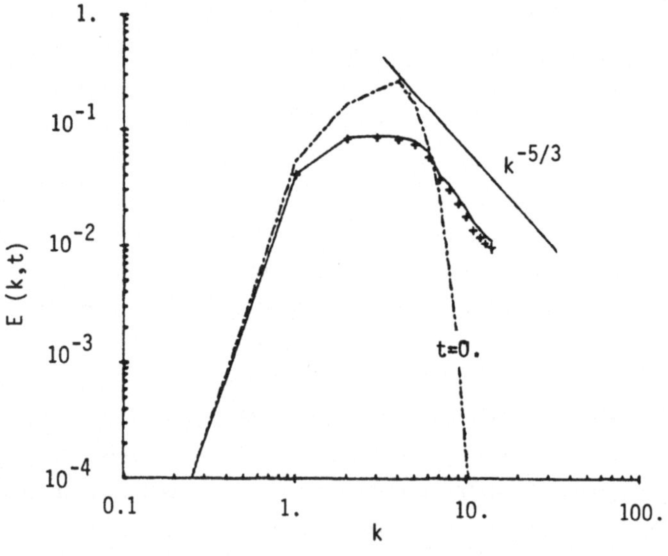

Fig. 6 : Energy spectra of freely decaying turbulence from a large eddy simulation (at $t = 12$ sec $\simeq$ 6,7 initial large eddy turn-over times) with
——— a "full coupling", as statistical calculation of small scales
+ + + the $\nu_t^+(k/k_c) \cdot (E(k_c,t)/k_c)^{1/2}$ eddy viscosity.

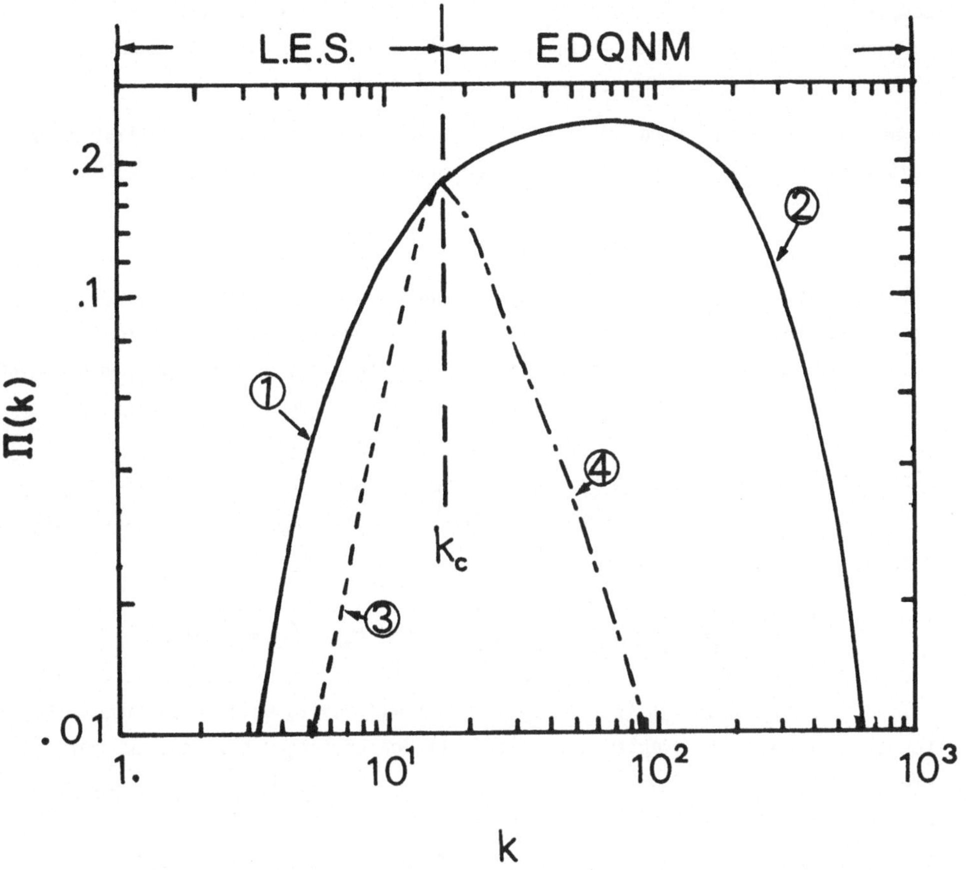

Fig. 7 : Energy flux $\Pi(k)$ from a large eddy simulation with a "full coupling" as SGS model (at t = 1.2).

(1) total flux from LES : $-\int_0^k T(k)dk$

(2) total flux from EDQNM : $\int_k^\infty T(k)dk$

(3) from SGS modelling : $-\int_0^k T^>(k)dk$

(4) from "large scales modelling" : $\int_k^\infty T^<(k)dk$

T(k) is the energy transfer at wavenumber k. The "large scales modelling" corresponds to the transfer of energy from large to small scales, in the EDQNM computation of the small scales.

viscosity (ν_t^+) overestimates the subgrid scale transfer as long as the spectrum is steeper than a $k^{-5/3}$ law. Discrepancies should be expected to reach the largest scales after a longer time as suggested in section 4.

The energy flux is plotted on fig. 7, from the evolution of a realization of the velocity field, using the "full coupling" procedure as the Subgrid Scale Model. We observe the continuity of the flux at the cutoff and how the subgrid scale transfers contribute to the total energy flux.

"Full coupling" is still conditioned by hypothesis, even if there is no more inertial range asumptions. Our approach was restricted to homogeneous isotropic turbulence. However, there have been successful attempts at extending it to nonisotropic homogeneous flows (Bertoglio, 1984) at the necessary cost of additional assumptions.

4. Predictability of large scales.

We consider two realizations of the velocity field $u^{(1)}$ and $u^{(2)}$ which are identical initially except in the small scales :

$$u_i^{(1)}(\vec{k},t_0) = u_i^{(2)}(\vec{k},t_0) \quad , \text{for } |\vec{k}| < k_e(t_0)$$
$$u_i^{(2)}(\vec{k},t_0) - u_i^{(1)}(\vec{k},t_0) = \delta_i(\vec{k},t_0) \text{ ,for } k_e(t_0) < |\vec{k}| < k_c$$

The initial error $\delta_i(k,t)$ is introduced in the small scales which are computed explicitly by the numerical simulation, that is at scales larger than the subgrid (small) scales. This initial error, which is random in direction and amplitude, preserves the energy distribution E(k) and the incompressibility condition.

The initial error will gradually contaminate the whole range of scales and consequently the predictability of the large scales is limited by this error growth. In real flows, this error can be generated by many causes which we do not attempt to analyse here : coarse grid of initial data, inaccuracies of a subgrid scale model, ...

Non linear terms of the flow equations have not only the property of transfering energy from large to small scales (the usual concept of cascade) but can also transfer errors from small to large scales. This has been studied in the framework of statistical closures by Leith (1971) and Leith and Kraichnan (1972) for stationary turbulence and extended to both stationary and freely-evolving turbulence by Metais et al. (1983) and Metais and Lesieur (1984). These two latter references emphasize how the error-growth behaves differently in two and three dimensions especially along the inertial ranges.

Let us consider the cross spectrum :

$$E_W(k) = 2\pi k^2 < u_i^{(1)}(\vec{k})\, u_i^{(2)}(-\vec{k}) >$$

and the error spectrum : $E_\Delta(k) = E(k) - E_W(k)$

which is a measure of the decorrelation between the two fields.

Statistical theories are not free of arbitrariness, and extra assumptions have to be introduced when extending their use from kinetic energy to error spectra (or cross-correlation spectra) especially with respect to the definition of the relaxation time which makes the closure possible. Also, it is of interest to study error growth through numerical simulations. Such a study has been done by Herring et al. (1973) for low Reynolds numbers and then with a strong damping effect from the molecular viscosity. We attempt the same calculations at high Reynolds number, which requires the use of a subgrid scale model. We consider that the two realizations $u^{(1)}$ and $u^{(2)}$ behave in the same way with respect to the subgrid scales.

The initial fields ($t = t_0$) are such that spectra exhibit a well established inertial range. In fig. 8, error-spectra are plotted for a numerical test where $k_e(t_0) = 10$. We observe that the error-spectrum $E_\Delta(k)$ behaves like k^4 in the largest scales, this result has been already mentioned in Metais et al. 1983 and Chollet 1983 b together with evolutions of statistical characteristics of the error growth. This k^4 law is in perfect agreement with numerical and analytical results obtained by Metais and Lesieur (1984) via statistical theory. Therefore no front error steeper than a k^4 law can propagate from small to large scales.

These results are of importance since it means that the field which is recovered from the numerical simulation is probably not the real field even if it looks like it. Atmospheric turbulence gives a typical example of predictability requirements; the meteorologist is interested in the complete details of a particular realization, when the climatologist can be satisfied in generating flows which have only some general properties of the real flows.

The comparison of two fields, as used in this study of predictability, can also be used as a method to test the validity and reliability of calculations, besides other classical procedures such as comparisons with experiments, recovery of inertial ranges, ... Bertoglio (1984) uses this error growth to test backtransfers through a cutoff scale, and also obtains very nice k^4 error spectra.

We have paid attention only to one of the possible origins of unpredictability : the error which would be initially located in small scales. Other phenomena can also yield unpredictability for example initial errors over the whole range of scales (a white noise affecting the velocity field in physical space would give a k^2 error spectrum).

When comparing the fields $u^{(1)}$ and $u^{(2)}$, statistical theories give access to spectra and evolutions of averaged quantities. The numerical simulations make possible a direct comparison of the velocity fields as illustrated by cross sections of the three dimensional fields in fig. 9. The fields under comparison have still strong similarities since the evolution time is not very long, nevertheless some discrepancies are already quite significant. This direct comparison between velocity fields is liable to further developments with the availability of new interactive graphics

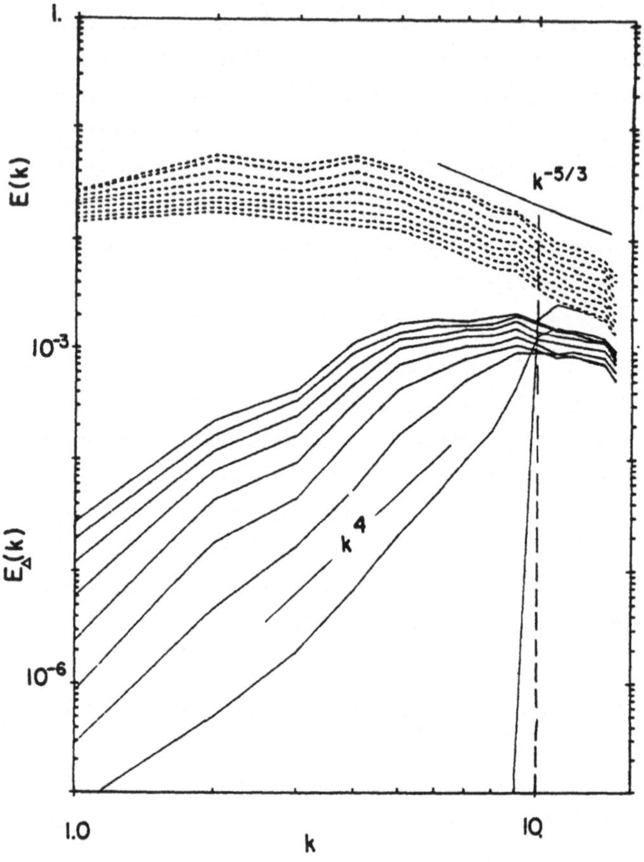

Fig. 8 : Evolution of energy spectra E(k,t) and error spectra $E_\Delta(k,t)$, at every 0,4 sec (the initial large eddy turn over time is 0,18 sec), calculated with a large eddy simulation.

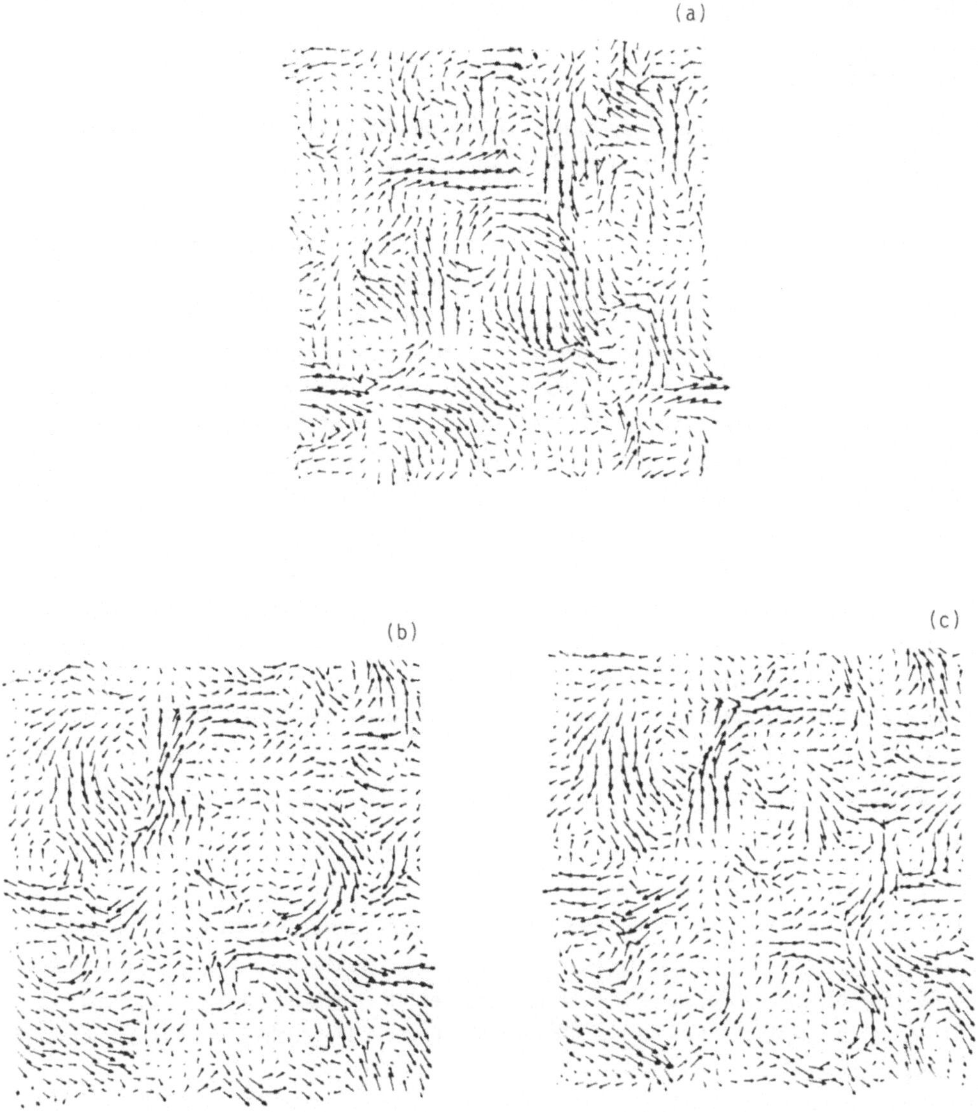

Fig. 9 : Plane cross-section of three dimensional velocity field : (a) initial field ; (b) $u^{(1)}(\vec{x},t)$; (c) $u^{(2)}(\vec{x},t)$ at t = 3 sec (same fields realizations as in fig. 8).

facilities.

5. Spectral gaps.

Spectral gaps seem to be of special interest when dealing with the method of homogeneization since they provide a clear separation between small scales and large scales. A well defined spectral gap means two inertial and two dissipative ranges which makes the full numerical simulation of the corresponding turbulent field far beyond the capabilities of available computers. Calculations of such spectra can be easily done in the frame of a statistical closure especially in stationnary turbulence with the constant rates of energy forcing $\mathcal{E}_1$ at the wave number k_1 (large scales) and $\mathcal{E}_2$ at k_2 (small scales). A numerical example already mentioned in Pouquet et al. 1983 is presented differently in fig. 10 in order to point out the dynamics of the energy transfer across the gap. Spectra are calculated (1) with all the terms of the transfer ; and (2) when omitting these terms in the transfer which can be referred to as non local, such that k interacts with p and q , $p \simeq q$ and k/p (or k/q) $<$ 0.19. This latter contribution to the transfer corresponds to a large part of the viscosity terms mentioned in section 2 when the limit k_c between large and small scales is located in the gap, around k_G . These eddy viscosity interactions are responsible for most of the transfer across the gap since the depth of the gap is observed to be considerably reduced when a large part of these interactions is omitted.

Statistical methods can be used to study the backtransfer of errors introduced in the smallest scales $k_e(t_0) > k_2$, then the error growth is observed to be blocked in the small scale end of the spectrum (around k_2).

Let us notice that, for the purpose of a direct simulation of the largest scales (around k_1), the "coupling procedure" of section 3 could take into account the distribution of energy beyond a cutoff wave number k_c located between k_1 and k_G. The usual eddy-viscosity (such as ν_t^+) could not handle such a small scale distribution of energy.

These spectral gaps seem to be features of three dimensional turbulence. In two dimensions, the energy forcing around k_2 would induce a backtransfer of energy, then filling up the gap. These results can hardly be compared to real situations which would be often characterized by three-dimensional small scales and quasi two-dimensional large scales.

Conclusions.

The two-point statistical closure (such as the EDQNM) gives a self consistent method to derive subgrid scale modeling. The eddy-viscosity or -diffusivity formalism is not free of arbitrariness even if these quantities are derived from statistical theories. "Coupling" LES to a complete statistical description of small scales gives access to any distribution of energy in the subgrid range. The small scale errors backscatter towards large scales as observed from both numerical simulations and statistical closure calculations.

Acknowledgements.

This work was supported by the I.N.A.G. under the A.T.P. Recherches Admosphériques and by the D.R.E.T. under contract n° 83/314. Computer resources were granted by the Scientific Committee of the Centre de Calcul Vectoriel pour la Recherche.

References.

Aupoix, B., 1984, Eddy viscosity subgrid scale models for homogeneous turbulence, Workshop on Macroscopic modelling of turbulent flows and fluid mixtures, Nice, Déc. 1984.

Bertoglio, J.P., 1984, A stochastic subgrid model for sheared turbulence, Workshop on Macroscopic modelling of turbulent flows and fluid mixtures, Nice, Dec. 1984.

Chollet, J.P., 1983 a, Statistical closure to derive a subgrid-scale modelling for large eddy simulations of three-dimensional turbulence, NCAR Technical Note TN-206 + STR, Boulder, Colorado.

Chollet, J.P., 1983 b, Two-point closures as a subgrid scale modelling for large eddy simulations, Turbulent shear flows IV, ed. Springer-Verlag.

Chollet J.P. and Lesieur, M., 1981, Parametrization of small scales of three-dimensional isotropic turbulence utilizing spectral closures, J. Atmos. Sci., 38, 2747-2757.

Chollet J.P. and Lesieur, M., 1982, Modélisation sous-maille des flux de quantité de mouvement et de chaleur en turbulence tridimensionnelle isotrope, La Météorologie, VIe série, 29 et 30, 183-191.

Chollet, J.P., 1984, Turbulence tridimensionnelle isotrope : modélisation statistique des petites échelles et simulation numérique des grandes échelles, Thèse de Doctorat ès Sciences, Université de Grenoble.

Herring, J.R., Riley J.J., Patterson, G.S. and Kraichnan, R.H., 1973, Growth of uncertainty in decaying isotropic turbulence, J. Atmos. Sci., 30, 997-1006.

Herring, J.R., Schertzer, D., Lesieur, M., Newman, G.R., Chollet, J.P. and Larchevêque, M., 1982, A comparative assessment of spectral closures as applied to passive scalar diffusion, J. Fluid Mech., 124, 411-437.

Kraichnan, R.H., 1976, Eddy-viscosity in two and three dimensions, J. Atmos. Sci., 33, 1521-1536.

Leith, C.E., 1971, Atmospheric predictability and two-dimensional turbulence, J. Atmos. Sci., 28, 145-161.

Leith, C.E. and Kraichnan, R.H., 1972, Predictability of turbulent flows, J. Atmos. Sci., 29, 1041-1058.

Lesieur, M. and Schertzer, D., 1978, Amortissement autosimilaire d'une turbulence à grand nombre de Reynolds,
J. Mécanique, 17, 610-646.

Leslie, D.C. and Quarini, G.L., 1979, The application of turbulence theory to the formulation of subgrid modelling procedures,
J. Fluid Mech., 91, 65-91.

Metais, O., Chollet, J.P. and Lesieur, M., 1983, Predictability of the large scales of freely evolving three and two-dimensional turbulence,
A.I.P. Conference Proceedings, n° 106, "Predictability of fluid motions", La Jolla Institute.

Metais, O. and Lesieur, M., 1984, Statistical precictability of decaying turbulence, submitted to J. Atm. Sci.

Pouquet, A., Frisch, U. and Chollet, J.P., 1983, Turbulence with a spectral gap, Phys. Fluids, 26, 877-880.

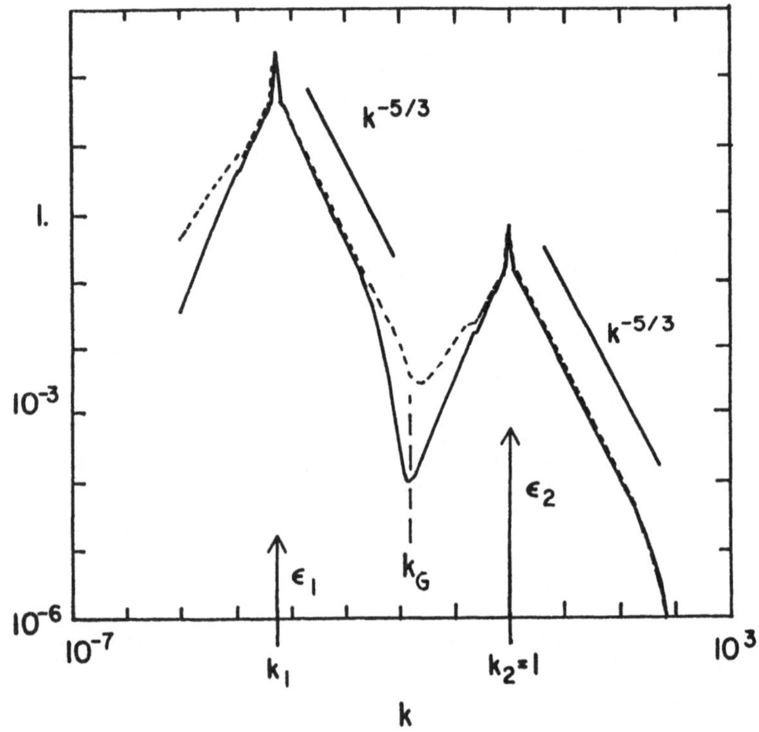

Fig. 10 : Energy spectrum with a gap, calculated with the EDQNM approximation by C. Montmory
——— with all the interactions of the energy transfer
------ without the eddy viscous non local interactions ($k_1 = 5.10^{-5}$; $k_2 = 1$; $\varepsilon_1/\varepsilon_2 = 10^{-7}$).

MODELLING OF THREE-DIMENSIONAL SHOCK WAVE TURBULENT BOUNDARY LAYER INTERACTIONS

Doyle D. Knight
Department of Mechanical and Aerospace Engineering
Rutgers University - The State University of New Jersey
New Brunswick, New Jersey 08903

I. INTRODUCTION

Over the past forty years, a significant effort has been concentrated towards the understanding of two-dimensional and three-dimensional shock wave - turbulent boundary layer interactions (denoted herein as "2-D" and "3-D turbulent interactions" for brevity). The motivation for this research arises from the presence of 2-D and 3-D turbulent interactions in a wide variety of practical configurations, including gas dynamic lasers (Christiansen et al 1975), transonic airfoils, supersonic inlets, nozzles, and deflected control surfaces and wing-fuselage junctures at transonic and supersonic speeds (Green 1970, Hankey and Holden 1975, Korkegi 1971).

The first systematic investigations of shock wave - turbulent boundary layer interaction were performed by Ackeret (1947) and Liepmann (1946) (see Adamson and Messiter 1980). In recent years, significant emphasis has been focused on both experimental and theoretical (in particular, computational) research in 2-D and 3-D interactions. A sample of recent experimental research includes the work of McCabe (1966), West and Korkegi (1972), Freeman and Korkegi (1975), Peake (1976), Oskam, Vas and Bogdonoff (1976, 1977), Roshko and Thomke (1976), Debieve et al (1982), Kubota and Stollery (1982), Zheltovodov (1982), Dolling and Bogdonoff (1983), McClure and Dolling (1983), Holden (1984), Muck and Smits (1984), and Settles and Teng (1984). A brief sample of recent theoretical research on 3-D turbulent interactions employing numerical solution of the 3-D Reynolds-averaged compressible Navier-Stokes equations includes the work of Hung and MacCormack (1978), Shang, Hankey and Petty (1979), Horstman and Hung (1979), Brosh et al (1983) (combining both experiment and computation), Knight (1983, 1984a, 1984b), and Horstman et al (1985). A survey of numerical solutions of the compressible Navier-Stokes has been given by Shang (1984). Recent analytical research on 3-D interactions includes the work of Inger (1984) and Stalker (1984).

The focus of the present paper is the examination of the particular 3-D turbulent interaction depicted in Fig. 1. The physical configuration, denoted as the "3-D sharp fin", is formed by a fin or wedge of angle α_g attached normal to a flat plate. An equilibrium supersonic turbulent boundary layer develops on the flat plate, and is subject to an oblique shock-boundary layer interaction due to the shock formed by the deflection α_g of the fin. The details of the flowfield structure are not fully understood, with a variety of flow models proposed (Token 1974, Korkegi 1976, Oskam, Vas and Bogdonoff 1976,1977, Kubota and Stollery 1982, Zheltovodov 1982).

The purpose of this paper is to examine the accuracy of numerical computations of the 3-D sharp fin flowfield using the Reynolds-averaged Navier-Stokes equations. The basic mode is a comparison of computed flows with a set of benchmark experiments at Mach 3 for two different Reynolds numbers. Detailed comparison is performed between experimental data and separate computational results of the author and Horstman (1984b) for the 3-D sharp fin at α_g = 10 deg in order to examine the accuracy of two different turbulence models. The computed surface pressure for the 3-D sharp fin at α_g = 20 deg by the author is compared with recent experimental data. The evaluation of the efficacy of the turbulence models is viewed as an integral part of an overall effort to obtain a better understanding of complex 3-D turbulent interactions.

II. METHOD OF SOLUTION

A. Governing Equations and Boundary Conditions

The governing equations are the full mean compressible 3-D Navier-Stokes equations using mass-averaged variables (Rubesin and Rose 1973) and strong conservation form (Pulliam and Steger 1980). The molecular dynamic viscosity is given by Sutherland's law. The molecular Prandtl number is 0.73 (air).

A three-dimensional coordinate transformation ($\xi(x,y,z)$, $\eta(x,y,z)$, $\zeta(x,y,z)$) is utilized to map the physical domain (denoted by the dotted lines in Fig. 1) into the unit cube in the transformed domain (Fig. 2) whose simple shape facilitates the coding of the numerical algorithm and the application of the boundary conditions. A wide variety of methods have been developed for numerical generation of boundary-fitted coordinates (Thompson and Warsi 1982). For the 3-D sharp fin configuration, the simple structure of the physical domain permits analytic specification of the transformation.

The effects of turbulence are modelled using the two-layer algebraic turbulent eddy viscosity model of Baldwin and Lomax (1978). The inner eddy viscosity ϵ_i is given by

$$\epsilon_i = \rho(\kappa \ell D)^2 \omega \tag{1}$$

where ρ is the density, κ is von Kármán's constant, ℓ is the mixing length, D is Van Driest's damping factor and ω is the magnitude of the mean vorticity. The mixing length is specified by the formula of Buleev (1963) as discussed in Gessner and Po (1976). For the 3-D sharp fin configuration, the expression is

$$\ell = 2\mathcal{Y}\mathcal{Z}/[\mathcal{Y} + \mathcal{Z} + (\mathcal{Y}^2 + \mathcal{Z}^2)^{1/2}] \tag{2}$$

where $\mathcal{Y}$ and $\mathcal{Z}$ are cartesian coordinates defined in Fig. 1. The Van Driest damping factor D is specified by

$$D = 1 - \exp(- \ell u_*/26\nu_w) \tag{3}$$

where $u_* = (\tau_w/\rho_w)^{1/2}$, τ_w is the wall shear stress, and ν_w is the wall kinematic viscosity.

The outer eddy viscosity ϵ_o is defined by

$$\epsilon_o = \rho k C_{cp} F_{wake} F_{Kleb} \tag{4}$$

where k and C_{cp} are constants. The outer function F_{wake} is

$$F_{wake} = \ell_{max} F_{max} \tag{5}$$

where

$$F_{max} = \max(\ell \omega D) \tag{6}$$

and ℓ_{max} is the value of ℓ where $\ell\omega D$ is a maximum. The Klebanoff intermittency function is

$$F_{Kleb} = \{1 + 5.5\ (C_{Kleb}\ell/\ell_{max})^6\}^{-1} \tag{7}$$

The constants k, C_{cp} and C_{Kleb} are assigned the values 0.0168, 2.08, and 0.3, respectively, based upon previous experience (Baldwin and Lomax 1978, Knight 1984a and 1984b). These constants have been recently examined in detail by York and Knight (1985). The turbulent Prandtl number is 0.9 .

The turbulent eddy viscosity is implemented by dividing the y-z plane into two regions by a line emanating from the corner (formed by the fin and flat plate) and oriented at approximately 45 deg. Within the region adjacent to the sharp fin, the inner and outer profiles of the eddy viscosity are obtained along y = constant lines, and the eddy viscosity is switched from ϵ_i to ϵ_o at the location where $\epsilon_i > \epsilon_o$. Similarily, within the region adjacent to the flat plate, the inner and outer profiles of ϵ are obtained along z = constant lines. Further details are provided in Knight (1984b). This method has been employed by Hung and MacCormack (1978).

B. Numerical Algorithm

The governing equations are solved by an efficient hybrid explicit-implicit numerical algorithm (Knight 1984b). The method employs the second-order explicit finite-difference algorithm of MacCormack (MacCormack 1971, Baldwin and MacCormack 1975) and the second-order implicit method ("Box Scheme") of Keller (1974). The implicit method of Keller is used in a thin layer (denoted the "computational sublayer") adjacent to the solid boundaries, where the large gradients require extremely fine grid spacing for accurate resolution. The Box Scheme is applied to the asymptotic form of the Navier-Stokes equations in a local cartesian coordinate system (x',y',z') as illustrated in Fig. 3. Based upon previous study (Knight 1981a, 1981b, 1981c, 1984b), the extent of this region is givn by

$$z_m'^{+} \leq 60 \quad (8)$$

where $z_m'^{+} = z_m' u_* / \nu_w$. Within the above limit, the computed solutions have been found to be essentially independent of the height z_m'. The computational sublayer typically constitutes a few percent of the local boundary layer thickness. The explicit algorithm of MacCormack is applied to the full Navier-Stokes equations in the remainder of the physical domain (denoted the "ordinary points"). The use of an implicit method in the region $z_m'^{+} \leq 60$ overcomes the extreme time step limitation encountered by application of an explicit algorithm to this portion of the flow. Further details are provided in Knight (1984b).

The boundary conditions are implemented in the numerical code in the conventional manner. At the upstream boundary ABHG (see Fig. 1), the flow variables are held fixed at conditions corresponding to an equilibrium flat plate turbulent boundary layer. On the solid boundaries ABCDEF (flat plate) and FEKL (fin), the velocity vector is set to zero, the surface temperature is specified, and the normal gradient of the static pressure is set to zero (Deiwert 1975, Knight 1981a). On the plane of symmetry AFLG, the normal component of the velocity is zero, and the normal derivatives of the remaining flow variables are set to zero. On the outer boundaries BCDJIH ($\eta = 1$) and HIJKLG ($\zeta = 1$), the zero gradient conditions $\partial/\partial\eta = 0$ and $\partial/\partial\zeta = 0$ are specified. These boundaries are located at a sufficient distant from the interaction to insure that the flow is locally two-dimensional. On the downstream boundary EDJK ($\xi = 1$), the conventional zero gradient condition $\partial/\partial\xi = 0$ is specified.

The hybrid algorithm has been successfully applied to numerous two- and three-dimensional flows exhibiting shock-boundary layer interaction and flow separation (Knight 1981a, 1981b, 1981c, 1983, 1984a, 1984b). The 3-D version was originally written in SL/1, a vector-processing language developed at NASA Langley Research Center for the CYBER 200 and CYBER 203 computers. The present version of the 3-D code is written in CYBER 200 FORTRAN. The explicit portion of the algorithm is highly vectorized with typical vector lengths of 1500 employed to date, and achieves an execution rate of approximately 100 MFlops (million floating point operations per second) on the VPS 32 at NASA Langley using a 32-bit word length. The VPS 32 is a vector-processing computer which is architecturally similar to the CYBER 205. The hybrid algorithm provides a substantial improvement in efficiency when compared to a vectorized version of MacCormack's explicit algorithm alone. Benchmark studies (Knight 1984b) have demonstrated the present method to be a factor of 16 to 21 times faster than a vectorized time-split operator version of MacCormack's explicit algorithm alone.

III. ACCURACY OF MODELLING OF 3-D SHARP FIN FLOWFIELD

A. Introduction

The flow geometry depicted in Fig. 1 has been extensively surveyed by Oskam, Vas and Bogdonoff (1976, 1977) and McClure and Dolling (1983) at Mach number $M_\infty \cong 3$ for a range of fin angles α_g and Reynolds numbers $Re_{\delta\infty}$, where δ_∞ is the boundary layer thickness on the flat plate in the vicinity of the leading edge of the fin. The data of Oskam et al includes surface pressure and heat transfer on the flat plate, and profiles of pitot pressure, yaw and pitch angles at $Re_{\delta\infty} \cong 9.0 \times 10^5$. The data of McClure and Dolling includes surface pressure, surface oil film photographs, and boundary layer profiles of pitot pressure p_p and yaw angle at **two** Reynolds numbers, specifically $Re_{\delta\infty} = 2.8 \times 10^5$ and 8.0×10^5.

Computations have been performed by Horstman and Hung (1979) and Knight (1984b) for the flow conditions of Oskam et al for two fin angles, namely, $\alpha_g = 3.7$ and 9.7 deg. The algebraic turbulent eddy viscosity models of Escudier (1965) and Baldwin and Lomax were employed by Horstman et al and the author, respectively. In general, both computations obtained good agreement with a variety of measured flow quantities, including surface pressure, skin friction, heat transfer, pitot pressure and yaw angle.

B. Comparison with Experimental Data of McClure and Dolling at $\alpha_g = 10$ deg

In the present paper, specific comparison is performed between the experimental data of McClure and Dolling, and the separate computations of the author and Horstman (1984b) for the 3-D sharp fin at $\alpha_g = 10$ deg at two values of the Reynolds number. The specific flow conditions of the experiments are indicated in Table 1, where $p_{t\infty}$ and

$T_{t\infty}$ are the upstream total pressure and total temperature, respectively.

Table 1. Flow Conditions for Experiments of McClure and Dolling (1983) for 3-D Sharp Fin for α_g = 10 deg

Case No.	δ_∞ (cm)	M_∞	$Re_{\delta\infty}$	$p_{t\infty}$ (kPa)	$T_{t\infty}$ (deg K)
1	0.45	2.91	2.75 x 10^5	690	276
2	1.29	2.93	8.0 x 10^5	690	271

The calculations of Horstman (1984b) solve the mean compressible 3-D Navier-Stokes equations using the explicit-implicit algorithm of MacCormack (1982), with the effects of turbulence incorporated using the two-equation k-ε model of Jones and Launder (1972) together with the wall function boundary conditions of Viegas and Rubesin (1983) (see also Horstman 1984a). **The present comparison, therefore, provides an opportunity to evaluate the efficacy of two different turbulence models (i.e., the Baldwin-Lomax and Jones-Launder models) for the 3-D sharp fin configuration.** The calculations of the author at $Re_{\delta\infty}$ = 2.8x10^5 and 9.3x10^5 have been previously compared to specific experimental profiles of McClure and Dolling, and Oskam, Vas and Bogdonoff in Knight (1984a, 1984b). The combined comparison with the computed results of Horstman (1984b) and the experimental data of McClure and Dolling is a unique feature of the present paper.

The flow conditions for the computations are presented in Tables 2 and 3. It is noted that there is a slight difference in Reynolds number for Case 2 between the author and McClure and Dolling. This is due to the fact that the freestream conditions for the author's computation were chosen to closely match the experimental test conditions of Oskam et al. For the computations by the author, the flat plate surface temperature was 280.6 deg K, and the fin surface temperature was 280.6 and 252.8 deg K, respectively, for Cases 1 and 2. These values were chosen on the basis of the experimental conditions, and are close to adiabatic conditions. For the computations by Horstman, the flat plate and fin surfaces were assumed adiabatic.

Table 2. Flow Conditions Calculations of Knight (1984a, 1984b) for 3-D Sharp Fin for α_g = 10 deg

Case No.	δ_∞ (cm)	M_∞	$Re_{\delta\infty}$	$p_{t\infty}$ (kPA)	$T_{t\infty}$ (deg K)
1	0.45	2.91	2.75×10^5	690	276
2	1.37	2.94	9.25×10^5	690	256

Table 3. Flow Conditions Calculations of Horstman (1984b) for 3-D Sharp Fin for α_g = 10 deg

Case No.	δ_∞ (cm)	M_∞	$Re_{\delta\infty}$	$p_{t\infty}$ (kPa)	$T_{t\infty}$ (deg K)
1	0.54	2.94	3.41×10^5	690	267
2	1.4	2.94	8.83×10^5	690	267

The numerical grids employed by the author for Cases 1 and 2 are detailed in Knight (1984a, 1984b). For Case 1, two separate grid systems (denoted Grid Nos. 1 and 2) were employed. These grids differed principally in the height z_m' of the computational sublayer adjacent to the flat plate. The purpose of performing these two separate computations was to investigate the sensitivity of the calculated solution to the height of the sublayer within the limit of Eq. (8).

The numerical grids utilized by Horstman (1984b) for Cases 1 and 2 employed a uniform streamwise spacing $\Delta x = 1.1\delta_\infty$ and $0.39\delta_\infty$, respectively. This is comparable to the values $\Delta x = 1.1\delta_\infty$ and $0.49\delta_\infty$ for Knight (1984b). The grid spacing in the y- and z-directions was stretched exponentially near the surface and followed by uniform spacing. The uniform spacing Δy_∞ in the y-direction was $1.1\delta_\infty$ and $0.39\delta_\infty$ for Case 1 and 2, respectively. The corresponding grid spacing for the author is $0.68\delta_\infty$ and $0.62\delta_\infty$, respectively. The uniform spacing Δz_∞ in the z-direction was equal to Δy_∞ for Horstman. The corresponding grid spacing for the author was $0.29\delta_\infty$ to $1.2\delta_\infty$ for Case 1, and $0.53\delta_\infty$ for Case 2. A total of 90,112 grid points each were used by Horstman for Cases 1 and 2. The author employed 106,316 points for Case 1 and 79,727 points for Case 2.

The location of the boundary layer measurements of McClure and Dolling are indicated in Fig. 4. Profiles of the pitot pressure p_p and yaw angle (defined as the angle of the mean velocity vector in the plane of the flat plate) were measured at a series of streamwise locations at a fixed spanwise location z = 6.4 cm and 12.1 cm for Cases 1 and 2, respectively. The profiles were measured within the boundary layer on the flat plate, both upstream and downstream of the shock location. The general orientation of the profile is shown in Fig. 5. The number of experimental profiles each for p_p and yaw was 21 and 22 for Cases 1 and 2, respectively.

The calculated and measured pitot pressure p_p is shown in Figs. 6 and 7 for Cases 1 and 2, respectively, at several locations both upstream and downstream of the 3-D turbulent interaction. The locations of the profiles are indicated by x_s/δ_o, where $x_s = x - x_{shk}$, x_{shk} is the location of the theoretical inviscid shock wave at the particular spanwise location z, and δ_o is the boundary layer thickness on the flat plate at $(x,z) = (x_{shk},z)$ in the absence of the fin. The shock location x_{shk} is z = 12.0 cm and 22.8 cm for Cases 1 and 2, respectively. The local undisturbed boundary layer thickness δ_o is 0.59 cm and 1.55 cm, respectively.

In Fig. 6a, the computed p_p profiles of the author for Case 1 at $x_s/\delta_o = -1.71$ using the two different grid systems (Grid Nos. 1 and 2) are seen to been in excellent agreement. The comparison between the author's profiles and the experimental data is good, with the computations accurately predicting the peak value of the "overshoot" in p_p located outside the boundary layer. Within the boundary layer, the author's computations show good agreement with the measurements except for $y/\delta_\infty < 0.5$ where the pitot pressure is modestly overestimated. The p_p profile of Horstman is observed to be in good agreement with experiment, accurately predict the peak value of the "overshoot" in pitot pressure. At $x_s/\delta_o = -0.63$ (Fig. 6b), the computed p_p profiles of the author using Grid Nos. 1 and 2 are again seen to be in excellent agreement. The profiles of the author and Horstman are seen to be in good agreement with experiment, with similar accurate prediction of the peak pitot pressure and an improved comparison within the boundary layer. At x_s/δ_o = 1.0 (Fig. 6c), a similar evaluation holds. Overall, it is evident that the computed results of the author for the two grid systems (Nos. 1 and 2) are in excellent agreement. In addition, it is shown for Case 1 that the computed results for p_p using both the Baldwin-Lomax and Jones-Launder turbulence models are in good agreement with experiment.

In Fig. 7a, the calculated and measured p_p for Case 2 at x_s/δ_o = -1.47 are displayed. Although achieving good agreement between theory and experiment within the boundary layer, both computations fail to accurately predict the peak pitot pressure outside the boundary layer, although the results of Horstman are seen to be in better agreement with the data. The reason for this discrepancy is not clear. In particular, it is noted that the relative grid spacing (in terms of the local flow conditions) for Case 2 for the author was smaller than the relative grid spacing for Case 1 as discussed previously. At x_s/δ_o = -0.37 (Fig. 7b), 1.27 (Fig. 7c) and 3.40 (Fig. 7d), the computed and measured profiles are seen to be in good agreement, with the exception of the region $y/\delta_\infty \leq 0.3$ where p_p is underestimated. Comparison of Figs. 6 and 7 indicate that the pitot pressure in this region is more accurately predicted for the Case 1. Overall, it is concluded that the computed results for p_p for Case 2 using both the Baldwin-Lomax and Jones-Launder turbulence models are in general good agreement with experiment, with the exception cited above.

In Figs. 8 and 9, the computed results of the author and Horstman for the yaw angle are compared with the experimental data for Cases 1 and 2, respectively. In Fig. 8a, the computed results of the author for Case 1 at x_s/δ_o = -1.71 using the two grid systems are again observed to be in excellent agreement. The author's profiles show generally good agreement with experiment, with a modest underestimate in the region $y/\delta_\infty < 0.4$. The profile of Horstman is in close agreement with the experiment for $y/\delta_\infty > 0.4$, although displaying a more significant underestimate of the yaw angle for $y/\delta_\infty < 0.4$. The predicted yaw angle at the surface differs markedly for the two turbulence models, with values of 34 deg and 24 deg using Baldwin-Lomax and Jones-Launder, respectively. In Fig. 8b (x_s/δ_o = -0.63), the author's computed profiles using the two grid systems are again observed to be in excellent agreement. The author's profile is in generally good agreement with experiment, although somewhat overpredicting the yaw angle in the region $y/\delta_\infty > 0.4$. The "overshoot" in yaw angle ($4 < y/\delta_\infty < 8$) is not predicted, however. The computed profile of Horstman is observed to be very similar to the author's, except for $y/\delta_\infty < 0.2$ where it displays a significantly lower yaw angle. The predicted surface yaw angles are 36 deg (Baldwin-Lomax) and 26 deg (Jones-Launder). The profile of Horstman likewise fails to exhibit the observed overshoot in yaw angle. In Fig. 8c ($x_s/\delta_o = 1.0$), the author's computed profiles using Grid Nos. 1 and 2 are in close agreement. The author's profile is in close agreement with experiment, although underpredicting the experimental data for $y/\delta_\infty < 0.4$. The profile of Horstman is in close agreement with experiment, except for $y/\delta_\infty < 0.4$, where it displays a more marked underprediction of yaw angle. The predicted surface yaw angles are 37 deg (Baldwin-Lomax) and 28 deg (Jones-Launder). Overall, the computed profiles of the author using Grid Nos. 1 and 2 are in excellent agreement. Furthermore, the predictions using the Baldwin-Lomax and Jones-Launder models are observed to be in good agreement with experiment with two exceptions. First, the measured overshoot in yaw angle (Fig. 8b) is not predicted using either model. Second, both models typically underpredict the yaw angle within the lower part of the boundary layer, although the Baldwin-Lomax model yields a more accurate profile for this case.

In Figs. 9a and 9b, the computed and experimental yaw angle profiles for Case 2 are displayed at $x_s/\delta_o = -1.47$ and -0.37. The calculated profiles are again observed to underpredict the measured yaw angle near the wall, and fail to display the observed overshoot outside the boundary layer. In Fig. 9c, the computed profiles of the author and Horstman at $x_s/\delta_o=1.27$ are in general agreement with experiment, although again underpredicting the yaw angle near the surface. The results using the Jones-Launder model provide a more accurate prediction of the yaw angle in the outer portion of the boundary layer, while the Baldwin-Lomax model yields better results near the surface. Finally, at $x_s/\delta_o=3.39$ (Fig. 9d), the computed profiles using the Jones Launder model are in excellent agreement with experiment. The results using the Baldwin-Lomax model underpredict the observed profile. The predicted values of the yaw angle at the surface using both models are in good agreement with experiment.

In summary, the computations using the Baldwin-Lomax and the Jones-Launder models display the following major features :

- The computed pitot pressure is relatively insensitive to the turbulence model. A similar observation was made by Settles, Horstman and McKenzie (1984) for a 3-D swept compression corner configuration for which separate computations were performed using the algebraic turbulent eddy viscosity model of Cebeci and Smith (1974) and the two-equation model of Jones and Launder (1972). The inner layer of the two-layer Cebeci-Smith model is essentially the same as the Baldwin-Lomax, while the outer layer formulations differ significantly.

- The computed pitot profiles are in generally good agreement with the experimental data. Similar agreement has been achieved by Knight (1984b) for the 3-D sharp fin data of Oskam et al (1976, 1977), and Settles, Horstman and McKenzie (1984) for the 3-D swept compression corner. A specific exception is the failure to accurately predict the overshoot in p_p at $x_s/\delta_o = -1.47$ for Case 2.

- The computed yaw angle is sensitive to the turbulence model. Differences as large as 30% in the calculated surface yaw angle are observed, with the Baldwin-Lomax model typically yielding higher yaw angles near the surface. This is consistent with the results of Settles, Horstman and McKenzie (1984) concerning the Cebeci-Smith and Jones-Launder models for the 3-D swept compression corner.

- The computed yaw angle is in general agreement with experiment, although two specific deficiencies are noted. First, the computed profiles generally underpredict the measured yaw angle in the lower portion of the boundary layer. The Baldwin-Lomax model typically provides a more accurate prediction of the yaw angle in the immediate vicinity of the surface. Within the inner portion of the boundary layer, the Baldwin-Lomax model provides a more accurate prediction for Case 1, while the Jones-Launder is more accurate in Case 2. Settles, Horstman and McKenzie (1984) observed that the Cebeci-Smith algebraic model overpredicted the surface yaw angle within the separated region of the 3-D swept compression corner, while the Jones-Launder model provided an accurate prediction. Second, the observed overshoot in yaw angle outside the boundary layer and upstream of the theoretical inviscid shock (Figs. 8b and 9a) is not predicted.

C. Comparison with Experimental Data of Goodwin at $\alpha_g = 20$ deg

Recently, Goodwin (1984) has examined the 3-D sharp fin for a range of fin angles up to approximately $\alpha_g = 20$ deg at Reynolds numbers up to $Re_{\delta\infty} = 8 \times 10^5$. In the present section, the results of the author for the 20 deg sharp fin are compared with the data of Goodwin for the surface pressure on the flat plate. The flow conditions for the computation and experiment are indicated in Table 4.

Table 4. Flow Conditions for Experiment of Goodwin (1984) and Computation for 3-D Sharp Fin at $\alpha_g = 20$ deg

Case	δ_∞ (cm)	M_∞	$Re_{\delta\infty}$	$p_{t\infty}$ (kPa)	$T_{t\infty}$ (deg K)
Experiment	1.27	2.94	8.0×10^5	690	265
Computation	1.27	2.94	8.6×10^5	690	256

The numerical grid was generated according to the method of Knight (1984b). A total of 20 streamwise grid planes ($N_x = 20$) were employed which were uniformly spaced by an amount $\Delta x = \delta_\infty$. Within each plane, the mesh points were distributed in the y- and z-directions using a combination of geometric stretching near the flat plate and fin, and uniform spacing outside the boundary layers. The number of **ordinary** grid points in the y- and z-directions are $N_y = 32$ and $N_z = 48$. In addition, eight grid points (NSL = 8) were employed locally within the computational sublayer. A separate refined grid was utilized in the sublayer region in the immediate vicinity of the corner formed by the flat plate and the fin. A total of 40,140 grids points were utilized. The height of the first grid point adjacent to the fin or flat plate was less than 3.0 wall units at all locations (i.e., $\Delta y_2^+ \leq 3.0$, where $\Delta y_2^+ = \Delta y_2 u_* / \nu_w$, and Δy_2 is the distance of the first row of points from the flat plate ; a similar result holds for Δz_2, which is the distance of the first row of points adjacent to the fin). Two separate computations were performed in order to examine the sensitivity of the computed solution to the height z'_m of the computational sublayer adjacent to the flat plate. In each case, the maximum grid spacing in the y-direction $\Delta y_\infty = 0.58\delta_\infty$ and $0.59\delta_\infty$. The height of the computational domain in the y-direction is $8\delta_\infty$ for both cases. The width of the computational domain increases linearly from $13.0\delta_\infty$ at $x = 0$ to $28.6\delta_\infty$ at the furthest downstream plane. Likewise, the maximum grid spacing in the z-direction Δz_∞ varies from $0.42\delta_\infty$ to $1.07\delta_\infty$. The computed results using the two separate grids were found to be insensitive to the value of z'_m within the limits of Eq. (8).

The initial condition for the computations was obtained by propogating the upstream flow profile (which matched the experimental upstream boundary layer) throughout the computational domain. The computed flowfield was judged converged to steady state after a physical time of approximately 3.5 t_c, where t_c is the time required for a fluid parcel to travel from the upstream to the downstream end of the mesh in the inviscid region. The computer time required was 2.0 hrs on the VPS 32 for each computation. This represents an estimated **factor** of 14 to 20 times faster than a time-split vectorized version of MacCormack's explicit algorithm alone.

In Fig. 10, the computed and experimental surface pressure on the flat plate is displayed at three spanwise locations, specifically, z = 3.7 cm (2.92 δ_∞), 6.25 cm (4.92 δ_∞), and 8.79 cm (6.92 δ_∞). The computed results provide an accurate estimate of the upstream propogation of the interaction. Also, the plateau pressure is predicted with reasonable accuracy. In some of the regions of pressure gradient, however, the computed profiles do not agree closely with the experiment (e.g., Fig. 10c). This discrepancy may be due to the streamwise grid spacing ($\Delta x = \delta_\infty$). Although this was observed to be sufficient for the α_g = 10 deg case (Knight 1984b) wherein computations using $\Delta x = 0.5\delta_\infty$ and 1.0 δ_∞ were found to be essentially identical, the value $\Delta x = \delta_\infty$ may be insufficient for the present case.

IV CONCLUSIONS

The efficacy of theoretical modelling of three-dimensional shock wave - turbulent boundary layer interactions has been examined for the 3-D sharp fin configuration. The Reynoldsaveraged compressible

Navier-Stokes equations, incorporating the algebraic turbulent eddy viscosity model of Baldwin and Lomax, are solved numerically using a hybrid explicit-implicit algorithm. The results are compared with separate computations of Horstman (employing the two-equation Jones-Launder turbulence model) and the experimental data of McClure and Dolling at α_g = 10 deg for two separate Reynolds numbers. The computed pitot pressure profiles using the two turbulence models are observed to be very similar. The p_p profiles are in good agreement with experiment, with the exception of observed overshoot in p_p upstream of the shock at $Re_{\delta\infty} = 8x10^5$. The computed yaw angle profiles are seen to be sensitive to the turbulence model. In the immediate vicinity of the surface, the calculated results using the Baldwin-Lomax model are observed to be in better agreement with the data. Within the inner portion of the boundary layer, both models tend to underpredict the yaw angle, with the Baldwin-Lomax model providing a more accurate profile at $Re_{\delta\infty} = 2.8x10^5$ and the Jones-Launder yielding a better result at $Re_{\delta\infty} = 8x10^5$. In the outer portion of the boundary layer, both models accurately predict the observed yaw angle profile. Both computations, however, do not predict the observed overshoot in yaw angle.

The hybrid algorithm of the author is also applied to the computation of the 3-D sharp fin at α_g = 20 deg at $Re_{\delta\infty} = 8.6x10^5$. The computed surface pressure is observed to be in good agreement with the experimental data of Goodwin in regards to upstream propogation and plateau pressure. In the regions of rapid pressure rise, the computed pressure does not agree closely with all of the experimental profiles. The discrepancy may be attributable to the streamwise grid spacing.

ACKNOWLEDGMENTS

This research is performed through the sponsorship of the Air Force Office of Scientific Research and NASA Langley Research Center under AFOSR Grant 82-0040 monitored by Dr. J. Wilson. The author wishes to acknowledge helpful discussions with S. Bogdonoff, D. Dolling,S. Goodwin, C. Horstman, W. McClure, G. Settles, J. Shang and L. Smits. Special appreciation is extended to C. Horstman for his permission to include comparison of his computed results for the 3-D sharp fin flowfield.

REFERENCES

Ackeret, J., Feldman, F., and Rott, N. 1947 Investigations of Compression Shocks and Boundary Layers in Gases Moving at High Speeds. NASA TM 1113.

Adamson, T., and Messiter, A. 1980 Analysis of Two-Dimensional Interactions Between Shock Waves and Boundary Layers. Annual Review of Fluid Mechanics, Vol. 12, pp. 103-108.

Baldwin, B., and Lomax, H. 1978 Thin Layer Approximation and Algebraic Model for Separated Flows. AIAA Paper No. 78-257.

Baldwin, B., and MacCormack, R. 1975 A Numerical Method for Solving the Navier-Stokes Equations with Application to Shock-Boundary Layer Interactions. AIAA Paper No. 75-1.

Brosh, A., Kussoy, M., and Hung, C. 1983 An Experimental and Numerical Investigation of the Impingement of an Oblique Shock Wave on a Body of Revolution. AIAA Paper No. 83-1757.

Buleev, N. 1963 Theoretical Model of the Mechanism of Turbulent Exchange in Fluid Flows. AERE Translation 957, Atomic Energy Research Estab., Hartwell, England.

Christiansen, W.,Russell, D., and Hertzberg, A. 1975 Flow Lasers. Annual Review of Fluid Mechanics, Vol. 7, pp. 115-140.

Debieve, J.-F., Gouin, H., and Gaviglio, J. 1982 Evolution of the Reynolds Stress Tensor in a Shock Wave-Turbulence Interaction. Indian J. Tech., Vol. 20, pp. 90-97.

Deiwert, G. 1975 Numerical Simulation of High Reynolds Number Transonic Flows. AIAA J., Vol. 13, pp. 1354-1359.

Dolling, D., and Bogdonoff, S. 1983 Upstream Influence in Sharp Fin-Induced Shock Wave Turbulent Boundary-Layer Interaction. AIAA J., Vol. 21, pp. 143-145.

Escudier, M. 1965 The Distribution of the Mixing Length in Turbulent Flows Near Walls. Report TWF/TN/1, Imperial Coll., Mech. Engr. Dept., London.

Freeman, L., and Korkegi, R. 1975 Experiments on the Interaction with a Turbulent Boundary Layer of a Skewed Shock. ARL TR 75-0182.

Gessner, F., and Po, J. 1976 A Reynolds Stress Model for Turbulent Corner Flows - Part II : Comparison Between Theory and Experiment. J. Fluids Engr., Trans. of ASME, Vol. 98, Series 1, pp. 269-277.

Goodwin, S. 1984 An Exploratory Investigation of Sharp-Fin Induced Shock Wave/Turbulent Boundary Layer Interactions at High Shock Strengths. MS Thesis, Dept. Aero. and Mech. Engr., Princeton U.

Green, J. 1970 Interactions Between Shock Waves and Turbulent Boundary Layers Prog. Aero. Sciences, Vol. 11, pp. 235-240.

Hankey, W., and Holden, M. 1975 Two-Dimensional Shock Wave-Boundary Layer Interactions in High Speed Flows. AGARDograph No. 203.

Holden, M. 1984 Experimental Studies of Quasi-Two-Dimensional and Three-Dimensional Viscous Interaction Regions Induced by Skewed-Shock and Swept-Shock Boundary Layer Interaction. AIAA Paper No. 84-1677.

Horstman, C., and Hung, C. 1979 Computation of Three-Dimensional Separated Flows at Supersonic Speeds. AIAA Paper No. 79-0002.

Horstman, C. 1984a A Computational Study of Complex Three-Dimensional Compressible Turbulent Flow Fields. AIAA Paper No. 84-1556.

Horstman, C. 1984b Private Communications : June 1984, July 1984, November 1984.

Horstman, C., Kussoy, M., and Lockman, W. 1985 Computation of Three-Dimensional Shock-Wave/Turbulent Boundary-Layer Interaction Flows. Third Symposium on Numerical and Physical Aspects of Aerodynamic Flows, Long Beach, California.

Hung, C., and MacCormack, R. 1978 Numerical Solution of Three-Dimensional Shock Wave and Turbulent Boundary Layer Interaction. AIAA J., Vol. 16, pp. 1090-1096.

Inger, G. 1984 Analytical Investigation of Swept Shock-Turbulent Boundary Layer Interaction in Supersonic Flow. AIAA Paper No. 84-1555.

Jones, W., and Launder, B. 1972 The Prediction of Laminarization with a Two-Equation Model of Turbulence. Int. J. Heat and Mass Transfer, Vol. 15, pp. 301-304.

Keller, H. 1974 Accurate Difference Methods for Nonlinear Two-Point Boundary Value Problems. SIAM J. Numerical Analysis, Vol. 11, pp. 305-320.

Knight, D. 1981a Improved Calculation of High Speed Inlet Flows : Part I. Numerical Algorithm. AIAA J., Vol. 19, pp. 34-41.

Knight, D. 1981b Improved Calculation of High Speed Inlet Flows : Part II. Results. AIAA J., Vol. 19, pp. 172-179.

Knight, D. 1981c Calculation of High-Speed Inlet Flows Using the Navier-Stokes Equations. J. Aircraft, Vol. 18, pp. 748-754.

Knight, D. 1983 Calculation of a Simulated 3-D High Speed Inlet Using the Navier-Stokes Equations. AIAA Paper No. 83-1165.

Knight, D. 1984a Numerical Simulation of 3D Shock Turbulent Boundary Layer Interaction Generated by a Sharp Fin. AIAA Paper No. 84-1559.

Knight, D. 1984b A Hybrid Explicit-Implicit Numerical Algorithm for the Three-Dimensional Compressible Navier-Stokes Equations. AIAA J., Vol. 22, pp. 1056-1063.

Korkegi, R. 1971 Survey of Viscous Interactions Associated with High Mach Number Flight. AIAA J., Vol. 9, pp. 771-784.

Korkegi, R. 1976 On the Structure of Three-Dimensional Shock-Induced

Separated Flow Regions. AIAA J., Vol. 14, pp. 597-600.

Kubota, H., and Stollery, J. 1982 An Experimental Study of the Interaction Between a Glancing Shock Wave and a Turbulent Boundary Layer. J. Fluid Mech., Vol. 116, pp. 431-458.

Liepmann, H. 1946 The Interaction Between Boundary Layer and Shock Waves in Transonic Flow. J. Aero. Sciences, Vol. 13, pp. 623-637.

MacCormack, R. 1971 Numerical Solution of the Interaction of a Shock Wave with a Laminar Boundary Layer. Lecture Notes in Physics, Vol. 8, pp. 151-163.

MacCormack, R. 1982 A Numerical Method for Solving the Equations of Compressible Viscous Flow. AIAA J., Vol. 20, pp. 1275-1281.

McCabe, A. 1966 The Three-Dimensional Interaction of a Shock Wave with a Turbulent Boundary Layer, The Aeronautical Quarterly, Vol. 17, pp. 231-252.

McClure, W., and Dolling, D. 1983 Flowfield Scaling in Sharp Fin-Induced Shock Wave Turbulent Boundary Layer Interaction, AIAA Paper No. 83-1754.

Muck, K., and Smits, A. 1984 Behavior of a Turbulent Boundary Layer Subjected to a Shock-Induced Separation. AIAA Paper No. 84-0097.

Oskam, B., Vas, I., and Bogdonoff, S. 1976 Mach 3 Oblique Shock Wave/Turbulent Boundary Layer Interactions in Three Dimensions. AIAA Paper No. 76-336.

Oskam, B., Vas, I., and Bogdonoff, S. 1977 An Experimental Study of Three-Dimensional Flow Fields in an Axial Corner at Mach 3. AIAA Paper No. 77-689.

Peake, D. 1976 Three Dimensional Swept Shock/Turbulent Boundary Layer Separations with Control by Air Injection. Aero. Report No. LR-592, National Research Council - Canada.

Pulliam, T. and Steger, J. 1980 Implicit Finite-Difference Simulations of Three-Dimensional Compressible Flow. AIAA J., Vol. 18, pp. 159-167.

Roshko, A., and Thomke, G. 1976 Flare Induced Interaction Lengths in Supersonic Turbulent Boundary Layers, AIAA J., Vol. 15, pp. 873-879.

Rubesin, M., and Rose, W. 1973 The Turbulent Mean-Flow, Reynolds-Stress and Heat-Flux Equations in Mass Averaged Dependent Variables. NASA TMX-62248.

Settles, G., Horstman, C., and McKenzie, T. 1984 Flowfield Scaling of a Swept Compression Corner Interaction - A Comparison of Experiment and Computation. AIAA Paper No. 84-0096.

Settles, G., and Teng, H. 1984 Cylindrical and Conical Flow Regimes of Three-Dimensional Shock/Boundary Layer Interactions. AIAA J., Vol. 22, pp. 194-200.

Shang, J., Hankey, W., and Petty, J. 1979 Three-Dimensional Supersonic Interacting Turbulent Flow Along a Corner. AIAA J., Vol.

17, pp. 7Ø6-713.

Shang, J. 1984 An Assessment of Numerical Solutions of the Compressible Navier-Stokes Equations. AIAA Paper No. 84-1549.

Stalker, R. 1984 A Characteristics Approach to Swept Shock-Wave Boundary Layer Interactions. AIAA J., Vol. 22, pp. 1626-1632.

Thompson, J., and Warsi, Z. 1982 Boundary-Fitted Coordinate Systems for Numerical Solution of Partial Differential Equations - A Review. J. Comp. Physics, Vol. 47, 1982, pp. 1-1Ø8.

Token, K. 1974 Heat Transfer Due to Shock Wave/Turbulent Boundary Layer Interactions on High Speed Weapons Systems. Air Force Flight Dynamics Lab Report AFFDL-TR-74-77.

Viegas, J., and Rubesin, M. 1983 Wall-Function Boundary Conditions in the Solution of the Navier-Stokes Equations for Complex Compressible Flows. AIAA Paper No. 83-1694.

West, J., and Korkegi, R. 1972 Supersonic Interaction in the Corner of Intersecting Wedges at High Reynolds Number. AIAA J., Vol. 1Ø, pp. 652-656.

York, B., and Knight, D. 1985 Calculation of a Class of Two-Dimensional Turbulent Boundary Layer Flows Using the Baldwin-Lomax Model. AIAA Paper No. 85-Ø226.

Zheltovodov, A. 1982 Regimes and Properties of 3-D Separation Flows Initiated by Skewed Compression Shocks. Zhur. Prk. Mekh. i. Tekh. Fiz., No. 3, pp. 116-123.

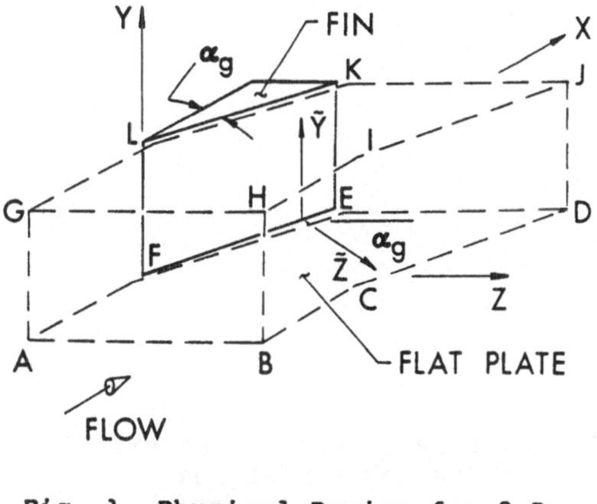

Fig. 1 Physical Region for 3-D Sharp Fin

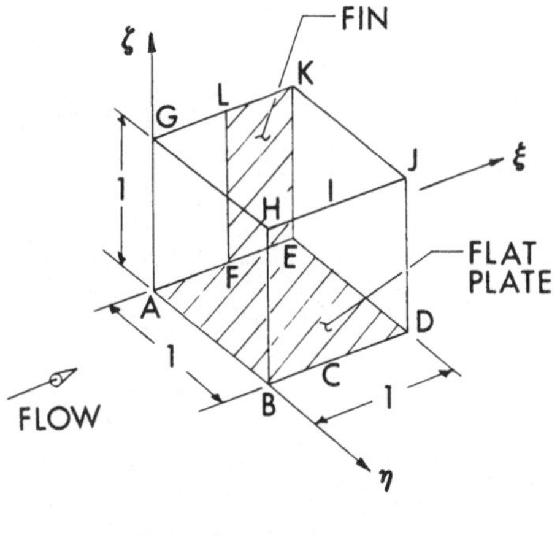

Fig. 2 Transformed Region for 3-D Sharp Fin

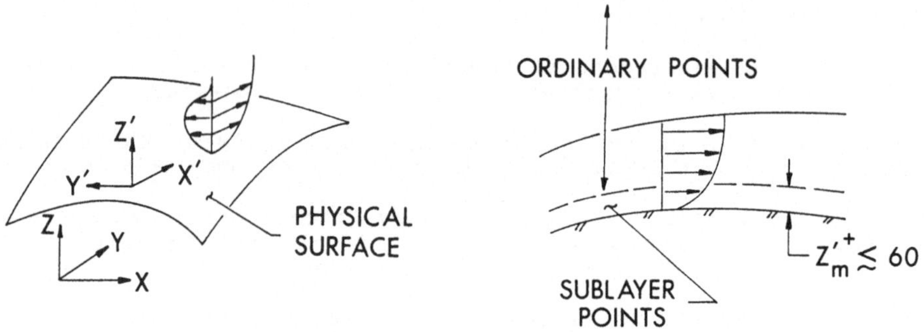

Fig. 3 Geometry of Hybrid Explicit-Implicit Algorithm

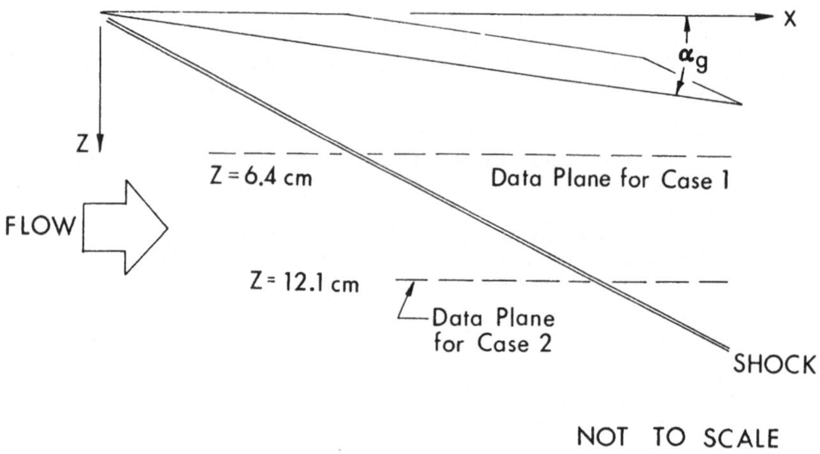

Fig. 4 Experimental Data Stations of McClure

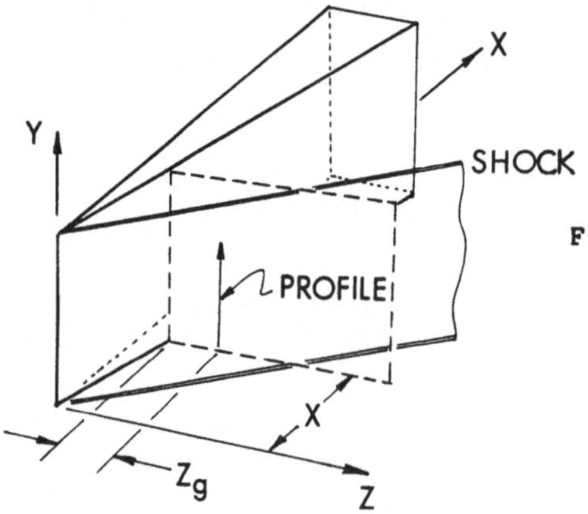

Fig. 5 Geometry of Pitot Pressure and Yaw Angle Profiles

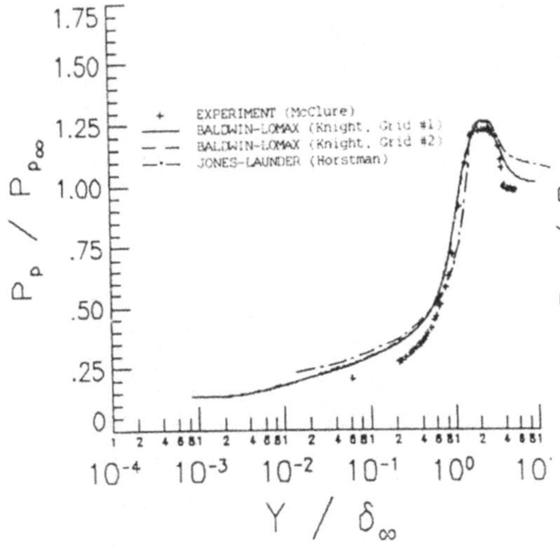

Fig. 6a Pitot Pressure at $x_s = -1.7\delta_o$ for Case No. 1

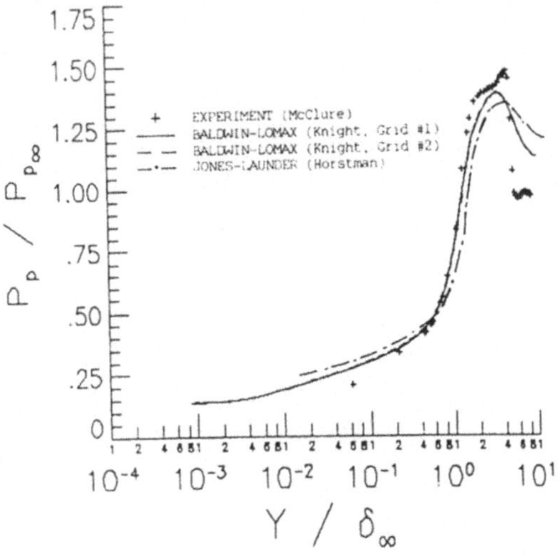

Fig. 6b Pitot Pressure at $x_s = -0.63\delta_o$ for Case No. 1

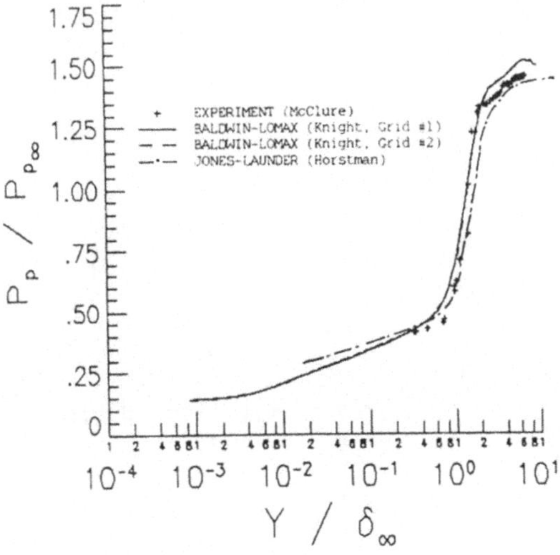

Fig. 6c Pitot Pressure at $x_s = 1.0\delta_o$ for Case No. 1

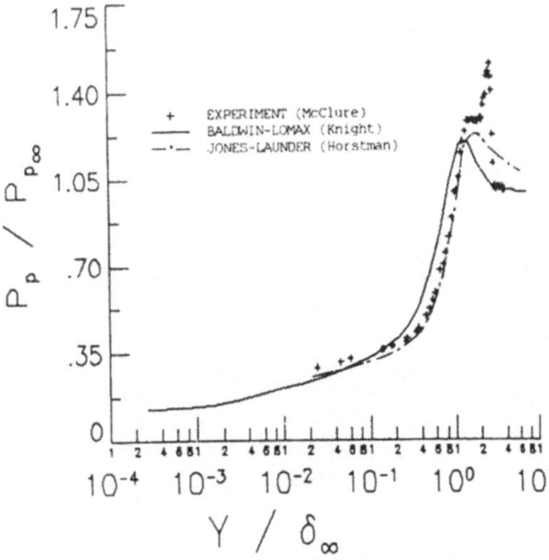

Fig. 7a Pitot Pressure at $x_s = -1.47\delta_o$ for Case No. 2

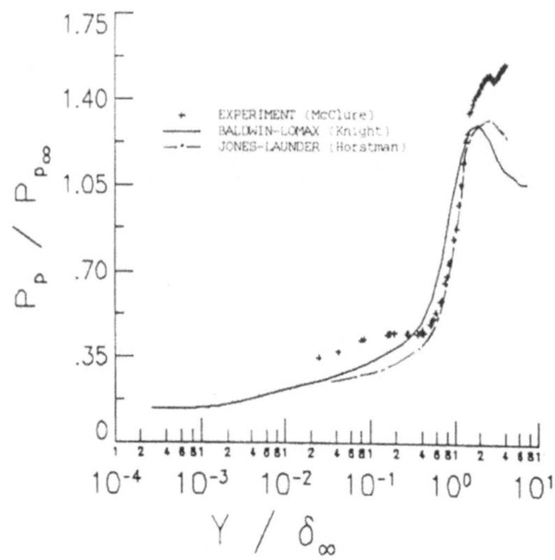

Fig. 7b Pitot Pressure at $x_s = -0.37\delta_o$ for Case No. 2

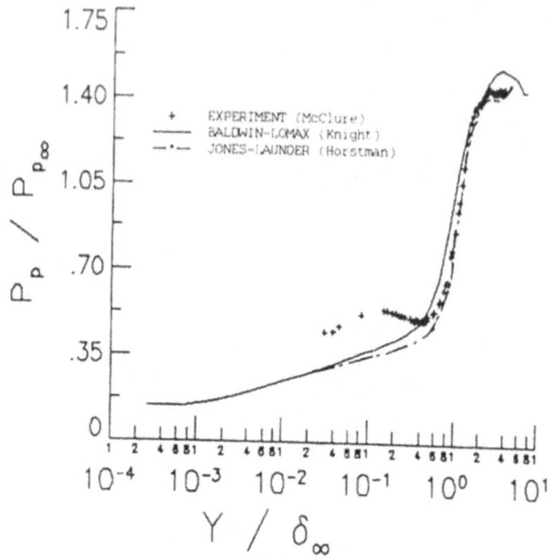

Fig. 7c Pitot Pressure at $x_s = 1.27\delta_o$ for Case No. 2

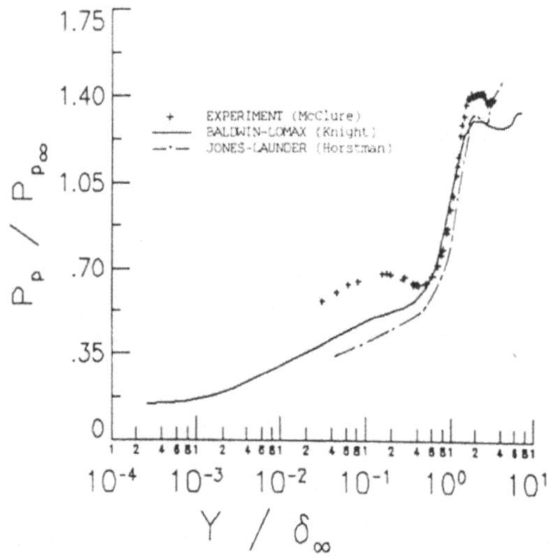

Fig. 7d Pitot Pressure at $x_s = 3.40\delta_o$ for Case No. 2

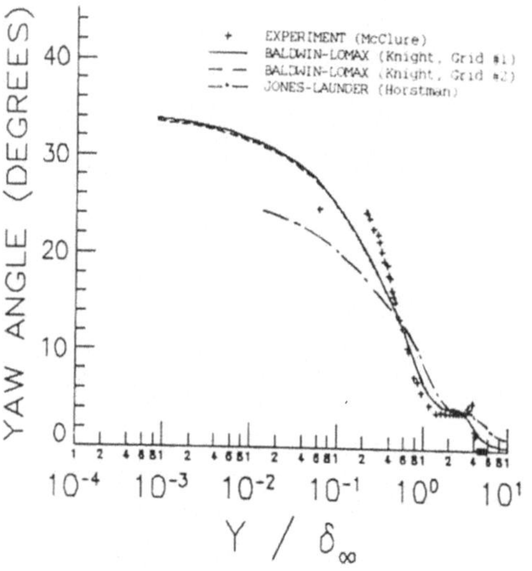

Fig. 8a Yaw Angle at $x_s = -1.70\delta_o$ for Case No. 1

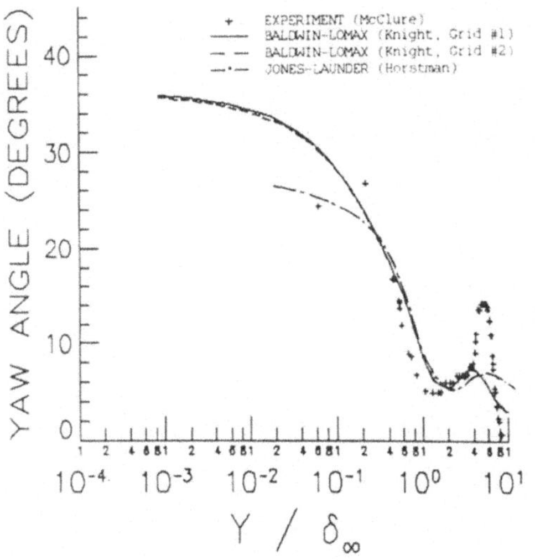

Fig. 8b Yaw Angle at $x_s = -0.63\delta_0$ for Case No. 1

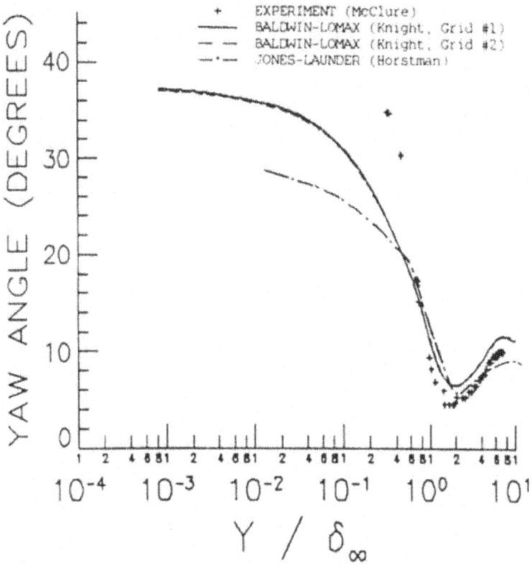

Fig. 8c Yaw Angle at $x_s = 1.0\delta_0$ for Case No. 1

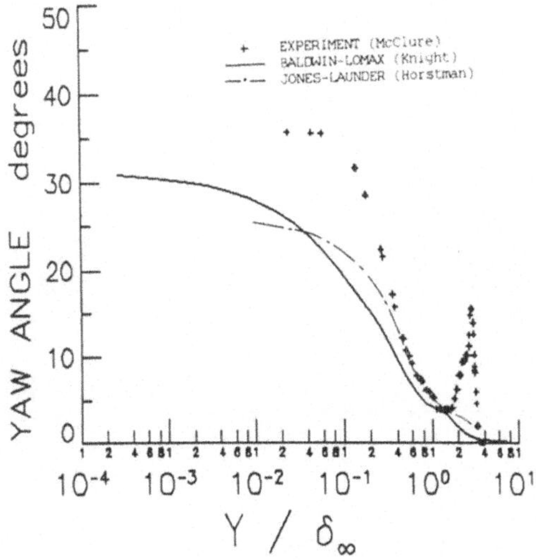

Fig. 9a Yaw Angle at $x_s = -1.47\delta_0$ for Case No. 2

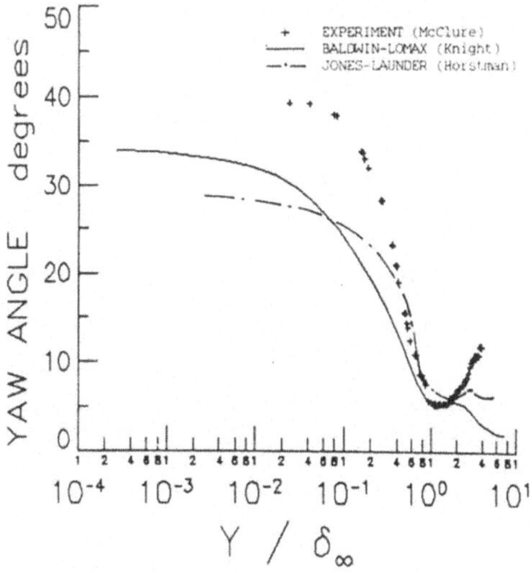

Fig. 9b Yaw Angle at $x_s = -0.37\delta_0$ for Case No. 2

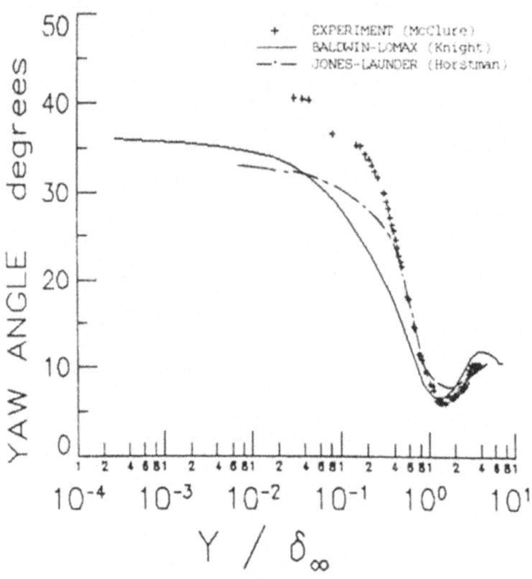

Fig. 9c Yaw Angle at $x_s = 1.27\delta_o$ for Case No. 2

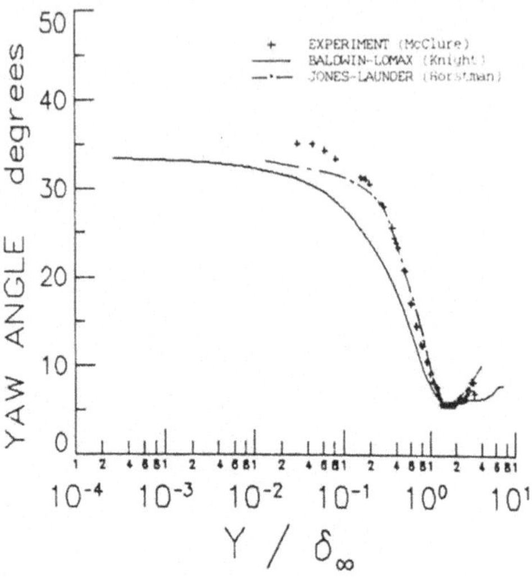

Fig. 9d Yaw Angle at $x_s = 3.40\delta_o$ for Case No. 2

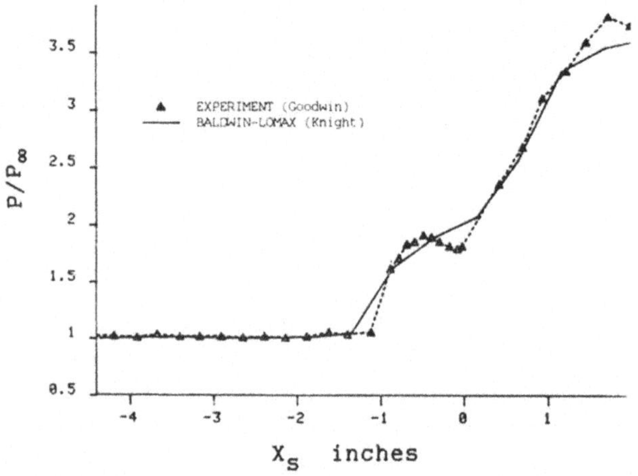

Fig. 10a Surface Pressure on Flat Plate at z = 3.71 cm for α_g = 20 deg

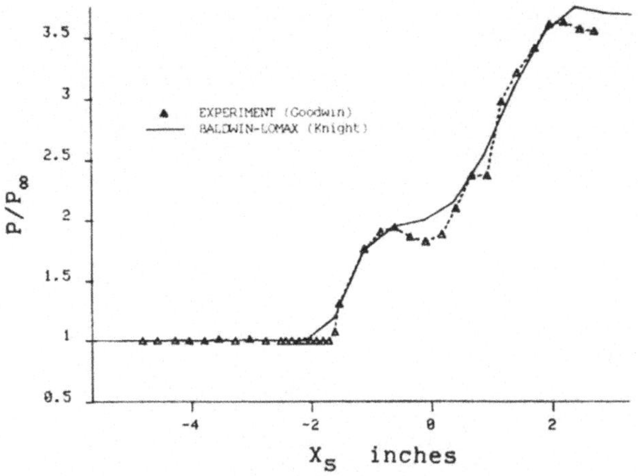

Fig. 10b Surface Pressure on Flat Plate at z = 6.25 cm for α_g = 20 deg

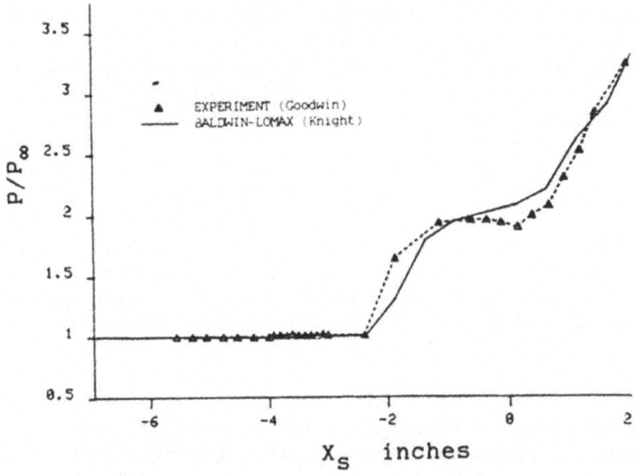

Fig. 10c Surface Pressure on Flat Plate at z = 8.79 cm for α_g = 20 deg

NUMERICAL AND THEORETICAL STUDY OF DIFFERENT FLOW REGIMES OCCURRING IN HORIZONTAL FLUID LAYERS, DIFFERENTIALLY HEATED

Bernard Roux, Patrick Bontoux and Daniel Henry
Institut de Mécanique des Fluides
1, rue Honnorat
13003 MARSEILLE, FRANCE

1. Introduction

This study is devoted to flow regimes of a low Prandtl number fluid in horizontal closed cavities with different end temperatures. It is connected to the problem of hydrodynamic perturbations induced by buoyancy forces in metallic melts (as pure Gallium) or in mixture (as As-Ga) during unidirectional solidification (horizontal Bridgman) processes where the fluid is submitted to horizontal temperature gradients.

Growth of metal or semiconductor crystals from melts is directly influenced by convective motions in the melts. Even for a small imposed temperature difference, i.e. for small Grashof number, the flow starts immediately although it may not immediately influence the temperature pattern. This weak flow is sufficient to induce inhomogeneous vertical segregation of impurities or dopants as shown by Nikitin et al. (1981). For higher temperature difference leading to Grashof number based on the melt depth of about 2.10^5, temperature oscillations can appear as shown in experiments by Hurle et al. (1974).

A survey of currently available literature on fluid motions in melt growth, covering Czochralski and Bridgman techniques, has been given from the point of view of fluid dynamicists by Pimputkar and Ostrach (1981). For the Bridgman growth they report on different interpretations of striations observed in crystal that have been attributed to a variety of cyclic phenomena which could occur in melt growth (freezing-remelting, temperature in the bulk, etc.). They point out, in addition, some contradictions between experimental results given by Hurle et al. (1974) and by Pamplin and Bolt (1976) concerning the dependence of oscillations frequency on melt depth.

Theoretical explanations of the onset of oscillatory instability have been suggested by Hart (1972, 1983) and Gill (1974). They studied the stability of disturbances developing in the direction normal to the plane of an unperturbed flow consisting in a single long convective cell, as given from asymptotic expansion by Birikh (1966) and Hart (1972).

A numerical simulation of the horizontal Bridgman growth of As-Ga has been recently carried out by Crochet et al. (1983) with a two-dimensional model taking into account a moving solid-liquid interface and a free upper horizontal surface. These authors report on oscillations obtained at $Ra = 3 \times 10^4$ and $Pr = 0.015$ ($Gr = 2.10^5$) for a Stefan number of 140 and for aspect ratios of the rectangular cavity ranging from 4 to 8.

The exact origin of the oscillations in molten metals is not yet well understood, neither the exact influence of the numerous parameters involved in Bridgman growth.

The goal of the present study is to contribute to a better analysis of flow regimes occurring in completely confined geometry with an imposed horizontal difference. Assuming that the origin of fluctuations is purely hydrodynamic, the present study is mainly oriented to the limit case where $Pr = 0$, in which the temperature field is completely frozen. This simplifies the problem and gives worse conditions for the onset of instabilities. In the same way, the rigid wall conditions lead to more severe limitations for this onset.

The paper begins with the determination of marginal stability threshold through a linear stability theory similar to the one proposed by Hart (1983) but carried out with an accurate method based on Chebyshev polynomials.

Two-dimensional numerical simulations are given afterwards for Gr, below and above the critical value given by this theory, for aspect ratios L/H (length/depth) of 10, and for higher Gr for L/H ranging from 3 to 4.

Finally, preliminary results of a three-dimensional simulation of motion in horizontal cylinders at Pr = 0 will be shown for L/H = 10.

2. The 2D model

A 2D flow of Boussinesq fluid is considered in the geometry shown in Fig.1. The governing equations (Navier Stokes and energy) are non-dimensionalized using scales $\Delta T/Az$, $\nu\, Gr_H/H$, H and H^2/ν, for temperature, velocity, length and time respectively.

The normalized equations written in vorticity and stream function formulation ,are:

$$\xi = \nabla^2\psi \quad ; \quad u = \frac{\partial\psi}{\partial y} \quad ; \quad v = -\frac{\partial\psi}{\partial x} \qquad (1)$$

$$\frac{\partial\xi}{\partial t} + Gr_H \left(u\frac{\partial\xi}{\partial x} + v\frac{\partial\xi}{\partial y}\right) = \nabla^2\xi + \frac{1}{8}\frac{\partial T}{\partial y} \qquad (2)$$

$$Pr\left[\frac{\partial T}{\partial t} + Gr_H\left(u\frac{\partial T}{\partial x} + v\frac{\partial T}{\partial y}\right)\right] = \nabla^2 T \qquad (3)$$

where $Az = \frac{L}{H}$ is the aspect ratio, $Gr_H = g\,\beta\gamma H^4/\nu^2$ is the Grashof number, $Pr = \nu/\chi$ is the Prandtl number and $\gamma = \Delta T/L$ is the horizontal temperature gradient externally imposed. Considering all the walls as rigid and impermeable, the end walls as isothermal and the horizontal walls as perfectly conducting or adiabatic, we have :

$$u = v = 0 \quad \text{on the boundaries} \qquad (4)$$

$$T(x, -L/2) = \frac{Az}{2} \quad ; \quad T(x, L/2) = \frac{Az}{2} \qquad (5)$$

$$T(H/2, y) = T(-H/2, y) = y/2 \quad \text{(Cond.)} \qquad (6)$$

$$\frac{\partial T}{\partial x}\left(\frac{H}{2}, y\right) = \frac{\partial T}{\partial x}\left(-\frac{H}{2}, y\right) = 0 \quad \text{(Adiab.)} \qquad (7)$$

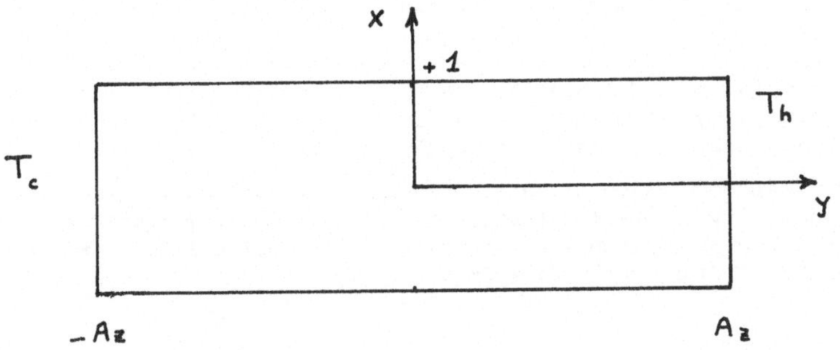

Fig. 1. Geometry of the problem

3. The 1D model and linear stability analysis

The system (1-3) admits a well known steady 1D asymptotic solution when $Az \to \infty$, as given previously by Birikh (1966) and Hart (1972). It becomes :

$$\hat{u}(x) = 0$$

$$\tilde{v}(x) = (x^3 - x) / 96 \qquad (8)$$

$$\tilde{T}(x, y) = y/2 + Pr\ Gr_H\ (3x^5 - 10x^3 + Cx) / 10920 \qquad (9)$$

where C = 7 for the conducting case (6) and C = 15 for the adiabatic case (7).

Following Hart (1983a) we consider the stability of small disturbances superimposed to this flow. Looking for disturbances in the (x,y) plane as developed in normal modes :

$$u' = u(x) e^{\lambda t} e^{iny},$$

the linearized differential system of perturbation becomes

$$\lambda\xi + Gr_H\ [u \frac{\partial\tilde{\xi}}{\partial x} + in\ \xi\tilde{v}] = (-n^2\xi + \frac{\partial^2\xi}{\partial x^2}) \quad \frac{in\ T}{8} \qquad (10)$$

$$\lambda Pr\ T + Pr\ Gr_H\ (u \frac{\partial\tilde{T}}{x} + in\ T\ \tilde{v} + \frac{v}{2}) = -n^2 T + \frac{\partial^2 T}{\partial x^2} \qquad (11)$$

$$\xi = -n^2\psi + \frac{\partial^2\psi}{\partial x^2} \ ; \quad u = in\psi \ ; \quad v = -\frac{\partial\psi}{\partial x} \qquad (12)$$

with the following conditions :

$$\psi(\pm 1) = 0 \qquad \frac{\partial\psi}{\partial x}(\pm 1) = 0$$

$$\frac{\partial T}{\partial x}(\pm 1) = 0$$

This stability problem is solved by a Tau method (Galerkin + boundary conditions) using Chebyshev polynomials as basis functions $u(x) = \sum_{n=o}^{N} a_n T_n(x)$.

The neutral stability curve ($\mathcal{Re}(\lambda) = 0$) so obtained agrees with the one given by Hart (1983a) for small Pr. The present results show an increasing difference when Pr is increased (Fig. 2). This could correspond to a round-off errors effect in Hart's results which were obtained on a Vax 11/750 with 6 1/2 digit arithmetic (Hart 1984).

A similar study has been carried out for transverse disturbances $u' = u(x)\ e^{\lambda t}\ e^{inz}$ but with a primitive variable formulation, which confirms the Hart's results, showing that the first instability corresponds to an oscillatory mode ($Im(\lambda) \neq 0$). But our results present a systematic deviation of $Gr_{H,C}^{osc}$ compared to Hart's results (Fig.2).

Nevertheless, the main conclusion of Hart's paper remains valid, in that sense that the two neutral stability curves for transverse and longitudinal disturbances intersect ; here near Pr = 0.05 in such a way that the first instability is monotonic for $0 \leqslant Pr \lesssim 0.05$ and oscillatory for $0.05 \lesssim Pr < 0.1$.

4. Numerical solution of the 2D model

The numerical technique used is described in detail in a previous paper by Roux et al (1978). It is used in the following form : 1/ a second order accurate central differenciation is used for spatial derivatives in the equations (2) and (3) ; 2/ a fourth order accurate compact Hermitian method is used for equation (1), and 3/ a second order accurate approximation of the boundary condition for vorticity is used with relaxation ; 4/ an alternating direction implicit (ADI) method is used to solve the finite difference form of equation (1-3) without internal iteration (false transient method).

The present code is used here mainly to analyse steady situations and to give qualitative insight of flow structure near the onset of oscillatory motions.

4.1. Long cavity (Az = 10) at Pr = 0

The present false transient code has been previously used with Nikitin et al, 1984, to study the onset of monotonic instabilities and the bifurcation to steady multicells structure already observed for moderate values of Pr (Pr < 0.1) and for adiabatic horizontal walls by Hart (1983b), and by Jones (1982) and Lee and Korpela (1983) for long vertical cavities.

The results given in the previous paper with Nikitin et al (1984) are repeated here in the case Pr = 0 for Az = 10 (Fig. 3), to give an example of shear flow instability generated on the boundary of two opposite flows. At $Gr_H = 7000$, below the theoretical $Gr_{H,c} = 7932$ given by Korpela et al (1973), this solution presents a one cell motion except small single cat's eyes near each end wall . At $Gr_H = 10000$, a smooth transition can be observed to a stable multicells solution which has been previously found by Lee and Korpela (1983) and by Hart (1983b) who presents this as an example of imperfect bifurcation. Here the correct solution is with 3 stable cells. At Gr = 50000 the observed solution corresponds to five equally strong and regularly spaced stable cells. This pattern is quite different from the one corresponding to shear flow instability.

Nikitin et al have also experienced that when the computation is made with a too coarse mesh size (98X9 for Az=10) a wrong but "converged" four rolls solution is obtained at $Gr = 10^4$. Furthermore, if this solution is used with interpolation as initial condition for a new calculation with a finer mesh, the four rolls solution is maintained. This wrong solution is very tricky because it corresponds to a wave length (2.5d) that would be in better agreement with the critical wave length given by linear stability theory (2.65d) carried out by Korpela et al (1974) for $Az = \infty$. The fact is that for Az = 10, the end walls play again an important role.

4.2. Moderately long cavity (Az = 4) at Pr = 0

The present code is used to confirm not yet published results obtained by Gresho and Sani (1984) with a finite element method carried out for fully transient Boussinesq equations at Pr = 0. One of the interesting features observed by these authors was that the solution for Az = 4 and $Gr_H = 50000$ appeared to have very stable oscillations (the pattern changing alternatively from 1 to 3 cells) during several ten cycles and then to present a sudden transient state before to converge to a steady two cells solution.Gresho and Sani noted in addition that during the intermediate transient phase the centro-symmetry of the solution was lost.

We repeated the computations in two ways. In the first one, the computation was started with an initial condition derived from asymptotic solution (8-9), then a stable oscillatory behaviour was obtained like in the beginning solution of Gresho and Sani's results (Fig. 4a). In a second way, the initial condition was perturbated asymmetrically multiplying the value of ψ (for I = 7 and J = 19) by a given factor. For a factor of 1.25 the stable oscillatory solution changes after only 3 cycles to transit to a two cells solution (Fig. 4b). For a factor of 1.5, the transition to a two cells solution is again more rapid (Fig. 4c). Most of the results presented hereafter concern iso-ψ lines for different IT (time) corresponding to successive maximum and minimum values of ψ at the center.

4.3. Moderately long cavity (Az = 4) for small but finite Pr

For Pr = 0.035, computations have been carried out for increasing values of Gr_H. For Gr_H = 10000 the solution converges rapidly (498 iterations) to a 1 cell solution. For Gr_H = 15000 the solution converges again to a 1 cell solution but with damped oscillations. At Gr = 20000 the solution presents stable oscillatory solutions for 8 cycles without damping, even when a perturbation with a factor 1.25 was used.

For Gr_H = 50000, we repeated the comparison made at Pr = 0 without perturbation or for perturbations with factors 1.25 and 1.5. In the first case we find again stable oscillatory solution from 1 to 3 cells alternatively (Fig.5a) . The damping effect of initial perturbations is not so strong than for Pr = 0 ; here the change for the factor 1.25 appears only after 6 cycles (Fig. 5b) and after 5 cycles for the factor 1.50

At Pr = 0.070, the solution for Gr_H = 10000 converges again more rapidly (386 iterations) to a steady one cell solution which is also reached for Gr_H = 20000 but after damped oscillations.

Additional results obtained for increasing Gr_H show that the solution is again much more stable than for the lower Pr. For Gr_H = 800000 the solution converges rapidly (173 itérations from the solution obtained at Gr_H = 750000) to a one steady cell.

This "stabilizing" effect of Pr , when this parameter is increased in the range $0 < Pr < 0.1$, is consistent with the results of marginal stability for transverse modes shown in Fig. 2, except for the spurious oscillations appearing for high Gr_H.

4.4. Real oscillating behaviour for higher Gr_H at small Pr.

Additional computations have been carried out at Az = 4, for higher Gr_H in order to observe real oscillatory behaviour.

At Pr = 0, the two cell solution mentioned before for Gr_H = 50000, has been also observed for Gr_H = 75000, 100000, 150000 and 200000 with increasing value of ψ_{max} (almost linearly) . But from Gr_H = 100000 oscillations appear, the amplitude of which increases with Gr_H (Fig. 6).

Using the last of these solutions as initial condition, computations have been made for Pr = 0.016 and Pr = 0.025 at Gr_H = 200000 which show that the solution is still oscillatory (Fig. 7). These oscillations have been registered for twelve characteristic values: ψ, u and v at the center of the cavity, ξ_w on the lower wall at mid-distance, v and T in four positions defined by J = 6 and I1 = 7, I2 = 16 , I3 = 31 and I4 = 46. Some significant samples (for v (I1,6) and v (I2,6)) presented on Fig.8 show that these oscillations are more pronounced for Pr = 0.016 with higher amplitude and modulation, than for Pr = 0.025 where the oscillations become regular; in that sense increasing Pr has also a stabilizing effect. But one can remark that the frequency of oscillations is higher for Pr = 0.025, and that the value of ψ_{max} decreases from 0.042 at Pr = 0, to 0.030 at Pr = 0.016 and 0.026 at Pr = 0.025, indicating that the maximum velocity is reduced with the same rate when Pr is increased.

These results show that oscillations can appear at Pr = 0, i.e. without temperature oscillation even from Gr_H = 100000; such oscillations are maintained at least at Gr_H = 200000 for values of Pr close to the one relevant for molten gallium.

4.5. Molten gallium (Pr = 0.02)

Numerical applications have been made for molten gallium (Pr = 0.02) for a range of values of Az between 2 and 3 relevant for comparison with experiments by Hurle et al (1974).

At Gr_H = 10000, the solution converges to a steady one cell.

At Gr_H = 50000, the solution still converges for Az = 2 and for Az = 2.5 but with damped oscillations for this last case, while for Az = 3, the solution presents stable oscillations (from 1 to 3 cells successively) after 1000 time steps, even with an initial perturbation on ψ by a factor 1.5. The frequency corresponds to 5 cycles for 800 time steps.

At Gr_H = 100000, the solution still converges but for Az = 2 only (Fig. 9a). For Az = 2.5, the flow shown in Fig. 9b presents 3 cells : a central one bigger and two secondary cells slightly oscillating (9 cycles for 800 time steps) with small amplitude. For Az = 3, stable oscillations are observed (Fig. 9c) with slightly higher frequency than for Gr_H = 50000 (6 cycles for 800 time steps).

At Gr = 150000, oscillations appear even for Az = 2 (Fig.10) the frequency of such oscillations is increasing when Az diminishes and when Gr_H increases.

These results give the expected tendency to have more stabilized flow when the cavity is more confined (i.e. when Az decreases for a fixed Ra_H).They show that the flow can become oscillatory for even smaller values of Ra_H than the ones for which experimental oscillations were reported by Hurle et al (1974).

In that case ($2 \leqslant Az \leqslant 3$) there is no evidence of spurious oscillatory flow as the one mentioned before (Fig. 4a and 5a) at Az = 4 for Pr = 0 and Pr = 0.035 when Gr_H= 50000.

5. 3D flow in horizontal cylinder at Pr = 0

Finally, we consider the flow in horizontal cylinder (differentially heated) in order to control the effect of viscosity damping in completely confined geometry. The governing 3D Boussinesq equations which are written in velocity and vorticity formulation, are solved by the finite difference method proposed by Leong and De Vahl Davis (1979). The code adapted for differentially (axially) heated cylinder has been described in detail in a previous paper by Smutek et al. (1984).

The preliminary results obtained for Pr = 0 and Az = 10 (here Az is length/diameter) for Gr_H = 7000 and 10000 show in Fig. 11, that the flow pattern does not present a multicell structure unlike the solution of 2D model presented in Fig. 3. Even for Gr_H = 20000 the one cell structure is still observed (Fig.11), indicating the strongly damping effect of the viscosity on the walls.

6. Conclusion

This paper focuses attention on some tricky numerical solutions. For long cavity such as Az = 10 an initialisation process with a coarse mesh size can induce a wrong number of cells. For moderate long cavity (Az = 4) spurious oscillations can be generated; they disappear when imposing some asymmetric disturbances.

As long as a 2D model is used to describe the onset of oscillatory motions in liquid metals the present computations show that stable oscillations can occur even for Ra number smaller than the one observed in the experiments by Hurle et al (1974). Nevertheless more accurate analysis would be necessary to give a definite insight on this problem as the confinement of the cavity in the third direction can strongly damp the perturbed motion, as it is shown for a closed cylinder at Az = 10.

Furthermore, a correct analysis of the real time dependent behaviour of the 2D oscillatory flows would require a truly transient version of the present code.

The 2D Boussinesq equations for Pr = 0 could be used as an interesting test problem to study complex unsteady convective flow occurring for higher values of Gr_H.

REFERENCES

R.V. Birikh, P.M.M., 1966, 30, 2, pp. 356-361
M.J. Crochet, F.T. Geyling and J.J. Van Schaftingen, 1983, J. of Crystal Growth 65, pp.166-172
A.E. Gill, 1974, J.F.M., 64, 3, pp. 577-583
P. Gresho and R. Sani, 1984, private communication
J. Hart, 1972, J.Atm. Sc., 29, pp. 687-697
J. Hart, 1983a, J.F.M., 132, pp. 271-281
J. Hart, 1983b, Int.J. H.M.T., 26, 7, pp. 1069-1074
J. Hart, 1984, private communication
D.T. Hurle, E. Jakeman and C.P. Johnson, 1974, J.F.M., Vol. 64, part 3, pp. 565-576
I.P. Jones, 1982, Harwell Rep^{t}. AERE-R 10416
S.A. Korpela, D. Gozum and C.B. Baxi, 1973, Int. J. Heat Mass Transfer, 16, pp.1683-1690
Y. Lee and S. Korpela, 1983, J.F.M., 126, pp. 91-121
S.S. Leong and G. De Vhal Davis, 1979, Num. Meth. in Thermal problems, Pmeridge Press, pp. 287-296
S.A. Nikitin, V.I. Polezhaev and A.I. Fedynskhin, 1981, J. Crystal Growth, 52, pp. 471-477
S.A. Nikitin, V.I. Polezhaev and B. Roux, 1984, submitted to Numerical Heat Transfer
R.R. Pamplin and G.H. Bolt, 1976, J. Phys. (App.Phys.) 9, pp. 145-150
S.M. Pimputkar and S. Ostrach, 1981, J. Crystal Growth, 55, pp. 614-646
B. Roux, P. Bontoux, T.P. Loc and O.Daube 1979, Lect. Notes in Math. , n°771, Springer Verlag, pp. 450-468
C. Smutek, B. Roux, P. Bontoux and G. De Vahl Davis, 1984, Vieweg Notes in Num.Fl. Mech., 7, pp. 338-345.

Acknowledgements

The authors are indebted to R. Sani and P. Gresho for suggesting a part of this work. They wish to thank the Centre Vectoriel pour la Recherche (CCVR), the Centre National Universitaire Sud de Calcul (CNUSC) and the Point d'Accès Informatique St.Charles for efficient technical assistance. The computation means were given by the Conseil Scientifique of CCVR and the research supported by Centre National d'Etudes Spatiales (Materials department); authors also express their thanks to Madame Ch. GUIBAUD for typing this manuscript.

c) $Gr_H = 10000$

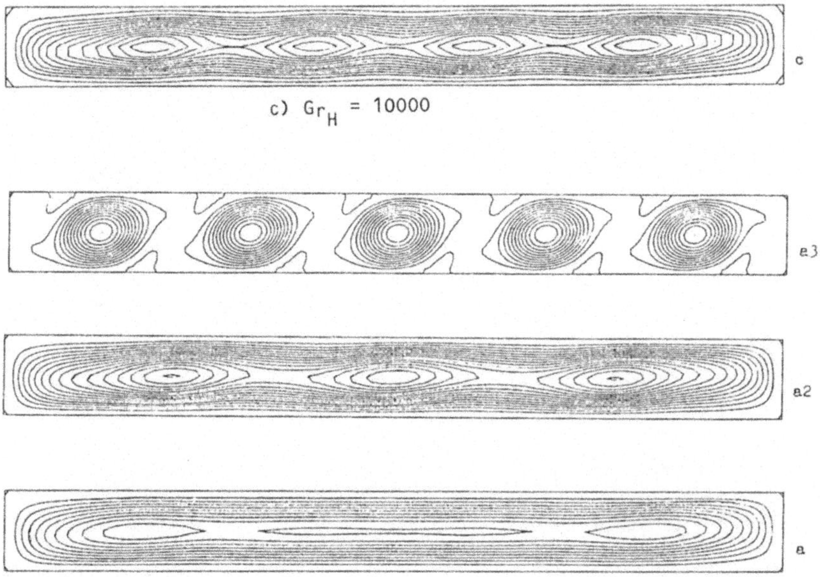

Fig. 3   ISO-ψ LINES ; Pr = 0 ; Az = 10   a) Gr_H = 7000 ; a2) Gr_H = 10000 ;

a3) Gr_H = 50000

—— Hart (1983)   present $\left\{\begin{array}{l} \bullet \text{ mon otonic (transverse)} \\ \triangle \text{ oscillatory (longitudinal)} \end{array}\right.$

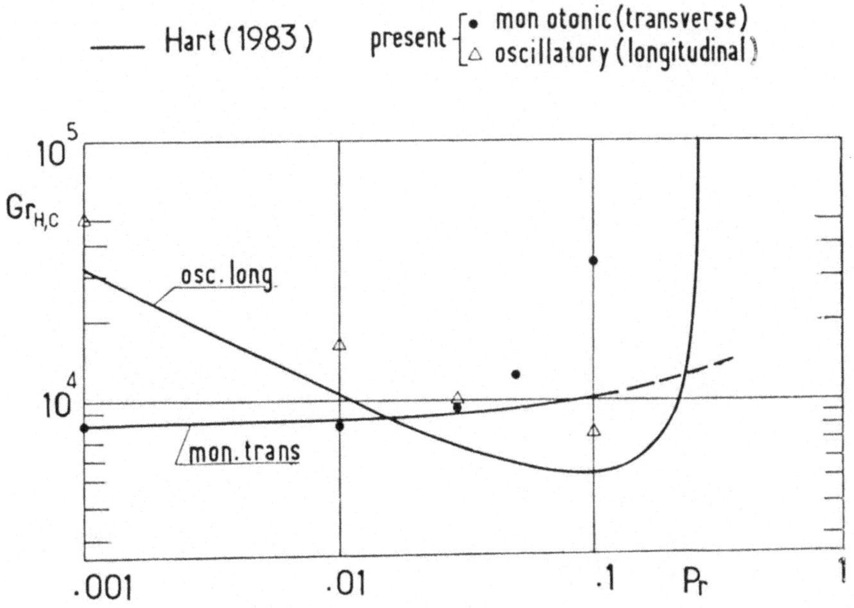

Fig. 2 NEUTRAL STABILITY LIMIT

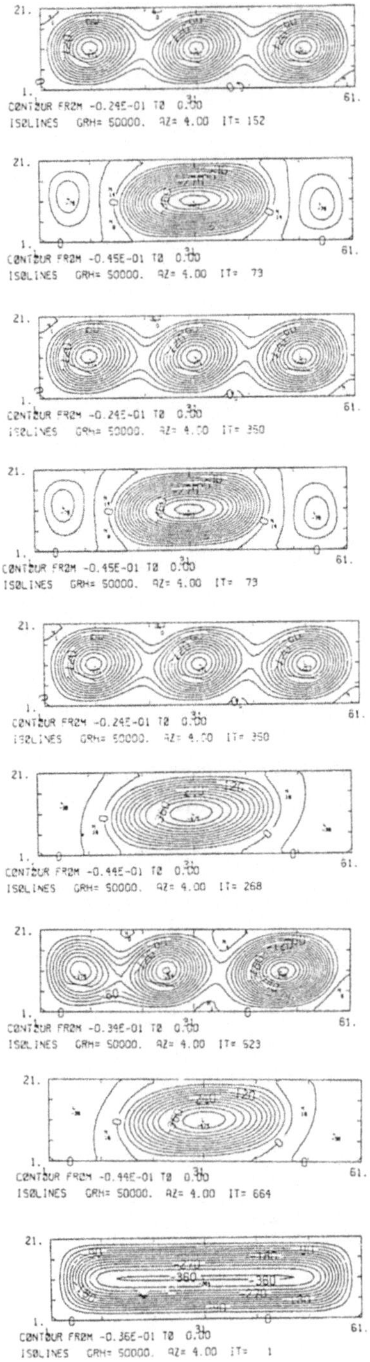

Fig. 4a PERTURB. PSI(7 .19) = 1.00

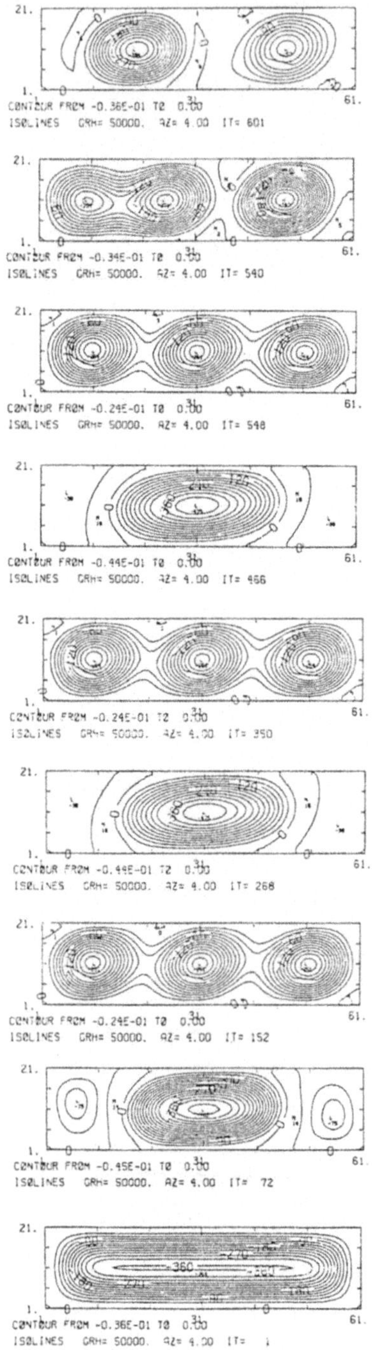

Fig. 4b PERTURB. PSI(7 .19) = 1.25

ISO-ψ LINES ; Pr = 0

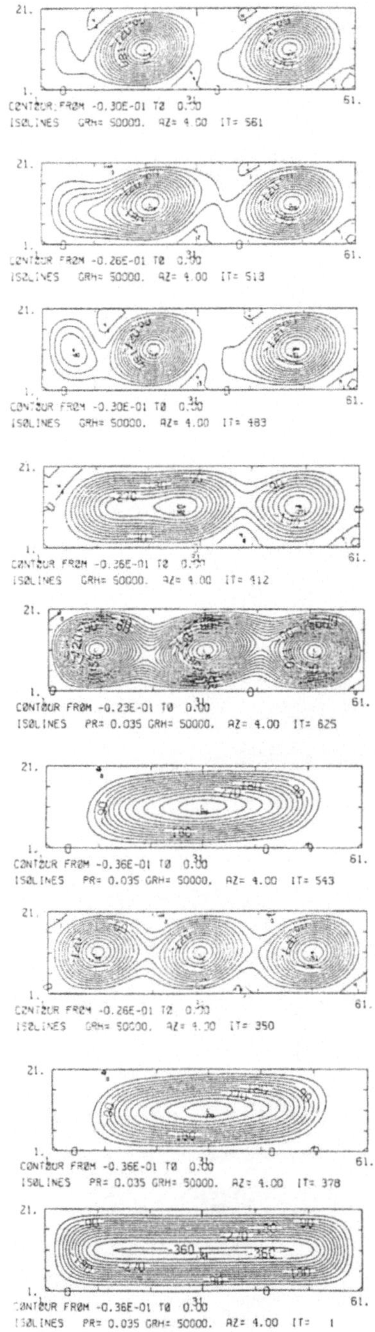

Fig. 4c PERTURB. PSI(7 ,19) = 1.50

Pr = 0

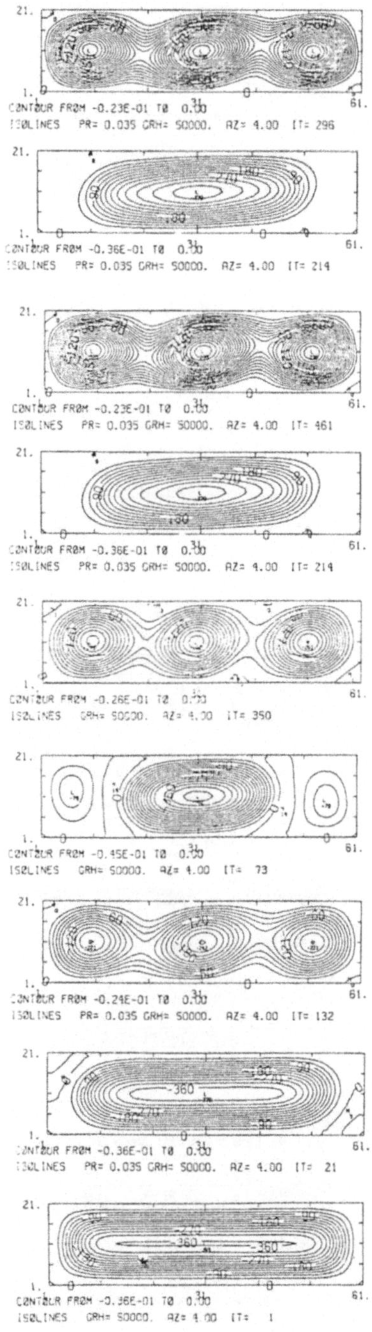

Fig. 5a PERTURB. PSI(7 ,19) = 1.00

Pr = 0,035

ISO-ψ LINES

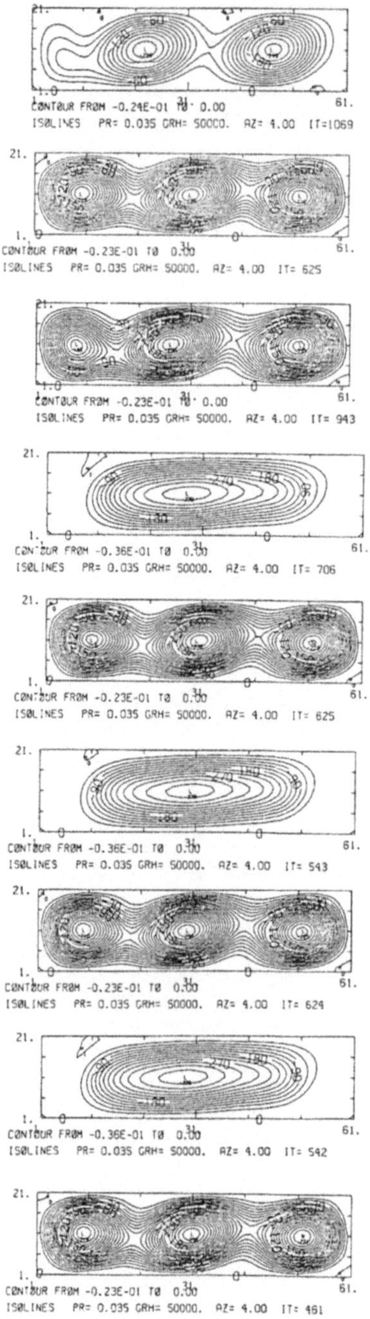

Fig. 5b

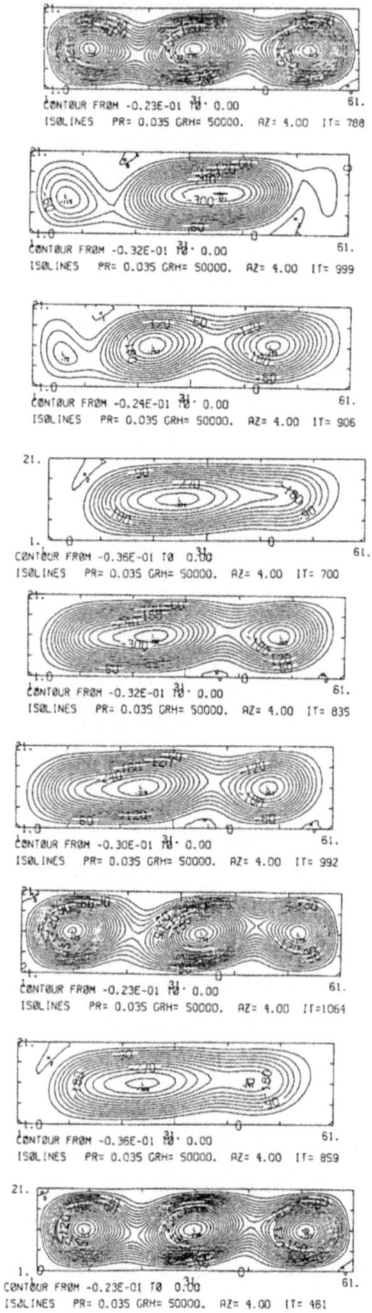

Fig. 5c

ISO-ψ LINES ; Pr = 0,035

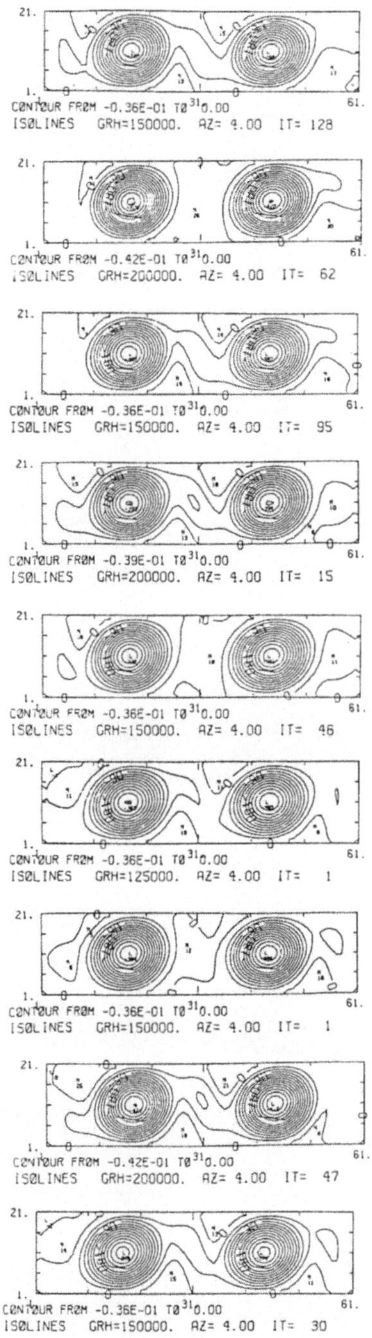

Fig. 6a

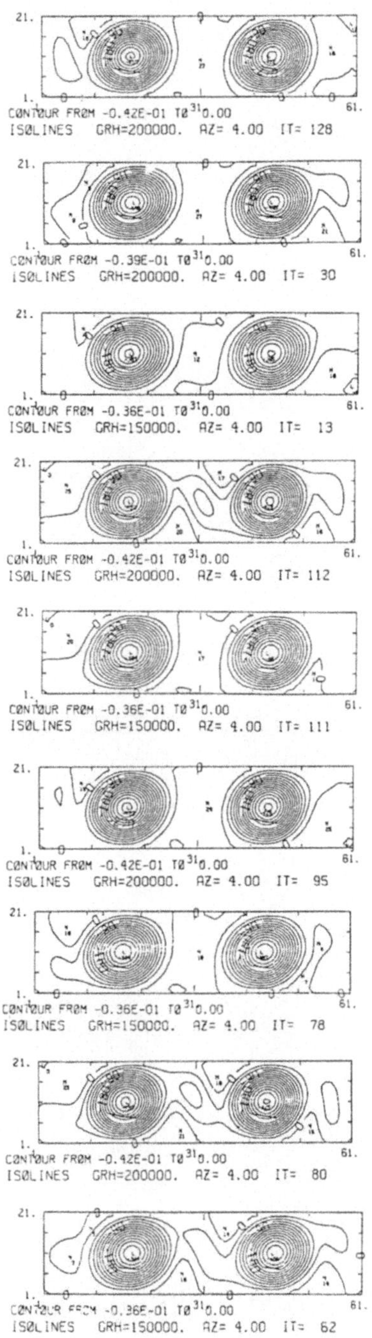

Fig. 6b

ISO-ψ LINES ; Pr = 0

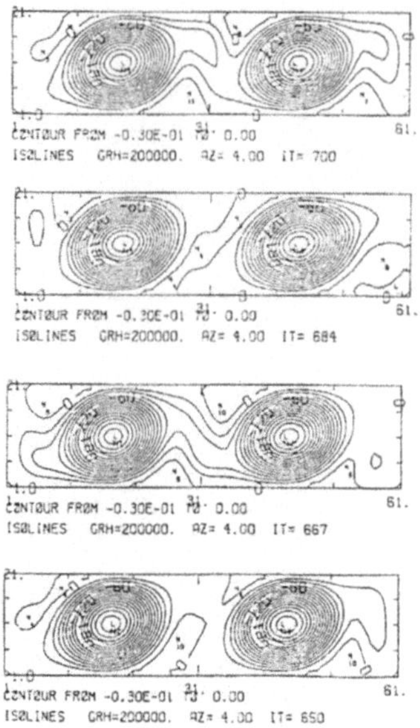

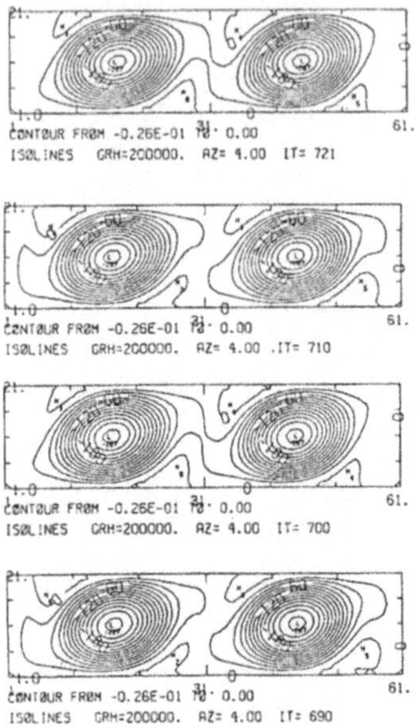

Fig. 7a ; Pr = 0,016

ISO-ψ LINES

Fig. 7b ; Pr = 0,025

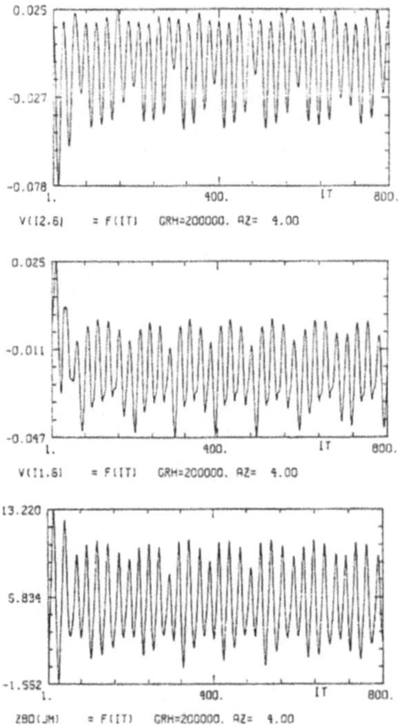

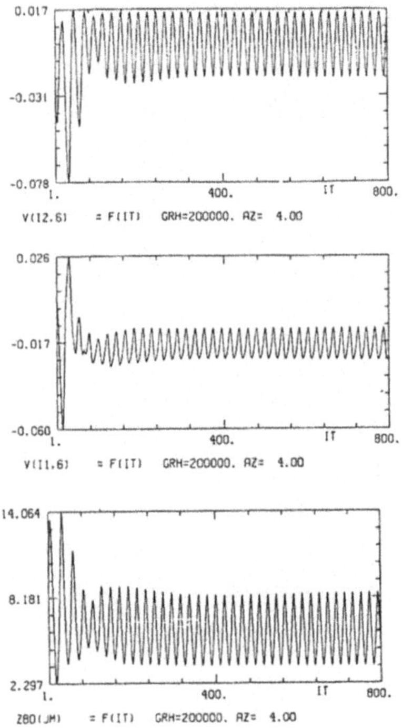

Fig. 8a Pr = 0,016

Fig. 8b ; Pr = 0,025

TIME EVOLUTION

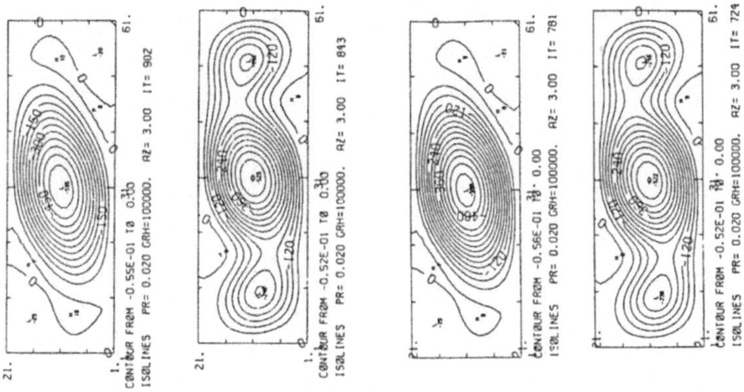

Fig. 9c Az = 3

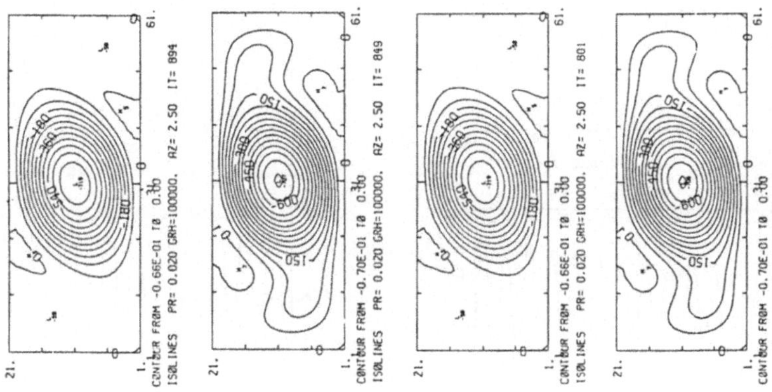

Fig. 9b Az = 2,5

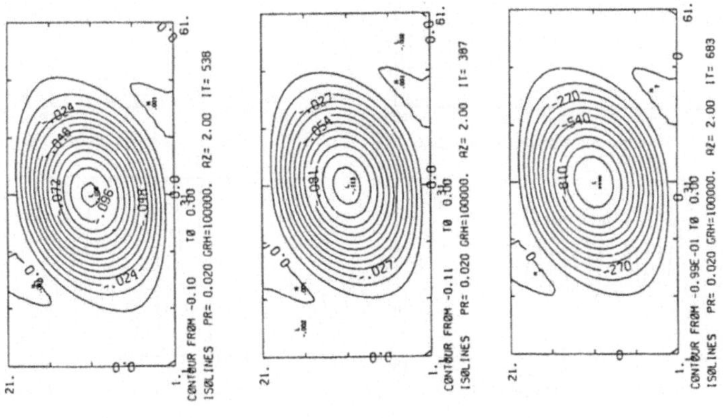

Fig. 9a Az = 2

ISO-ψ LINES ; Pr = 0,02

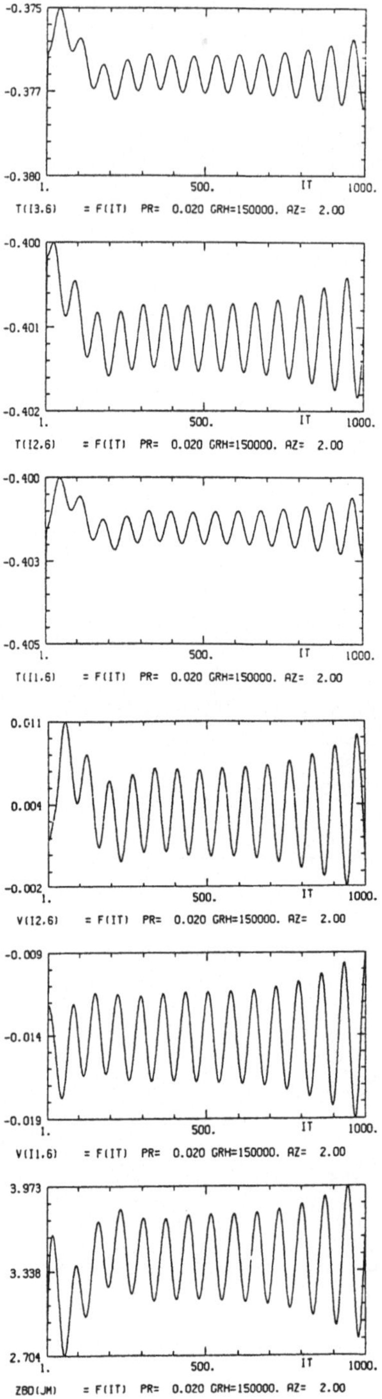

Fig. 10a Az = 2

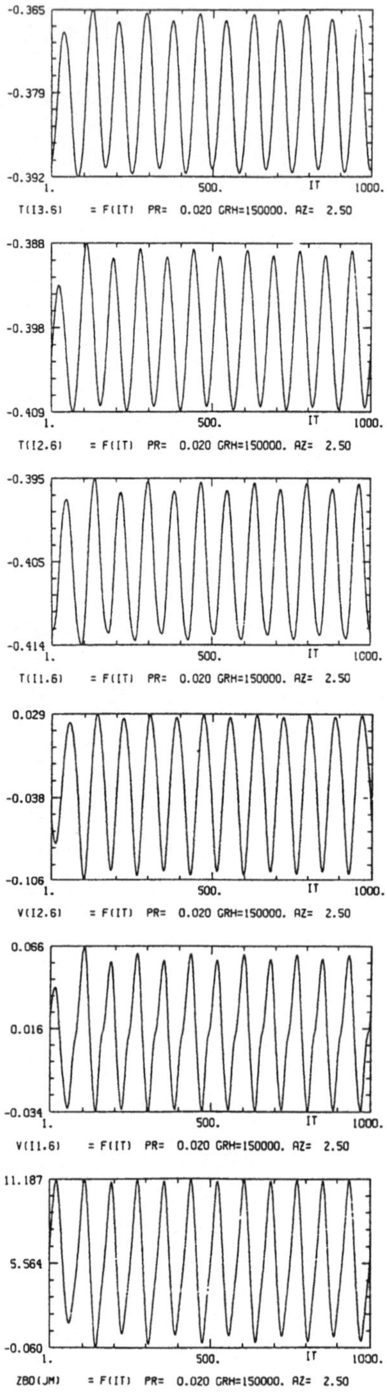

Fig. 10b Az = 2,5

TIME EVOLUTION ; Pr = 0,02

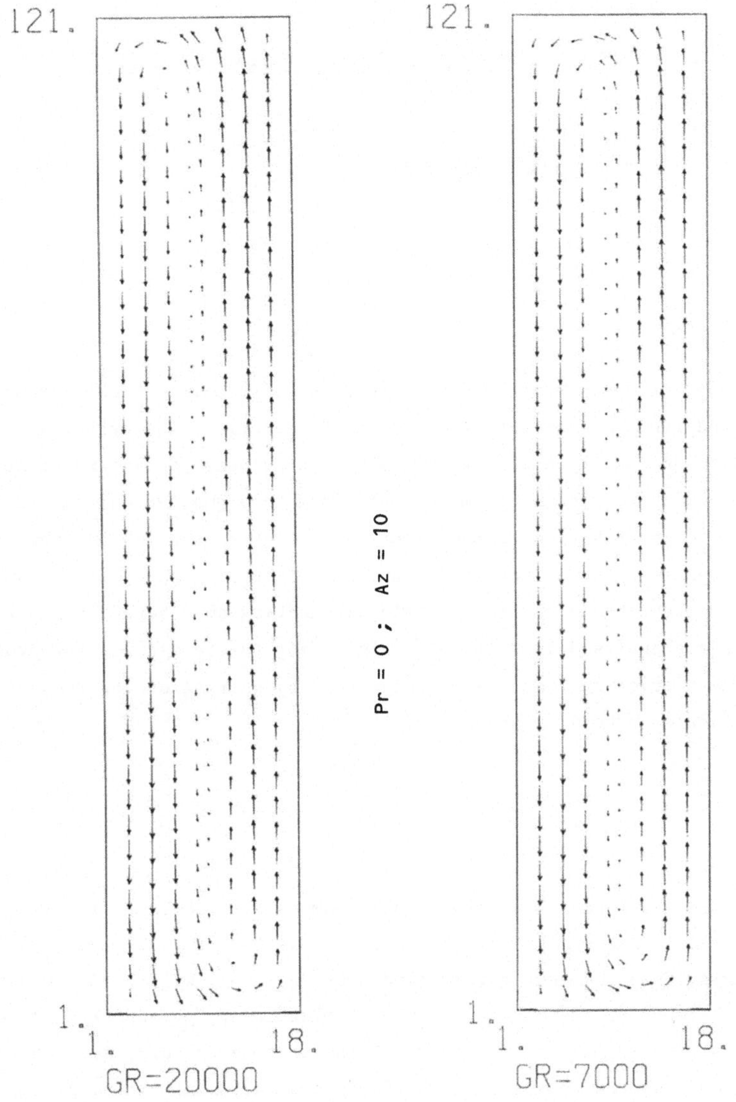

Pr = 0 ; Az = 10

Fig. 11 VELOCITY FIELD IN SYMMETRY PLANE
OF CIRCULAR CYLINDER ;

ROTATING TURBULENCE EVOLVING FREELY FROM AN INITIAL QUASI 2D STATE

M.Mory and E.J.Hopfinger
Institut de Mécanique de Grenoble,BP68
38402 Saint Martin d'Hères cédex/France

Abstract:

Recent experiments demonstrated the existence of quasi-twodimensional turbulence in a boundary-forced fluid system subjected to strong rotation. The principal results are briefly recalled and an inertial wave mechanism is proposed as an explaination for the observed sudden transition from 3D turbulence to a quasi-twodimensional turbulent flow. The main contribution of the paper is however concerned with the freely evolving state, obtained when the forcing is suddenly stopped. Experiments show an increase, in time, of the turbulence length scale, indicating an inverse energy flux. These observations are analysed in terms of a similarity theory derived for evolving turbulence with Ekman friction. The scale increase is by pairing of vortices of like sign and by large scale unsteady meandering motions.

1 -Introduction

The action of Coriolis force in a rotationally dominated turbulent flow (low Rossby number $R_o=u/2\Omega l$ flow) tends to orient the vorticity vector of turbulence parallel to the rotation axis. This tendency toward twodimensionality justifies the use of a twodimensional turbulence model to describe the statistics of "geostrophic" turbulence (Rhines (1)). In contrast with the importance of the phenomenon, only few laboratory experiments have been devoted to rotationally dominated turbulent flows. Among them, a first set of experiments (Gough and Lynden-Bell(2),Bretherton and Turner(3),McEwan (5)) was motivated by the study of the problem of angular momentum mixing rather than by the structure of turbulence itself. Experiments aimed at characterizing the structure of turbulence were conducted by Colin de Verdière(4),Hopfinger Browand & Gagne (6) (refered to as HBG) and Dickinson and Long (7). In these experiments the flow was boundary driven, with a continuous forcing provided either by an array of sources and sinks (5) or by a grid oscillation (6)(7). In the oscillating grid experiments the forcing was strong so that two different turbulent flows coexisted: close to the forcing position the turbulence was nearly 3D isotropic whereas at greater distance the flow state was quasi-twodimensional. HBG give a local Rossby number criterion for transition from one flow state to another. We attempt to explain this transition by a change from 3D turbulence to an inertial wave transport of energy.

Certain features (the strong vorticity concentration for example) in the experiment of HBG are probably the result of the existence of a 3D turbulent layer adjacent to the rotationally dominated turbulence. It is therefore of interest to study the freely evolving turbulence dynamics obtained when the forcing is suddenly stopped. The 3D turbulence close to the forcing plane dissipates rapidly and it is then two-dimensionalised by the rotational constraints. The quasi-two-dimensional turbulence can then evolve freely under the action of Ekman friction. Prelimary results on this problem were reported in Hopfinger Griffiths and Mory (8). Here, we give more extensive experimental results and analyse them in the context of a similarity theory developed for freely evolving turbulence with Ekman friction. It should be noted that this situation differs from that of Ibbetson and Tritton (9) and Wigeland and Nagib (10) in the initial state of turbulence. It has, however, close similarity with the numerical calculations by Roy and Dang (11) concerning the evolution of a two-dimensional turbulent flow perturbed by small scale 3D turbulence in the presence of rotation.

The article is organized as follows: the experimental conditions are given in section 2. In section 3 we recall the steady state turbulence properties and point out the importance of inertial waves. The theoretical aspects of freely evolving turbulence are treated in section 4 and in section 5 the experimental results are discussed in the context of this theory and of observations on vortex pairing.

2-Experimental conditions

2.1-Experimental apparatus

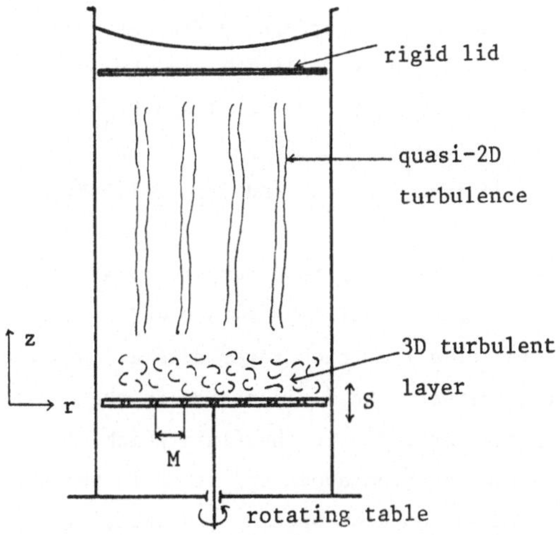

Figure 1- Schematic representation of the apparatus and the flow field.

Experiments were conducted in the experimental apparatus previously used by HBG which is shown schematically in figure 1; a brief description is again given here. This apparatus consists in a transparent cylindrical tank 40cm in diameter and 80cm in depth which is mounted on a rotating table. The rotation rate is adjustable from zero to 2π rad.s^{-1}. The turbulent velocity field is generated by a grid placed perpendicular to the axis of rotation, oscillating on the z axis around its mean position. This grid consists of 1cm square bars with a mesh M=5cm. In the experiments the stroke S (2x amplitude) of the oscillation was fixed at 4cm and its frequency n was set at 3.3Hz and 6.6Hz. In order to avoid perturbations of the free surface which would affect the quality of the visualisation (see §2.3) a rigid lid was placed at the top at 55cm above the mean position of the grid. The visualisation technique was improved: a laser light sheet was used and the photographic technique permitted to get the instantaneous velocity vector at a great number of points in the plane illuminated by the light sheet.

2.2-Experimental conditions

For the sake of comparison it was desirable to choose the same experimental conditions as in HBG. These have been shown by HBG to depend only on the grid Rossby number defined by

$$R_{og} = \frac{\pi n}{\Omega} \tag{1}$$

For values of R_{o_g}=3.3 and 6.6 the steady turbulence characteristics (when the oscillation of the grid is maintained) have been discussed by Hopfinger, Griffiths and Mory (8),Hopfinger, Mory and Gagne (12) and in particular by Mory (13). Since the experiments presented here are conducted by first establishing a steady state and then suddenly bringing the grid to rest at time t=0, the initial conditions to the unforced state are given by the steady state pattern. Different combinations of n and Ω can be used to give the same values of R_{o_g}. In the time evolving experiments reported here the values were: R_{o_g}=6.6 (n=6.6Hz, $\Omega=\pi$ rad.s^{-1}) and R_{o_g}=3.3 (n=3.3Hz, $\Omega=\pi$ rad.s^{-1}).

2.3-Visualisations and data collection technique

Visualisations were performed in planes perpendicular to the axis of rotation which were respectively located at the positions z=15cm and z=31cm above the mean position of the grid. Illumination of a 1cm thick layer of fluid was accomplished by expanding a laser beam into a light sheet. The fluid motion in that layer was obtained from the trace of neutrally buoyant particles in the fluid which diffused the light. Exposures were taken by a camera looking from above and fixed in the rotating frame. The use of a shutter, programed by a sequence of opening and shutting instructions permitted us to determine the velocity vector on the exposures (the end of each streak is marked by a shorter streak). Duration between two exposures was 2s. In the unforced experiments the exposure duration was changed during the experiment to adjust to the decay of the turbulent velocity by Ekman friction.

The image analysis consisted first in following the motion of the vortices in time (from one exposure to the next). We mainly looked for the relative displacements of the vortices and the merging events of vortices which sometimes occur. Digitalisation of the particle streaks on each exposure provided an estimate of the flow pattern from which turbulent velocities, spatial correlation functions and integral length scales were computed. In the steady state (initial conditions to the freely evolving situation) spectra were obtained from the computation of particle pair diffusion (see subsection 3.3).

3-Initial quasi-twodimensional steady state

3.1-Observations-Turbulent velocity and integral length scale

A streak line photograph of the turbulent flow field is shown in figure 2. It is taken in a cross section of the tank at 15cm above the grid midplane for $Ro_g=3.3$. It gives a view of the steady state pattern. A striking observation is the concentration of vorticity into a number of intense vortices (vorticity being one to two orders of magnitude greater than the background vorticity) as it was reported by HBG. These vortices are roughly parallel to the axis of rotation of the tank. The strength, scale and number of the vortices are shown to depend on the grid Rossby number Ro_g, with an increase in strength and number of the vortices and a decrease of their scale as the Coriolis frequency ($f=2\Omega$) increases for a fixed forcing (6). The conditions under which concentrated vortices appear are not well understood. Some experimental data on their cinematics have recently been obtained (13).

The existence of vortices with the vorticity vector roughly parallel to the rotation axis is an indication of a predominant 2D flow field. A more quantitative evidence of the quasi-2D structure of the flow is,however, given by the measurements (using an LDA technique) of the azimuthal rms velocity u at different distances z from the grid midplane.The variation of u is reproduced from (8) in figure 3 for two experimental conditions ($Ro_g=3.3$ and 6.6).A clear change in the variation of u with z occurs at the position $Z_T(Ro_g)$ beyond which the decay of turbulence is drastically reduced. At a greater distance from the turbulence generating plane, when $z > Z_R$, the profiles are nearly invariant with z. In the region closest to the grid ($z < Z_T$) the results fit, with good accuracy, the decay law of non rotating oscillating grid turbulence (14). Within experimental accuracy the turbulence in the layer $z < Z_T$ is three dimensional. A criterion for transition from 3D turbulence to rotationally dominated turbulence is deduced from figure 3. At Z_T the Rossby number u/fl,constructed with the local r.m.s turbulent velocity u and the integral length scale l,is found to be close to 0.2 for all experimental conditions.

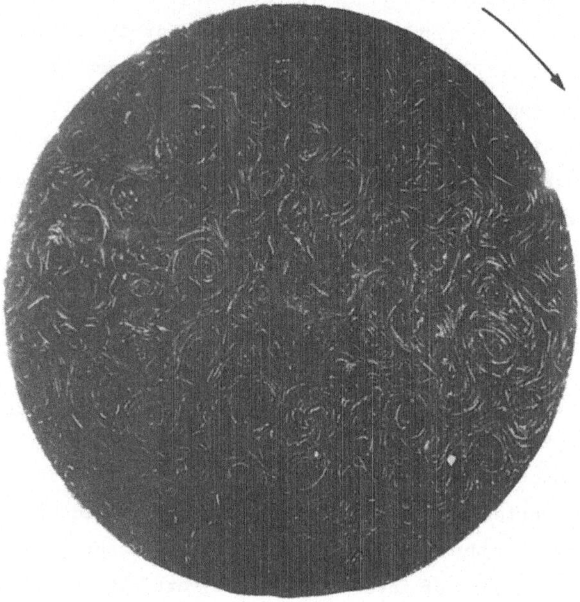

Figure 2-Streak line photograph of the steady turbulent flow field in a plane perpendicular to the axis of rotation. Experimental conditions are $\Omega = \pi \, rad.s^{-1}$, $n=3.3Hz$, $Ro_g=3.3$ Rotation is indicated by the black arrow.

Measurements of the integral length scale , plotted as a function of z in figure 4, were done by two techniques, first by computation of the time correlation function of a hot film probe velocity signal and second by computing the spatial correlation function from digitalised images as described in subsection 2.3. The method was rather tedious and this explains why only few points are plotted in figure 4. The straight lines in figure 4 represent the linear increase in length scale which is measured for non oscillating grid turbulence (14)(15). The length scale of turbulence subjected to rotation tends to saturate far away from the grid midplane.

3.2-How is turbulence two-dimensicnalised by rotation

The mechanism by which 3D homogeneous turbulence is twodimensionalised by rotational constrainsts is not well understood. Numerical simulations of turbulence subjected to rotation (Roy and Dang (11)) clearly show the formation of quasi-2D structures. However, the effect of rotation is subtle since it comes in trough the third order moment equation (Cambon(16)), indicating that the 3D cascade is modified by the generation of inertial waves.

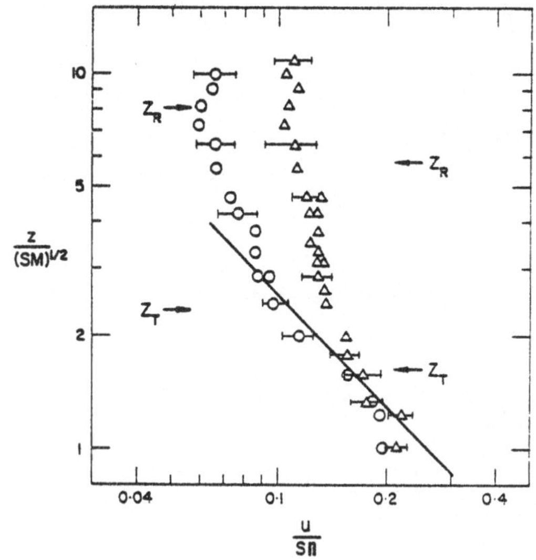

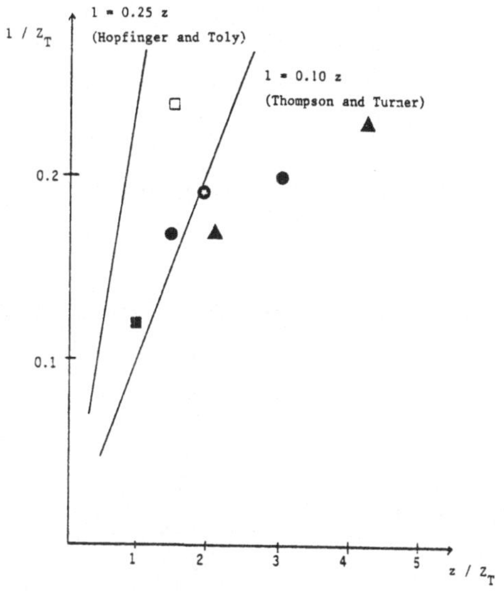

Figure 3-Variation of the r.m.s azimuthal turbulent velocity,non dimensionalised by the stroke S and frequency n,with distance from the grid midplane,non dimensionalised by S and by the grid mesh M=5cm.
Experimental conditions: △,Ro_g=3.3; ○,Ro_g=6.6.
—represents the decay of a 3D isotropic turbulence.
(figure reproduced from Hopfinger,Griffiths and Mory (8))

Figure 4-Variation of the integral length scale(non dimensionalised by the distance Z_T from the grid midplane where transition occurs from 3D turbulence to rotationally dominated turbulence)with distance from the grid midplane (non dimensionalised by Z_T)
Experimental conditions: ■,□ n=6.6Hz,Ω=0
●,○ Ro_g=6.6
▲Ro_g=3.3
—refers to the variation in length scale for a 3D isotropic turbulence.

This change from 3D turbulence to an inertial wave mechanism is schematically drawn in figure 5 where characteristic length scales L_I, 1 and L_K are plotted as functions of the distance z to the grid midplane. 1 and L_K are respectively the integral length scale (see figure 4) and the Kolmogorov viscous length scale ($L_K=(\nu^3/\varepsilon)^{1/4}$) for non rotating oscillating grid turbulence. These scales increase linearly with z as it is shown by Hopfinger and Toly (14). The length scale

$$L_I = \left(\frac{u^3}{\ell f^3}\right)^{1/2} \tag{2}$$

is the equivalent in rotating flows of the Ozmidov scale in stratified fluid. The scale L_I, given by equation (2), is characteristic of the wave length, inertial waves would be required to have for energy evacuation from a turbulent eddy in a time equal to the eddy turn-over time. The dissipation rate by way of the 3D turbulent cascade is

(3) $$\varepsilon \sim u^3/\ell$$

and the energy advection rate by the inertial waves is

(4) $$\varepsilon_I \sim v^2 C_g / L_I$$

Inertial waves have a group velocity $C_g \lesssim f\, L_I$ (taking $L_I = \lambda/2\pi$). The experiments by Dickinson and Long (7) and measurements of the time required for vortex generation when the grid is suddenly started, reported by HBG, suggest that $v^2 \lesssim (f\, L_I)^2$. Using these expressions in (4) and equating (3) to (4) gives expression (2).

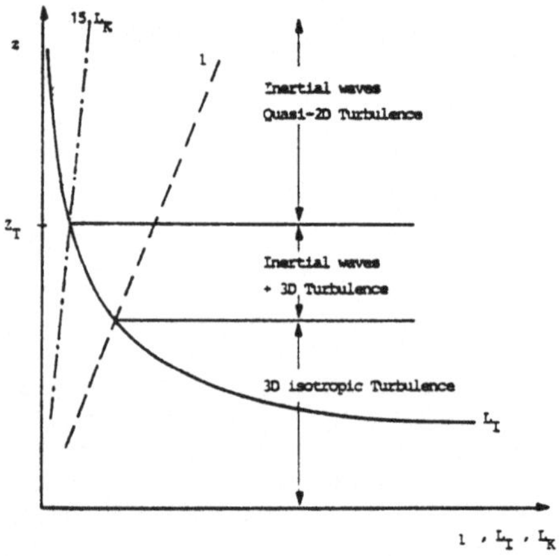

Figure 5- Schematic variation of the length scales l, L_I and L_K with distance to the grid miplane z. Regions corresponding to the different flow regimes are indicated.

Figure 5 shows that, close to the grid, the required wave length for evacuating the turbulent kinetic energy by an inertial wave mechanism is much greater than the integral length scale. And, since the wave length is imposed by the turbulence scale, the waves have practically no effect on the energetics in the region near the grid. The required wave length drops however very rapidly with distance (figure 5) and, when $L_I < l$, the energy radiation by waves is more efficient than the non linear direct energy transfer to small scales. The different regimes are summarized as follows :

- when $l < L_I$, advection of energy by inertial waves is small and turbulence is little influenced by rotation
- when $l > m\, L_K$, 3D turbulence cascade is replaced by more efficient energy advection along a vertical by inertial waves; turbulence is quasi-2D
- in the intermediate range $L_I < l < m\, L_K$, both 3D turbulence and inertial waves are at work.

The multiplication factor m of L_K has been taken equal to 15 in figure 5 by analogy with stratified flows (Stillinger, Helland and Van Atta (17)). The positions of the transitions from one regime to another are given by the intersections of the functions

$l(z)$,15 L_K and L_I.

Ekman friction is principally responsible for energy dissipation in the fluid layer and a steady state is reached when Ekman dissipation balances the advection of energy by the inertial waves. The time scale of energy flux by inertial waves is

(5) $$\tau_I \sim \frac{H}{f l_T}$$

where H is the fluid layer depth and l_T the turbulence length scale at the transition (where $l \sim L_I$). This time must be shorter than the Ekman dissipation time scale

(6) $$\tau \sim \frac{H}{2(\kappa f)^{1/2}}$$

This gives $l_T \gtrsim 2(\kappa/f)^{1/2}$, which means that l_T must exceed the Ekman layer thickness.

3.3-Energy spectra and particle dispersion laws

Since twodimensionalisation of the turbulence is achieved by a blockage in energy transfer to the small scales it should show up in the spectral energy density. Such measurements were attempted with a hot film probe but the results, though indicating a change in the spectral slope under the effect of rotation (12),(13), were not very convincing. Probe vibrations and probe drifts were the main difficulties encountered.

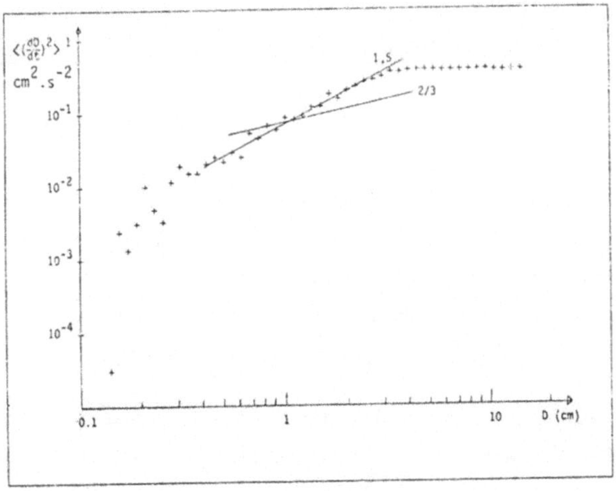

Figure 6-Rate of dispersion of particle pairs $\langle(\frac{dD}{dt})^2\rangle$ as a function of distance D between particles.
Experimental conditions: $R_{o_g}=3.3$, $\Omega=\pi$ rad.s^{-1}, n=3.3Hz.
Visualisation is performed in a plane located 15cm above the grid midplane. The straight lines with slopes 2/3 and 1.5 refer to energy spectral distributions varying like $k^{-5/3}$ and $k^{-2.5}$ respectively.

Inertial range laws of energy spectral densities can also be infered from particle pair dispersion measurements (Morel and Larchevêque (18)). This technique was adapted to laboratory flows by Griffiths and Hopfinger (19) and its validity has been further substantiated by Mory (13) and Mory and Hopfinger (23). The relationship between the dispersion law and the spectral density distribution is established as follows (within the range $1 < n < 3$, n being the exponent in the energy spectra)

(7a) $$< \left(\frac{dD}{dt}\right)^2 > \sim D^{n-1}$$

(7b) $$E(k) \sim k^{-n}$$

where D is the separation between two particles.

Figure 6 gives the dispersion law for the experimental conditions shown in figure 2 ($n=3.3$Hz, $\Omega=\pi$ rad.s^{-1}, $z=15$cm). In the range $0.4\text{cm} < D < 3\text{cm}$ the dispersion law is well approximated by a slope of 1.5 giving a spectral law in $k^{-2.5}$. Note that Griffiths and Hopfinger (19) found the same slope (using the same method) in a baroclinic turbulence. Buzyna, Pfeffer and Kung (20)also find energy spectra in $k^{-2.5}$ in the low Rossby number regime of a baroclinic turbulence in an annulus. Even though the measured slope is less steep than the classical phenomenological spectral law of turbulence ($E(k) \sim k^{-3}$) the general idea of existence of a reduced energy transfer to the small scales, valid for a 2D turbulent flow, still applies to the turbulent field with $k^{-2.5}$.

4-Similarity theory for freely evolving,rotationally dominated turbulence

As it was first shown by Batchelor (21), the hypothesis of similarity of the energy spectrum provides a useful model to describe the time evolution of 2D turbulence. Following the general idea of Batchelor we assume that the time evolving energy spectrum only depends on the wave number k, the time dependent turbulent velocity u(t) and the integral length scale l(t) (contrary to Batchelor's approach, the time is not used explicitly). The energy spectrum can thus be written in the form

(8) $$E(k,t) = u^2(t)\, \ell(t)\, g\{k\,\ell(t)\}$$

where g is a dimensionless function which determines the shape of the energy spectrum. For an inviscid motion subjected to a strong rotational constraint, as well as for a 2D flow, the conservation of enstrophy is an extra property of the motion in addition to energy conservation. The entrophy ξ^2 is related to the dimensional quantities u(t) and l(t) by

(9) $$\xi^2 = \int_0^\infty k^2 E(k,t)\, dk = B \frac{u^2(t)}{\ell^2(t)}$$

where B is a constant defined by

(10) $$B = \int_0^\infty x^2 g(x)\, dx$$

In real flows, dissipation has to be taken into account. It is acting both on energy and enstrophy in two ways: by molecular or eddy viscous dissipation and by

Ekman friction due to the Ekman boundary layers on the end walls. Therefore, the energy and enstrophy balance equations are written

$$\frac{du^2}{dt} = -\frac{2}{\tau}u^2 - 2\nu\xi^2 \tag{11}$$

$$\frac{d\xi^2}{dt} = -\frac{2}{\tau}\xi^2 - 2\nu\zeta(t) \tag{12}$$

In equations (11) and (12) the first terms on the right hand side give the Ekman dissipation and the second terms represent the eddy dissipation. τ is the Ekman characteristic time which is taken in this model as a constant, depending only on the initial conditions of the experiment. Due to the constraint of rotation, the energy flux to small scales is considerably reduced so that the viscous dissipation can be neglected compared to the energy dissipation by Ekman friction. The solution of equation (11) is then simply + :

$$u^2(t) = u^2(0)\, e^{-2t/\tau} \tag{13}$$

Conversely, enstrophy dissipation by viscosity remains large since regions of strong vorticity are to be expected in the flow. A likely scaling of $\zeta(t)$ is assumed as follows (A is a constant)

$$2\nu\zeta(t) = A\frac{u^3(t)}{\ell^3(t)} \tag{14}$$

Equation (12), after combining with equations (9) and (14) leads to (with T=A/2B) :

$$\frac{d\ell}{dt} = T\,u(t) \tag{15}$$

It is interesting that equation (15) is formally identical to the result of Batchelor's similarity theory for turbulence. Ekman friction does not appear explicitly in (15). It is taken into account through the decay of energy caused by Ekman friction. Integration of equation (15) gives the time evolution law for the integral length scale:

$$\ell(t) = \ell(0) + T\,u(0)\,\tau\,(1 - e^{-t/\tau}) \tag{16}$$

During the first stage of the decay, Ekman friction has little influence on the growth rate of the integral length scale. This length scale doubles in a time which is T^{-1} the initial eddy turn-over time $l(0)/u(0)$. At later time the effect of friction becomes significant. An important outcome of the present model is the existence of a maximum integral length scale which is reached at the end of the evolution:

$$\ell_{max} = \ell(0) + T\,u(0)\,\tau \tag{17}$$

An estimate of the time t_d at which Ekman friction becomes prevailing is obtained by equating the Ekman characteristic time τ to the instantaneous eddy turn-over time $l(t_d)/u(t_d)$. Since $\tau \gg l(0)/u(0)$ and $T \ll 1$ for any experimental conditions which were

---

\+ An exact analytical solution to equations (11) and (12) is found, due to the simple form of ζ (eq.14), without neglecting $2\nu\xi^2$ in equation (11). It leads, however, to an untrealistic behavior as $t \to \infty$ (negative energy). It is questionable whether it is appropriate to use simultaneously the similarity assumption and to take into account the eddy dissipation of energy.

considered, the time t_d has a simple expression:

(18) $$t_d \simeq -\tau \ln T$$

5-Experimental results on freely evolving turbulence

5.1-Qualitative observations

In figure 7 photographs of the evolving flow are shown for four time steps during a single decay experiment (Ro_g=3.3;position of the visualisation z=15cm). A photograph of the initial steady state(t=0) was reproduced in figure 2. Images 7a,b,c,d show the flow pattern at times $5\Delta t, 7\Delta t, 9\Delta t$ and $14\Delta t$ (Δt=2s) after the grid was brought to rest. Comparison of the flow structure at different times clearly indicates an increase in eddy spacing and eddy size and, consequently,a decrease in the number of vortices. This occurs through different kinds of mechanisms among which are pairing of vortices of like sign. Such events appear in figure 7, with vortices A and B in figure 7a pairing in C in figure 7b. Some of the properties of pairing observed in numerical experiments by McWilliams (22) are discernable in laboratory experiments : in particular pairing of vortices generally occurs when one of the two vortices is weaker. This is the case in figure 7a where vortex B is weaker then vortex A. When the two vortices roughly have an equal strength, interaction does generally not lead to merging of the two. Nevertheless, interaction is often dramatic: it produces shearing deformations of the vortices by which the vortices may loose a significant part of their vorticity. For example, three vortices of like sign (designed by A,B and C) are interacting in figure 7c whithout merging. However the vortices B and C have been weakened in figure 7d. Pairing occurs for cyclonic vortices as well as for anticylonic vortices but, cyclonic eddies, at later times, tend to disappear. The final state of the evolution consists in a few anticyclonic vortices driving an unsteady large scale meandering anticyclonic circulation. As the time increases, the effect of the sidewalls can no longer be neblected because the lateral boundary layer reaches an appreciable thickness. Furthermore, some of the vortices are dissipated during the evolution after being trapped inside this boundary layer. Whether or not interactions between vortices of opposite sign are present, in any case neither they are significantly disturbing nor do they lead to the formation of dipole modons as those observed by McWilliams.

5.2-Time evolution of turbulent velocity and integral length scale

The decay of the turbulent velocity u (non-dimensionalised with the turbulent velocity at time t=0) is plotted in figure 8 as a function of the non-dimensional time $t\Omega/2\pi$. The results are given for four cases i.e for the two studied experimental conditions with two laws representing the decay at the positions z=15cm and 31cm above the grid midplane.A semi-logarithmic scale is used for u which permits, by fitting a

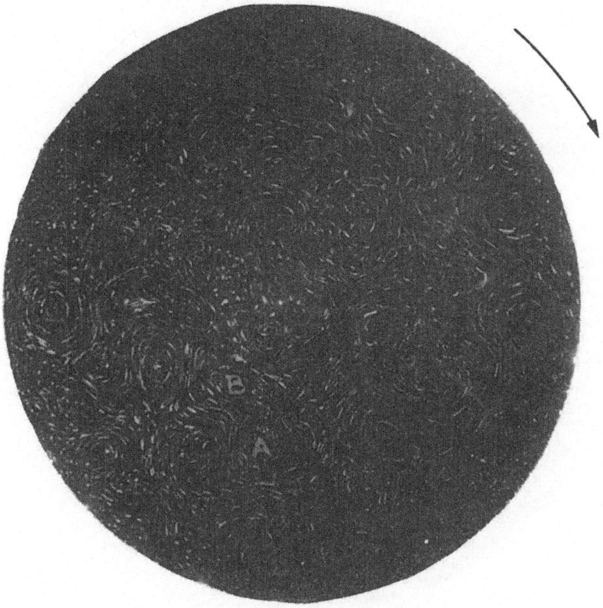

Figure 7a

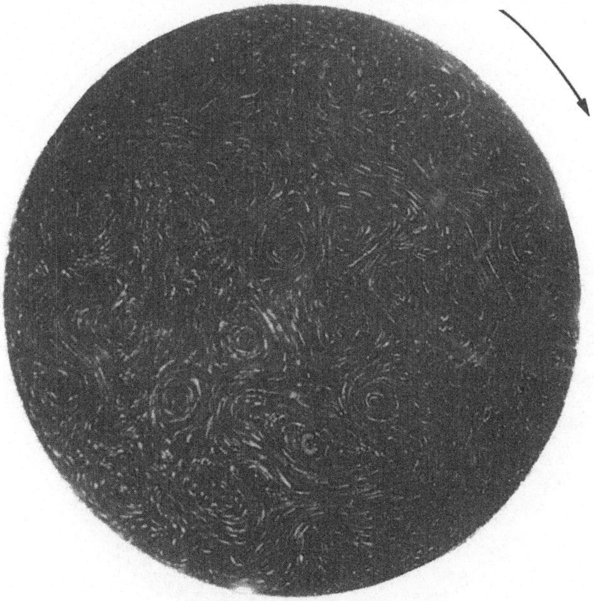

Figure 7b

Figure 7- Photographs of the flow field at different times after the grid was brought to rest (7a:5Δt;7b:7Δt;7c:9Δt and 7d:14Δt; with Δt=2s).

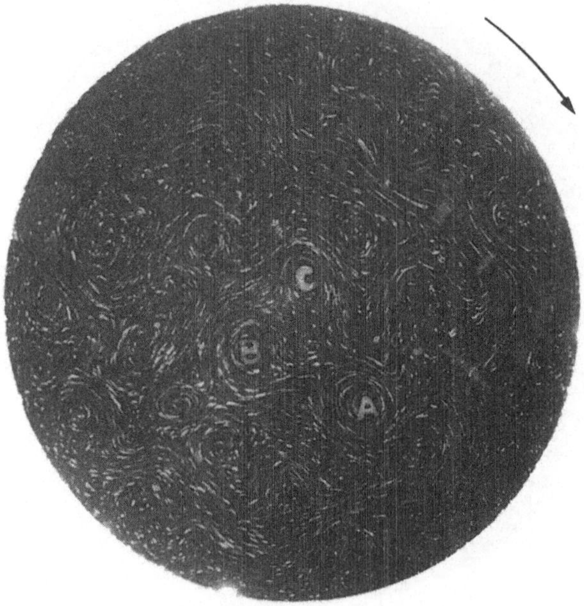

Figure 7c

Figure 7d

straight line to the data, an estimate of the Ekman characteristic time

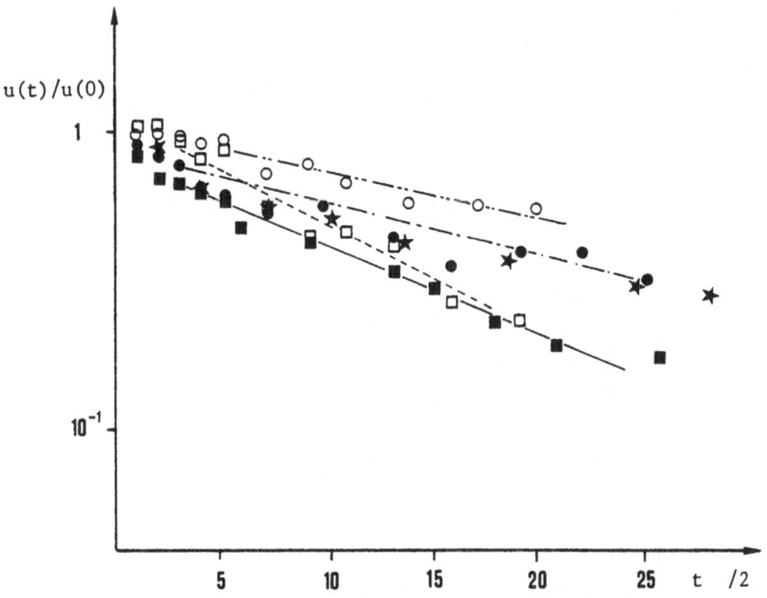

Figure 8-Evolution with time of the r.m.s velocity in freely evolving turbulence. Straight lines correspond to an exponential decay law fitting the data for the different conditions.
Symbols: □, - - -, R_{o_g}=3.3, z=31cm; ■, ——, R_{o_g}=3.3, z=15cm;
○, -··-, R_{o_g}=6.6, z=31cm; ●, -·-, R_{o_g}=6.6, z=15cm;
★ are experimental data reproduced from a previous experiment (8).
For all experiments $\Omega = \pi$ rad.s^{-1}, except for ($\Omega = \pi$ rad.s^{-1}).

In figure 9 are plotted the integral length scale l(t) (made dimensionless with the integral length scale at time t=0) as a function of the non-dimensional time $t\Omega/2\pi$. Using the same symbols as in figure 8, the results refer to the four cases studied. Figure 9 indicates a clear increase of the integral length scale with time. For each case the theoretical evolution law, given by equation (16) is adjusted to the experimental points giving a value of T. The relative error is about 10% for all experiments. This large error is due to the insufficient convergence of the correlation function which was computed from a set of about 150 particles.

We present in table 1 different quantities deduced from figure 8 and 9 among which are the Ekman characteristic time and the dimensionless rate T of increase of the length scale. Also given in table 1 are the values of the r.m.s velocity u and the integral length scale l at time t=0, the theoretical maximum length scale l_{max} (eq.17) and the relative error $\left\langle \left| \frac{l(t)-l_{th}(t)}{l_{th}(t)} \right| \right\rangle$ between the measured integral length scale and the theoretical prediction (equation 16). The scale l_{max} is calculated from equation (17) with the quantities reported in table 1. For all cases the calculated value of l_{max} is greater than the measured length scale during the decay.

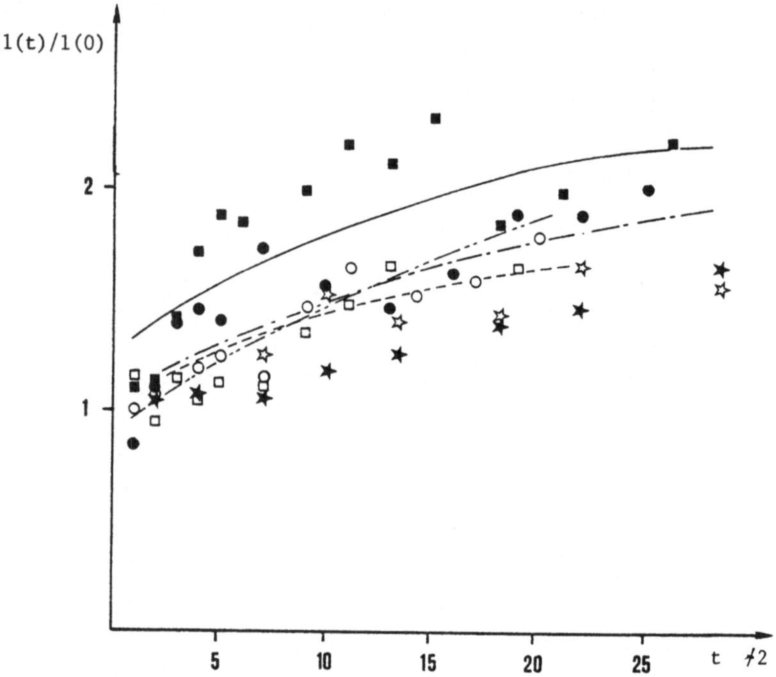

Figure 9 -Evolution with time of the integral length scale in freely evolving turbulence. Curves represent the theoretical law, given by eq.16 fitting the data.
Symbols are the same as those in figure 8. ★and ☆ are the data from the experiment of Hopfinger, Griffiths and Mory (8). ★and ☆ are respectively measurements of the eddy size and the eddy spacing.

		T	$\tau\Omega/2\pi$	$l(0)$ cm	$u(0)$ $cm.s^{-1}$	l_{max} cm	$\langle\left\|\frac{l-l_{th}}{l_{th}}\right\|\rangle$	$\frac{2\ l(0)}{Tu(0)}$
$Ro_g=6.6$	z=15cm	0.026	23.5	1.86	2.22	5.3	9%	1.37
	z=31cm	0.038	29	2.05	1.73	4.4	9%	1.07
$Ro_g=3.3$	z=15cm	0.031	15.5	1.31	1.75	3.3	11%	1.56
	z=31cm	0.032	12.5	1.71	1.58	3.3	12%	2.71

Table 1

For the sake of comparison we also included in figures 8 and 9 the results of a previous experiment by Hopfinger, Griffiths and Mory (8) obtained for slightly different conditions ($\Omega=2\pi$ rad.s^{-1},n=6.6Hz,Ro_g=3.3). For this experiment two different length scale were measured: the eddy spacing and the eddy diameter. The evolution of the turbulent velocity and lengths shows a behavior similar to the one observed in the present study. Numerical values of and T, measured by Hopfinger, Griffiths and Mory (8) ($\tau\, 2\Omega/2\pi \simeq 29$; $T \simeq 3.1\times 10^{-2}$) are in fairly good agreement with the present, more extensive measurements.

5.3-Inverse energy flux to large scales

A main conclusion emerging from the experimental data is the very slow decay of turbulence. The characteristic time of this decay is 12 to 30 times the period of rotation of the tank and 25 to 60 times the eddy turn-over time. In other words, only 15 to 30% of the energy is dissipated after 10 eddy turn-over times. By comparison 95% of the energy would be dissipated during the same time in a freely evolving 3D isotropic turbulent flow. Using the relationship for Ekman friction $\tau = H/2(f\,\kappa)^{1/2}$ (with H being the depth of the tank), the eddy viscosity in the Ekman layer can be calculated. For the experimental conditions $R_{o_g}=3.3$ and $R_{o_g}=6.6$ we respectively get $\kappa = 0.07\ cm^2.s^{-1}$ and $\kappa = 0.02 cm^2.s^{-1}$. These low values of the eddy viscosity agree well with the laboratory experiments of Colin de Verdière (5). Estimates of the characteristic dimensionless ratio $\kappa/u(0)l(0)$ are respectively $2\ 10^{-2}$ and $5\ 10^{-3}$ for $R_{o_g}=3.3$ and $R_{o_g}=6.6$ and are therefore close to the value $1\text{-}2\ 10^{-2}$ measured by Colin de Verdière.

Our estimates of the dimensionless growth rate T of the length scale are in good agreement with the numerical results of Rhines (1). The value of T depends on the definition of the length scale. The length scale considered by Rhines is nearly equivalent to the integral length scale used here.

In interpreting the experimental results, the question arises whether or not the increase in length scale is indeed indicative of an effective energy flux to the large scales. An answer to this question requires to consider the time evolution of the energy spectrum in order to compare the energy flux to the large scales with Ekman dissipation. By using the similarity law (eq.8) for the spectrum, the energy transfer $\mathcal{T}(k)$ to wave number k is easily expressed since the Ekman dissipation is scale independent Thereby the existence of a non zero $\mathcal{T}(k)$ is a consequence only of the time evolution of the integral length scale. The expression for the transfer is then

$$\mathcal{T}(k) = T u^3(t)\left[g\{kl(t)\} + kl(t)\, g'\{kl(t)\}\right] \tag{18}$$

With respect to the largest energy containing eddies ($kl \ll 1$) we have $g'(kl) > 0$ and the energy transfer to the wave number k is found to be positive. Nevertheless this effect of energy cascade to large scales competes with the Ekman dissipation. In order to determine which of the two effects is prevailing the time derivative of the energy spectrum at wave number k is examined

$$\frac{\partial}{\partial t} E(k,t) = T u^3(t)\left[\left(1 - \frac{2l(t)}{\tau T u(t)}\right) g(kl) + kl\, g'(kl)\right] \tag{19}$$

For the largest scales ($kl < 1$) the sign of this quantity mainly depends on the order of magnitude of the dimensionless ratio $2l(t)/\tau T u(t)$. The values for this quantity, estimated at time t=0, are given in table 1 for the four experimental conditions. It turns out that the numerical values are in all cases slightly larger than one. The two effects are therefore of the same order of magnitude. Due to Ekman friction one should therefore not expect the large scales of the motion to significantly increase their energy. Although they gain energy from smaller scales by non linear interactions, most of this energy is immediately dissipated.

The strong effect of Ekman friction is not surprising if the principal features of the evolving flow pattern are recalled (§5.1). On the one hand the existence of merging events is a typical signature of an inverse cascade of energy. On the other hand it has been noticed that the energy decreases in the vortices.

6-Conclusion and further discussions

The constraint fo rotation tends to change the structure of turbulence toward a quasi-twodimensional turbulent flow field. This is most clearly demonstrated by the experiments on freely evolving turbulence presented in this paper. The direct energy cascade is strongly reduced in the presence of rotation and, in the experiments, energy dissipation is mostly by Ekman friction which is scale independent. Following the general idea of Batchelor's similarity theory of freely evolving 2D trubulence, a model is proposed which incorporates Ekman friction. In interpreting the experimental data on hand of this model, the rate of inverse energy flux to the large scales is obtained which is in good agreement with the rate obtained from numerical simulations (1). The rate of energy dissipation by Ekman friction is equivalent in order of magnitude to the rate of energy transfer to the large scales. Despite the dissipation, it is conceivable to interpret this energy flux to the large scales as an "inverse energy cascade" phenomenon. This occurs through pairing events of vortices of like sign in a way similar to what was observed in numerical simulations carried out by McWilliams. An important consequence of Ekman dissipation, demonstrated by the model, is the existence of a maximum length scale which is reached at the end of the evolution. Theoretical estimates of the maximum length scale for the experimental conditions studied are not inconsistent with observations.

The similarity model of freely evolving turbulence (presented in section 4) should carry over to the steady forced turbulence described in section 2. A blockage of the inverse energy cascade is observed in the steady state experiment, and the question is whether this blockage is predicted by similarity theories. The derivation of a similarity model for the evolution toward a forced steady state is more difficult than it is for the freely evolving case: the energy and enstrophy injection terms (respectively designed by ε and β) must be added in equations (11) and (12). The expression for the rate of change of the length scale is then:

$$\frac{d\ell}{dt} = T\,u(t) + \frac{1}{2}\frac{\ell}{u^2}\,\varepsilon - \frac{\beta}{2B}\frac{\ell^3}{u^2} \qquad (20)$$

When it is assumed that the energy and enstrophy injection occurs at the same rate, such that

$$\beta \sim \frac{\varepsilon B}{\ell^2} \qquad (21)$$

equation (20) is identical to the freely evolving case, given by equation (15). Since the turbulent velocity remains finite and non zero for large times, the present model does not give a bound on the scale of the turbulence. These conclusions are not satisfying for explaining the blockage of the inverse cascade in steady forced turbulence. A more complete model would be required to explain the evolution toward a steady state.

Ackowledgements

The authors wish to thank D.J.Tritton for his contribution in clarifying the action of Ekman friction in the similarity model. The work was financially supported by the Centre National d'Exploitation des Océans under contract n° 84/3276.

References

(1) Rhines,P.B.-"Geostrophic Turbulence",Ann.Rev.Fluid Mech,1979,11,401-441.

(2) Gough D.O and Lynden-Bell D.-"Vorticity expulsion by turbulence: astrophysical implications of an Alka-Seltzer experiment",J.Fluid Mech.,1968,32,437-447.

(3) Bretherton F.P. and Turner J.S.-"On the mixing of angular momentum in a stirred rotating fluid",J.Fluid Mech.,1968,32,449-464.

(4) Colin de Verdière A.-"Quasi geostrophique turbulence in a rotating homogeneous fluid",Geophys.Astrophys.Fluid Dyn.,1980,15,213-251.

(5) McEwan A.D.-"Angular momentum diffusion and the initiation of cyclones",Nature, 1976,260,126-128.

(6) Hopfinger E.J.,Browand F.K. and Gagne Y.-"Turbulence and waves in a rotating tank", J.Fluid Mech. ,1982,125,505-534.

(7) Dickinson S.C and Long R.R.,-"Oscillating grid turbulence including effects of rotation",J.Fluid Mech.,1982,126,315-333.

(8) Hopfinger E.J.,Griffiths R.W. and Mory M.-"The structure of turbulence in homogeneous and stratified rotating fluids",J.Mech.Theor.Appl.,1983,Numéro spécial,21-44.

(9) Ibbetson A. and Tritton D.J.-"Experiments on turbulence in a rotating fluid", J.Fluid Mech.,1975,68,639-672.

(10) Wigeland R.A. and Nagib H.M.-"Grid generated turbulence with and whithout rotation about the streamwise direction",IIT fluids an heat transfer report R78-1,Illinois Inst. of Tech.,Chicago,Illinois,1978.

(11) Roy P. and Dang K.-"Numerical simulations of homogeneous turbulence subject to strong rotation",10th EGS Annual meeting,Louvain la Neuve,Belgium,1984.

(12) Hopfinger E.J.,Mory M. and Gagne Y.-"Two dimensionalisation by rotational constraints of homogeneous turbulence",IUTAM Symposium on Turbulence and Chaotic Phenomena in Fluids,1983,Kyoto,Japan.

(13) Mory M.-"Turbulence dans un fluide soumis à forte rotation",Thèse de Docteur Ingénieur, Inst. National Polytech. de Grenoble,1984.

(14) Hopfinger E.J. and Toly A.J.-"Spatially decaying turbulence and its relation to mixing accross density interfaces",J.Fluid Mech.,1976,78,155-175.

(15) Thompson S.M. and Turner J.S.-"Mixing accross an interface due to turbulence generated by an oscillating grid",J.Fluid Mech.,1975,67,349-368.

(16) Cambon C.-"Departure from isotropy of an homogeneous turbulence submitted to rotation",10th EGS Annual Meeting,Louvain la Neuve,Belgium,1984.

(17) Stillinger D.C.,Helland K.N. and Van Atta C.W.-"Experiments on the transition of homogeneous turbulence to internal waves in a stratified fluid",J.Fluid Mech.,1983,131,91-122.

(18) Morel P. and Larchevêque M.-"Relative dispersion of constant level balloons in the 200mb general circulation",J.Atmos.Sci.,1974,31,2189-2196.

(19) Griffiths R.W. and Hopfinger E.J.-"The structure of mesoscale turbulence and ocean spreading at ocean fronts",Deep-Sea Res.,1984,31,245-269.

(20) Buzina G.,Pfeffer R.L. and Kung R.-"Transition to geostrophic turbulence in a rotating differentially heated annulus of fluid",J.Fluid Mech.,1984,145,377-403.

(21) Batchelor G.K.-"Computation of the energy spectrum in homogeneous two-dimensional turbulence",Phys. of Fluids,1969,Suppl. II,12,233-238.

(22) McWilliams J.C.-"The emergence of isolated coherent vortices in turbulent flow", J.Fluid Mech.,1984,146,21-43.

QUASI-GEOSTROPHIC TURBULENCE AND THE MESOSCALE VARIABILITY

by Jackson R. Herring

Mesoscale Research Section
Atmospheric Analysis and Prediction Division
National Center for Atmospheric Research
Boulder, CO, Box 3000 80307

1. Introduction

We discuss here the quasi-geostrophic turbulence theory of Charney (1971), and its possible application to the mesoscale variability of the earth's atmosphere. The Charney theory consists of a two-dimensional advection of the three-dimensional potential vorticity field, whose vertical stratification is dynamically determined. Charney argued that the quasi-geostrophic dynamics provided an equipartitioning of kinetic and potential energy and was able to provide observational support for his ideas in the small-scale end of the planetary wave spectrum. This equipartitioning was deduced on the basis of general statistical mechanics arguments--strictly applicable only to inviscid flows: the actual problem of interest is highly dissipative of both enstrophy (squared vorticity) and kinetic energy. It is then of some interest to investigate such putative equipartitioning, especially since the dynamics (two-dimensional advection) is patently anisotropic. Charney used the inviscid constraints of energy and enstrophy to deduce a k^{-3} spectrum for the total (kinetic plus potential) energy.

More recently, observations of the large scales of mesoscale variability (i.e., scales of motion which lie between the small thunderstorm scale (~1 km) and 500 km which show very nearly a $k^{-5/3}$ range for the kinetic energy spectrum (see Nastrom <u>et al.</u>, 1984, or Lilly and Petersen, 1983, for a review). Indeed, the closeness of the data to $k^{-5/3}$ (as opposed to a k^{-2} or $k^{-3/2}$ range) led Gage (1979) to hypothesize that this range of scales was essentially an inverse cascading two-dimensional turbulence of the sort proposed by Kraichnan (1967). The energy source (at the small-scale end of the spectrum) was thought by Lilly (1983) to be decaying convecting clouds and thunderstorm anvil outflows. However, Gage (1984) argues for breaking internal waves for the source. Fig. 1 shows the near tropopause GASP aircraft data, as given by Nastrom <u>et al.</u> (1984). Note that the k^{-3} planetary scale range spectrum merges with the $k^{-5/3}$ mesoscale spectrum near $(2\pi/k)=500$ km. If the dynamics is in fact quasi-geostrophic turbulence, there needs to be a sink at ~500 km.

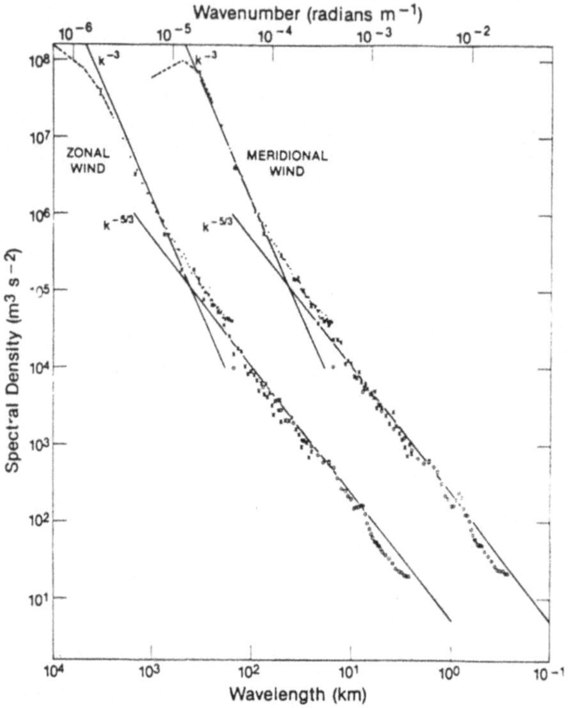

Fig. 1. Kinetic energy spectrum near the tropopause from the GASP aircraft data. After Nastrom et al. (1984).

These observations--if they prove to be typical--argue against a spectral gap in the mesoscale as proposed by Fiedler and Panofsky (1971). Also, we may conjecture on the basis of general arguments (Pouquet et al., 1983) that a spectral gap should not be expected. Indeed, these authors point out that only if the small-scale energy considerably exceeds that of the large (planetary) scales should there be a gap. Such is not the case in the earth's atmosphere.

Although quasi-geostrophic turbulence arguments seem plausible, we should stress that a wave dynamical explanation has also been proposed (Van Zandt, 1982) and is still a contender. In this case, the spectrum may be explained à la Garrett and Monk (1972), as for the oceanographic mesoscale kinetic energy spectrum. However, in that case, to argue for a $k^{-5/3}$ range instead of a k^{-2} or $k^{-3/2}$ seems strained, at least from the perspective of statistical turbulence theory. Other, more concrete explanations have been offered by Gage and Nastrom (1984). Their point here is simply that the observed vertical variability is in much better accord with quasi-geostrophic theory than with (at least the linear version of) wave theory.

Reality probably lies somewhere between wave theory (at small k) and quasi-geostrophic (or an appropriate generalization) theory at large k. Such a perspective has been stressed by Holloway (1980) in his theoretical proposals to treat combined waves and turbulence. In any case, one perplexing problem remains with respect to quasi-geostrophic theory: what is the dynamics whereby the flow becomes (nearly, as observed) homogeneous in the vertical? The resolution of this problem would probably be more interesting than a complete understanding of homogeneous quasi-geostrophic turbulence. But such issues lie outside the scope of the present paper.

The quasi-geostrophic theory assumes for its validity that the rotation rate, f, is large compared to inertial terms. Quantitatively, this takes the form $Ro = (u'/LF) << 1$, where u' and L are typical speed and length scale of a fluid parcel. Ro is the Rossby number; for planetary scales we estimate (from Fig. 1) its value to be $\gtrsim 0.01$. Note that for $E(k) \sim k^{-3}$, f is constant. Then, if E(k) becomes less steep (as for $k^{-5/3}$), Ro increases as $k \rightarrow \infty$ as $k^{3/4}$, for a $k^{-5/3}$ range. This means that the quasi-geostrophic theory eventually becomes invalid at large k; indeed, for the present problem, we estimate that at most 1.5 octaves inferior (in scale) to k = 500 km constitute its domain of validity.

In this paper, we shall focus on the basic dynamics of homogeneous turbulence. Key dynamical issues considered are: (1) how does this (dissipative) flow become isotropic (in the sense of Charney) in the face of an anisotropic dynamics, (2) what are the spectral laws of inertial ranges ($k^{-5/3}$, k^{-3}), and how might concepts like negative eddy viscosity (Kraichnan, 1976) generalize to quasi-geostrophic flows; and (3) what are the special anisotropic features of such turbulence if driven by a spiked random forcing (the prototype of the mesoscale problem), and can the $k^{-5/3}$ range also be considered isotropic as was suggested by Charney for the k^{-3} large scale problem?

2. Formulation of the Problem

In quasi-geostrophic theory (see, i.e., Pedlosky, 1975, for a more complete discussion) the equation of motion for the (potential) vorticity field, Q, is:

$$(\partial/\partial t + \underline{u}\cdot\nabla)Q = \nu_n \nabla^n Q \quad , \qquad (2.1)$$

where,

$$\underline{u} = (-\partial\psi/\partial y \; , \partial\psi/\partial x) \quad , \quad Q = -\nabla^2\psi \quad , \qquad (2.2)$$

and

$$\nabla = \hat{i}\,\partial/\partial x + \hat{j}\,\partial/\partial y + \hat{k}\,\partial/\partial z \quad . \tag{2.3}$$

The right hand side of (2.1) is a (fictitious) dissipation introduced to remove Q at very small scales, beyond the region of real interest. We recall that the relation of ($\underline{u}$, Q) to the Navier-Stokes equations,

$$\psi = (1/\rho_o f)\,\delta p \quad ,$$

where ρ_o is the (constant and stable) density stratification, f the earth's rotation rate, assumed large compared to (u'/L), where u' and L are a typical velocity and length scales. Further, δp is the variation of the pressure about its combined gravitationally stratified and geostrophic form. In (2.2), z is a scaled coordinate, not the physical, z'. To bring out its meaning more clearly, we re-write (2.1) in physical units:

$$Q = \{\partial^2/\partial x^2+\partial^2/\partial y^2+(f/N)^2\partial^2/\partial z^2\}\psi \quad ,$$

where N is the Brunt-Väisälä frequency,

$$N^2 = -(g/\rho_o)\,d\rho_o/dz \quad . \tag{2.4}$$

Equations (2.1)-(2.4) assume a stable and constant stratification ($N^2 > 0$), in which ρ_o decreases linearly with z'. The density fluctuation (from ρ_o) is given by:

$$\delta\rho = -(\rho_o f/g)(\partial\psi/\partial z') \quad .$$

At this level of approximation, the vertical velocity field is,

$$w = -(f/N^2)(\partial/\partial t+\underline{u}\cdot\nabla)\,\partial\psi/\partial z' \quad ,$$

not at all related in a simple way to ψ itself. Here, we do not consider at all problems in which f and N vary with (x,y,z).

Equation (2.1) has only two inviscid quadratic constants of motion,

$$E = (1/2)\int d\underline{x}\ \{\underline{u}^2(x,t)+(\partial\psi/\partial z)^2\} \quad , \tag{2.5}$$

$$E = (1/2)\int d\underline{x}\ Q \quad . \tag{2.6}$$

Here, we remind that $(1/2)(\partial\psi/\partial z)^2$ is (in scaled z-coordinates) the potential energy density.

For homogeneous flows, it is convenient to resolve Q(x,t) into its Fourier modes = $\sum \exp(i\underline{k}\cdot\underline{x})Q(\underline{k},t)$, and characterize the simplest statistical aspects of the flow by;

$$U(\underline{k},t) = \langle Q(\underline{k},t)Q(-\underline{k},t)\rangle \quad . \tag{2.7}$$

We use a simple (EDMQN) closure (Orszag, 1974), Pouquet et al. (1975), Herring (1980) to investigate U(k,t):

$$\dot{U}(k)=\int_\Delta d\underline{p}[(\underline{p}\times\underline{q})\cdot\hat{\underline{n}}]\; B(k,p,q)U(\underline{q})\{U(\underline{p})-U(\underline{k})\} \quad . \tag{2.8}$$

Here, $\underline{k} = \underline{p} + \underline{q}$, and

$$B(k,p,q) = [(q^2-p^2)(q^2-k^2)/(k^2p^2q^4)]\;\theta(k,p,q) \quad , \tag{2.9}$$

The factor θ(k,p,q) is the triple-moment relaxation time, and the remainder of (2.9) stems entirely from the convective derivative in (2.1). The relaxation factor, θ(k,p,q) is computed here simply via the EDMQN prescription:

$$\theta(k,p,q) = \nu_n(k^n+p^n+q^n) + \sqrt{\int_0^k k'^2dk'U(k')} + \sqrt{\int_0^p p'^2dp'U(p')} + \sqrt{\int_0^q q'^2dq'U(q')} \tag{2.10}$$

where ν_n is given in (2.1). The prescription (2.10) is admittedly crude, but a proper algorithm (via a Galilean invariant scheme) is too formidable.

The three dimensional energy spectrum (kinetic plus potential) is given in terms of U(k) as:

$$E(\underline{k}) = 2\pi\; U(\underline{k})=2\pi\;[U_{KE}(\underline{k})+U_{PE}(\underline{k})] \tag{2.11}$$

$$U_{KE}(\underline{k}) = \sin^2\theta\; U(\underline{k}) \quad , \quad U_{PE}(\underline{k}) = \cos^2\theta\; U(\underline{k}) \tag{2.11'}$$

U_{KE} ($\underline{k}$) and U_{PE} ($\underline{k}$) are the kinetic and potential energies respectively. In what follows, we employ U instead of E($\underline{k}$), for simplicity.

3. How Does U(k) Become Isotropic?

Charney's idea requires that U(k) become isotropic in an (enstrophy) inertial range. How is this possible in view of the anisotropy of (2.1)? First, as already noted, since these equations have isotropic inviscid constants of motion (2.5) and (2.6), there should be a tendency towards isotropy contained in any realistic closure. We employ two approaches to discuss this issue; (a) a linearization in the extent of anisotropy combined with a diffusion approximation suitable for large k, and (b) an angular modal expansion.

(a) Linearization for Weak Anisotropy.

As a first approach to see how such a tendency is reflected in (2.9) at large k, we make a "diffusion" approximation, in which either (p,q) lies in the energy containing range of scales, and k is at a much larger inertial range-k. Supposing for example q << (k,p) we expand

$$\int d\underline{p} U(\underline{p}) U(\underline{q}=\underline{k}-\underline{p}) B(\underline{k},\underline{p},\underline{q})$$

about q = 0 and retain only leading order terms. The present problem requires some attention to details of the angular integration. Taking care to make the approximation symmetric in (p,q), we find:

$$\dot{U}(\underline{k}) = k^2[(1-\mu^2)\{\partial^2/\partial k^2+(4/k)\partial/\partial k\}+L(\mu)/k^2]U(\underline{k})\Omega \quad , \qquad (3.1)$$

where,

$$L(\mu) = (1-\mu^2)\ \partial^2/\partial\mu^2-4\mu(1-\mu^2)\partial/\partial\mu \quad , \qquad (3.2)$$

and,

$$\Omega^2=\int d\underline{p}U(p,t) \quad . \qquad (3.3)$$

It is of some interest to inquire as to the analytic character of (3.1). Let C satisfy $L(\mu)C_\lambda(\mu) = \lambda(1-\mu^2)C_\lambda(\mu)$. Then, single valuedness requires C_λ to be polynomial, and we find $LC_n(\mu) = -n(n+3)(1-\mu^2)C_n(\mu)$, with, $C_n(\mu)$ Gegenbauer polynomials of degree (3/2). These are orthonormal of weight $(1-\mu^2)$ (Mangus and Oberhetinger, 1947, p. 76 <u>et seq.</u>). We may construct the solution of (3.1) via the expansion

$$U(\underline{k},t) = \sum C_n(\mu)U_n(k,t) \quad , \qquad (3.4)$$

where $U_n\ (k,t)$ satisfies;

$$\sum_m c_{nm}\ \dot{U}_m(k,t)=\{\partial^2/\partial\xi^2-3\partial/\partial\xi-n(n+3)\}U(\xi=\ln(k),t)\Omega \quad , \qquad (3.5)$$

with,

$$c_{nm}=\int_{-1}^{1} d\mu C_n(\mu)C_m(\mu) \quad .$$

We may readily investigate the problem in which (3.1) is driven by an impulsive anisotropic force of the form:

$$F(k,\mu) = (1-\mu^2) \quad A_n\delta(k-k_o)C_n(\mu) \quad . \qquad (3.6)$$

The time-independent solution to (3.5) is then;

$$U(k,\mu) = \sum_n A_nC_n(\mu)/k^{n+3}, k>k_o \quad . \qquad (3.7)$$

Thus, in the enstrophy inertial range, the anisotropy characterized by $C_n\ (\mu)$ decays at a rate k^{-n} faster than the energy spectrum (k^{-3}).

A peculiarity of quasi-geostrophic turbulence is that if the forcing, $F(k,\mu)$ is less isotropic than $(1-\mu^2)$ the dynamics cannot transfer energy out of the forcing region, thereby allowing an accumulation of energy at k_o. This simply corresponds to the fact that the horizontal advection is unable to disperse energy concentrated at $\mu = 1$. Indeed, $U(\underline{k})\ \delta(\mu^2-1)$ is an exact, time independent solution to (2.1).

(b) Angular Modal Expansion for Strong Nonlinearity.

The above analysis is linear, and cannot be extended to small k, which is the region where we expect $k^{-5/3}$. One way to proceed is to introduce a modal expansion of $U(k,t)$ in terms of a suitable set. Some time ago (Herring, 1980), we described the results of a Legendre expansion for decaying turbulence. Our analysis of Sec. 3(a) suggests that the Gegenbauer functions $C_n(\mu)$ may be more suitable. At low order, these are equivalent, and we shall here sketch only briefly the results of a Galerkin expansion of $U(k,t)$ of the form:

$$U(k) = \sum U_n(k)P_n(\underline{k}/|k|) \quad , \qquad (3.8)$$

as it pertains to stationary turbulence.

We recall the structure of the equations for $(U_0(k), U_2(k),\ldots)$ (Herring, 1980). Here, we record only the $(P_0(\mu), P_2(\mu))$ components for simplicity:

$$\dot{U}_0(k,t) = T(U_0,U_0)-(1/5)[T(U_0,U_2)+T(U_2,U_0)]+\tilde{T}(U_2,U_2)+\ldots,$$

$$U_2(k,t) = -(1/5)T(U_0,U_0)+R(U_0,U_2)-(1/7)T(U_2,U_2)+\ldots. \qquad (3.9)$$

Here, $T(U_0,U_2)$ is the angular average (over 4Π) of the right hand side of (2.8) for isotropic functions (U_0,U_2). The functional form $\tilde{T}$ is more complicated, but both it and T satisfy energy and enstrophy conservation for arbitrary $(U_0,U_2, \ldots)$. The term $R(U_0,U_2)$ is not conservative, and is in fact responsible for the isotropization ($U_2 \to 0$), if there is no anisotropic driving (its "diffusion"-limit form is the anisotropic part of (3.1)-(3.2)). We shall not record them in any more detail here. The "diffusion" limit forms of T and R are:

$$T(U_0,U_0) \to (4\Pi/45)[\int dpp^2U_0(p)](1/k^2)\partial/\partial k(k^4\theta(k,p,p)U_0(k)) \quad , \quad (3.10a)$$

$$R(U\ ,U\ ) \to (136\Pi/4410)[\int dpp^2U_2(p)][k^2\partial^2U_0/\partial k^2+(32/17)k\partial U_0/\partial k]\theta(k,p,p)$$

$$-(5/7)\eta(k)U_2(k) \qquad (3.10b)$$

Here, the eddy-viscosity $\eta(k)$ entering (3.10) has the form,

$$\eta(k) = \int dpdq\tilde{B}(k,p,q)U_0(q)/q^2 \quad , \qquad (3.11)$$

$$\tilde{B} = (4\pi/3)\theta pq(p^2-q^2)(k^2-q^2)\sin^2(p,q)/k^3 \quad .$$

Note that $U_0(k) \sim k^{-3}$ yields $T_0(U_0,U_0)=0$, an enstrophy inertial range.

The return to isotropy, $R(U_2,U_0)$ (3.10b), is somewhat involved, but its physical interpretation is simple. The first term is a production of anisotropy by a straining of isotropic eddies of scale k by larger-scaled anisotropy. The second term is a destruction of anisotropy (U_2), which is in general much larger. A closer examination of the exact terms shows in fact that the negative term in (3.10b) is nearly cancelled by a large positive contribution from the region $(p/k) \sim 1$. The net effect is to reduce the drain term from $\eta(k)$ down to the eddy circulation time at scale $k \sim [\int^k dpp^2E(p)]^{1/2} \ll \eta(k)$.

(c) Some Numerical Results.

We now present some numerical result illustrating the points made above. First, we should comment on the form of the eddy-viscosity (3.11), which is somewhat different from that encountered in other problems. Consider for example the limit k--> 0:

$$\eta(k) \to -(16\Pi/45)k^2\int_0^\infty p^2 dp U_0(p)\theta(k,p,p)/p^2 \qquad (3.12)$$

Thus, at $k \to 0$, the eddy viscosity for this problem is always negative, even in the absence of a low-k cutoff (recall that in two-dimensional turbulence a negative eddy viscosity occurs only if a low-wave number cut-off is applied). To see what this means in practice, we consider a forced problem in which both forcing and dissipation are ~ $\sin^2\theta$. Then from Sec. 3(a), the steady-state spectrum $U_0(k)$ is isotropic. The numerical solution to such a problem is shown in Fig. 2, with forcing and damping parameters as given in the figure caption. Also given is $\eta(k)$, with the scaling to the right. Note that $\eta(k) < 0$ only below the peak energy wave number where Rayleigh friction is significant in maintaining a balance between inverse cascade and dissipation.

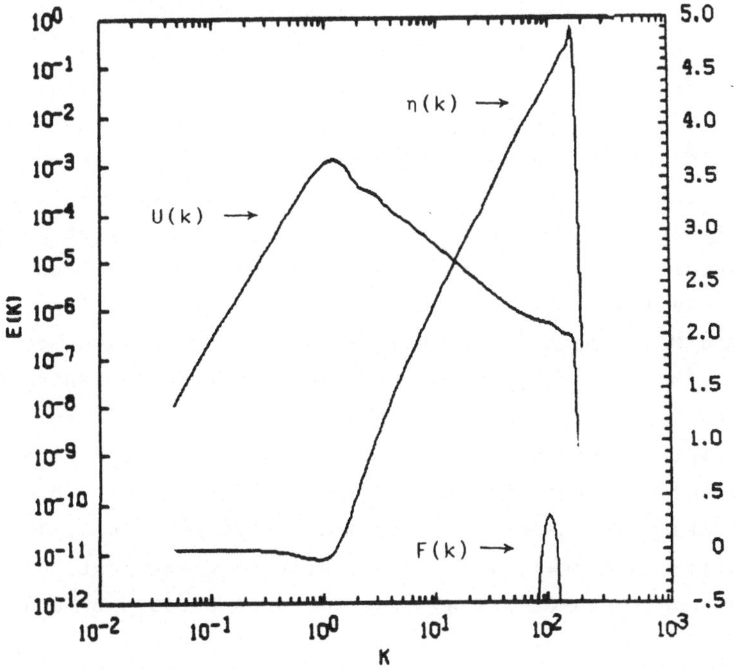

Fig. 2. Total energy spectrum, U(k) for stationary (forced) isotropic quasi-geostrophic turbulence, with an anisotropic forcing

~ $(k-k_F)\sin^2\theta$, $k_F = 100$. Both forcing and dissipation have the angular form $\sin^2\theta$. (The curve labeled "$\eta(k)$" is actually $\mathrm{Log}(1+100\ \eta(k))$ (scale to the right). Notice $\eta(k)<0$ for $k<k_P$, where k_P is the energy-peak wave number. Dynamically, $\eta(k)$ $(k<k_P)$ plays an insignificant role in determining the shape of $U(k)$. An hyperviscosity ~ $(k-k_H)^4$ is applied for $k>k_H$ ($k_H = 150$) and prevents any significant k^{-3} range at large k.

4. Concluding Comments

Our brief exploration of the quasi-geostrophic theory presented here suggests that within a limited context the theory provides for an isotropic inverse cascading $k^{-5/3}$ energy range at scales larger than the forcing wave number. The general considerations of Sec.3(a) suggest an isotropic statistics at scales smaller than the forcing wave number, k_F. However this approximate analysis may not be extended to scales inferior to k_F. The problem is that the "diffusion" analysis misses--apparently--some essential transfer processes, since we may verify directly that the isotropic form of (3.1) with (3.3) modified to include only straining scales larger than k_F does not yield a $k^{-5/3}$ range, contrary to what happens in two and three dimensional flows. We are then thrown back on the more complete modal analysis of Sec.3(b) and the rather specialized numerical calculation of Sec.3(c). The former is here simply cited and briefly discussed as providing an appropriate mechanism for isotropization. The latter numerical calculation provides--for a particular forcing and dissipation--evidence for the isotropic $k^{-5/3}$ law, and a context in which to discuss the negative eddy viscosity, special to quasi-geostrophic turbulence. The latter may be a vestige of baroclinic instability expressed primarily at large scales.

We should stress that the closure framework described here may be limited as to its validity, even for randomly forced flows. Recently, Herring and McWilliams (1984) have presented comparisons of closure and numerical simulations which indicate that inverse cascading two-dimensional turbulence, driven by a random forcing at small scales develops an appreciable amount of isolated vortices, which tend to strongly inhibit the inverse cascade necessary for the $k^{-5/3}$ range. Preliminary numerical study of (nearly) homogeneous flows (Lien-Hua, 1984, private communication; and McWilliams, 1984, private communication) also suggests a strongly non-Gaussian flow, but does produce evidence for isotropization.

REFERENCES

Charney, J .G., 1971. J. Atmos Sci., 28, 1087-1095.

Fiedler, F., and H. A. Panofsky, 1970. Bul. Am. Met. Soc., 51, 1114-1119.

Gage, K. S., 1979. J. Atmos. Sci., 36, 1950-1954.

_____, G. D. Nastrom, 1984. Position paper for Second Workshop on Technical Aspects of MST Radar, Urbana, IL, 21-25 May, 1984.

Garratt, C., and W. Monk, 1975. Ann. Rev. Fluid Mech., 11, 339-369.

Herring, J. R., 1980. J. Atmos. Sci., 37, 969.

_____, and J. C. McWilliams, 1984. Preprint, to appear in J. Fluid Mech.

Holloway, G., 1980. J. Geophys. and Astrophys. Sci., 11, 271-287.

Hoyer, J. E., and R. Sadourney, 1982. J. Atmos. Sci., 39, 707.

Kraichnan, R. H., 1967. Phy. Fluids, 10, 1417.

Lilly, D. K., 1983. J. Atmos. Sci., 40, 749-761.

Magnus, W., and F. Oberhettinger, 1949. Formulas and Theorems for the Functions of Mathematical Physics. Chelsea Publishing Co., New York, 170 pp.

Nastrom, G. D., K. S. Gage, and W. H. Jasperson, 1984. Nature, 310, 36.

Orszag, S. A., 1974. Statistical Theory of Turbulence: Les Houches Summer School on Physics. Gordon and Breach.

Pedlosky, J., 1979. Geophysical Fluid Dynamics. Springer-Verlag, 624 pp.

Pouquet, A., U. Frisch, and J.-P. Chollet, 1983. Phys. Fluids, 26, 877.

SMALL-SCALE ATMOSPHERIC TURBULENCE

AND ITS INTERACTION WITH LARGER-SCALE FLOWS

J.C. André

Centre National de Recherches Météorologiques (DMN/EERM)
42 Avenue G. Coriolis
31057 Toulouse Cedex - FRANCE

Abstract

The properties of small-scale atmospheric turbulence and its influence on larger-scale flows are briefly described, with particular reference to the effects of stratification. The questions for which no satisfactory answers are presently available are reviewed.

1. Introduction

The aim of this presentation is to show how small-scale atmospheric motions are parameterized in atmospheric and meteorological numerical models. Most of these parameterization schemes are based on a simplifying assumption, namely that small-scale and large-scale atmospheric motions are segregated in spectral space, or in other words that the interaction processes can be simply described by a more or less sophisticated eddy-viscosity formulation. Even with such a simple assumption it will be shown that important questions remain to be answered, related to either the time variability of surface boundary-layer quantities or the existence of shear-generated dissipative layers in the upper part of the atmosphere.

Specific features of small-scale atmospheric turbulence will also be discussed, with particular reference to the case where there exist density stratification, either unstable or stable. In the former case some indication will be given on how to deal with the non-local turbulent transfer due to medium-scale convective eddies. The latter case will be addressed in the light of recent developments, indicating that stratification-generated gravity waves may interact with three-dimensional turbulence to give rise to larger- and larger-scale two-dimensional eddies, and consequently leading to question the validity of the spectral gap hypothesis.

2. The spectrum of atmospheric motions

Energy which feeds atmospheric motions ultimately comes from solar radiation. The radiative energy undergoes many transformations, specially at the earth surface, before it is transformed from internal (or more precisely available potential) energy into kinetic energy through various instability mechanisms (see e.g. Charney, 1973). It is sufficient to say here that the rate at which kinetic energy is generated and dissipated in the atmosphere is approximately 2 to 3 Wm^{-2}, while the total kinetic energy amounts to around 100 W hr m^{-2} (see e.g. André, 1978, for more detail on atmospheric energetics). This means that kinetic energy is recycled within 2 days or, equivalently, that only two days would be needed to totally slow down the atmosphere if energy production was stopped. Such a small time duration indicates that parameterization schemes for dissipation due to turbulence are of great importance for meteorological purposes, specially for weather prediction on a scale of a few days to a week.

According to "standard wisdom", atmospheric motions can be separated into two spectral ranges. The large scale motions, with typical horizontal scale $k_L^{-1} \sim 1000$ km and typical velocity $v_L \sim 10$ ms^{-1}, correspond for example to the almost horizontal eddies which affect our mid-latitude weather (storm systems, high and low pressure elements,...). The small scale motions, with typical size $k_s \sim 100$m and typical velocity $v_s \sim 1$ ms^{-1}, correspond to the familiar turbulence, due for example to wind shear/or convective activity. Typical atmospheric spectra have been assembled (see e.g. Vinnichenko, 1970), which are consistent with the idea that intermediate scales are not energetic. If this is to be true, as it was and is still frequently assumed, one can apply the simple arguments developed by Pouquet et al. (1983) to estimate the effective Reynolds number R_L of large-scale motions corresponding to the eddy viscosity $v_s k_s^{-1}$ due to small-scale turbulence

$$R_L = \frac{v_L}{k_L} \cdot \frac{k_s}{v_s} \qquad (1)$$

This Reynolds number is obviously very large (of the order of 10^5 if one uses the above estimates for v's and k's) and leads to a dissipative length scale for large-scale two-dimensional motions given by standard Kolmogorov phenomenology (e.g. Kraichnan, 1967) :

$$k_D = R_L \, k_s \qquad (2)$$

From the above estimates it is then clear that k_D would be much smaller than k_s, thus confirming the idea that small-scale and large-scale atmospheric motions would be spectrally segregated and that their interaction could be described by an eddy-viscosity assumption.

It however remains to determine where this dissipation of energy takes place in the atmosphere. There are two possible locations, one close to the earth surface, in the atmospheric boundary layer, and one in the upper atmosphere where there exist a strong wind shear associated with the so-called jet stream. Some estimates, either experimental (Hollingsworth, 1977) or numerical (Miyakoda and Sirutis, 1977), indicate that both regions may contribute to the total dissipation.

3. Small-scale turbulence in the atmospheric surface layer

We shall first address the problem of small-scale eddy fluxes in the lower part of the atmosphere, in the so-called surface boundary layer, where momentum is lost to the earth surface and where heating and cooling of the underlying surface due to radiative effects influence the termal stratification of the atmosphere. The various surface properties are supposed to be known, so that we concentrate here only on turbulence properties and vertical structure of eddy fluxes. More detail about the parameterization of the atmospheric boundary layer can be found for example in André (1983).

The well-kown assumptions which lead to the formulation of the universal law of the wall, or logarithmic law (see e.g. Monin and Yaglom, 1971) can be generalized to the case of the atmospheric surface layer where temperature cannot be considered anymore as a passive scalar. The corresponding equations, above a possible viscous layer, read then

$$\frac{\partial}{\partial z}\overline{u'w'} = 0 \quad ; \quad \frac{\partial}{\partial z}\overline{w'\theta'} = 0 \qquad (3)$$

where z is the vertical (upward) coordinate, θ is the temperature and u and w are respectively horizontal and vertical velocity components. Primes denote turbulent fluctuations and overbars indicate averaging.

While in the more usual case of a neutral logarithmic layer there is only one governing parameter u_*, equal to the square-root of the constant momentum flux

$$\overline{u'w'} = -u_*^2 \quad , \quad (4)$$

we have here to consider an additional parameter Q_0, equal to the constant heat flux,

$$\overline{w'\theta'} = Q_0 \quad . \quad (5)$$

From u_*, Q_0, the local altitude z and the buoyancy parameter β measuring the intensity of stratification effects ($\beta = \alpha g$ where α is the coefficient of thermal expansion and g is the gravity acceleration), Monin-Obukhov (1954) simularity theory predicts that mean profiles are given by universal functions of a dimensionless stability parameter ζ

$$\frac{\partial \bar{u}}{\partial z} = \frac{u_*}{kz} \varphi_u(\zeta) \; ; \; \frac{\partial \bar{\theta}}{\partial z} = -\frac{Q_0}{u_* k z} \varphi_\theta(\zeta) \quad (6)$$

$$\zeta = -\frac{k \beta Q_0 z}{u_*^3} \quad (7)$$

where the von Karman's constant k is included for convenience. In the case of a neutrally stratified boundary layer, one has $Q_0 = \zeta = 0$ so that $\varphi_u(0) = 1$. When the boundary layer is stratified, either stably (i.e. cooled at the bottom or $Q_0 \leqslant 0$) or unstably (i.e. heated at the bottom or $Q_0 \geqslant 0$), various field experiments have allowed for a relatively unambiguous determination of universal functions, like (Businger et al., 1971)

$$\varphi_\theta(\zeta) = \begin{cases} 0.74\,(1-9\zeta)^{-1/2} & \text{for } \zeta \leqslant 0 \\ 0.74 + 4.7\,\zeta & \text{for } \zeta \geqslant 0 \end{cases} \quad (8)$$

which is represented in Fig 1. The fact that φ_θ remains positive for all values of the stability parameter ζ simply means that stable stratification always corresponds to a positive temperature gradient $\partial\bar{\theta}/\partial z \geq 0$, and that unstable stratification always corresponds to $\partial\bar{\theta}/\partial z \leq 0$, which is obvious if one considers only short time- or space-intervals.

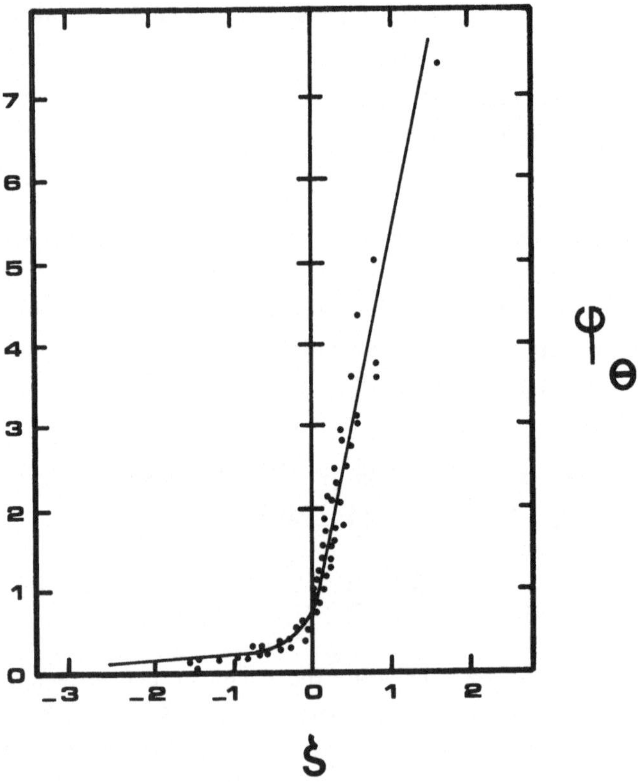

Figure 1 : Non-dimensional temperature gradient as a function of stability for small-scale averaging

Unfortunately meteorologists may have to deal with either larger-scale spatial averages (the grid size of a numerical weather prediction model is typically 100 km!) or diurnal averages. In such a case it may well happen that unstable regions are characterized by large upward heat fluxes (large positive Q_0) but relatively small negative temperature gradients $\partial\bar{\theta}/\partial z$, since convectively driven turbulence tends to very efficiently mix flow properties, while stable regions correspond to small downward heat fluxes (small negative Q_0) but fairly large positive temperature gradients $\partial\bar{\theta}/\partial z$, since stable stratification tends to strongly inhibit turbulence. On the average it turns out that globally unstable situations, in the sense $\langle Q_0\rangle > 0$ where $\langle\cdot\rangle$ denotes large-scale averaging, are associated with positive temperature gradient $\langle\partial\bar{\theta}/\partial z\rangle$, as can be seen in Fig. 2 where is shown the large-scale averaged function Ψ_θ defined by (Mahrt et al., 1985)

$$\left\langle\frac{\partial\bar{\theta}}{\partial z}\right\rangle = -\frac{\langle Q_0\rangle}{\langle u_*\rangle k z}\,\Psi_\theta\left(-\frac{k\beta\langle Q_0\rangle z}{\langle u_*\rangle^3}\right) \qquad (9)$$

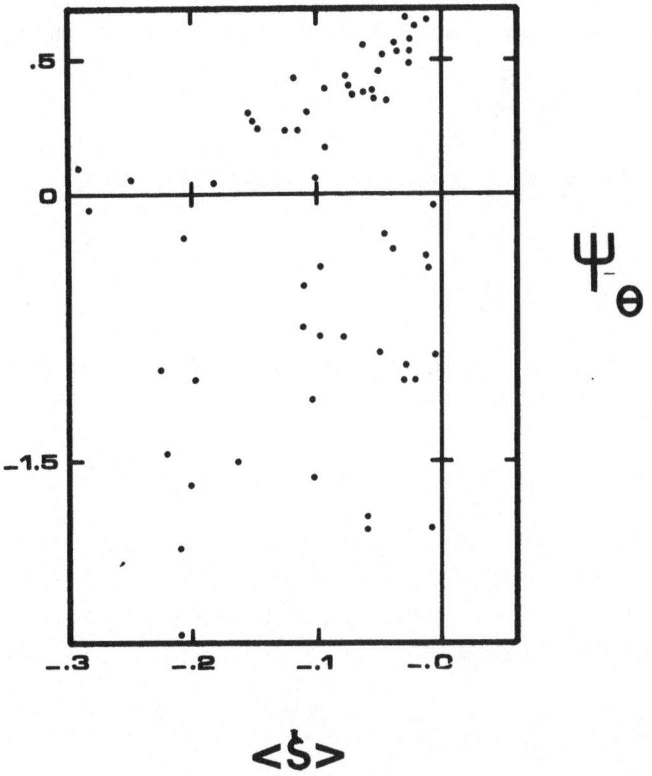

Figure 2 : Non-dimensional temperature gradient as a function of stability for large-scale averaging

The only way to solve such a paradox is to allow for counter-gradient heat fluxes, e.g. (Mahrt et al., 1985)

$$\langle Q_o \rangle = - K_e \left(\langle \frac{\partial \bar{\theta}}{\partial z} \rangle - \gamma \right) \tag{10}$$

where γ is a yet-to-be-determined counter-gradient correction and K_e an eddy diffusivity.

4. Small-scale atmospheric turbulence and eddy-kinetic energy dissipation

All the above developments explicitely or implicitely assume that there exists an eddy diffusivity, as for example in (10), or in (6) and (5) which amount to

$$\overline{w'\theta'} = - K_e \frac{\partial \bar{\theta}}{\partial z} \quad ; \quad K_e = \frac{k u_* z}{\varphi_\theta(\zeta)} \tag{11}$$

The prescription of this eddy diffusivity requires the knowledge of two independant parameters, for example a velocity-scale V and a length-scale L_k, as respectively u_* and z in (11) :

$$K_e = L_k V \tag{12}$$

A general way to determine the velocity scale V, which reduces to u_* when specialized to the particular case of the surface layer, is to relate it to the eddy-kinetic energy $\bar{e}$:

$$V^2 = \bar{e} = \frac{1}{2} \overline{u'_i u'_i} \tag{13}$$

Eddy-kinetic energy is then usually computed from a simplified form of its rate equation (see e.g. André, 1983), which allows for an influence of local shear and local thermal stratification :

$$\frac{D\bar{e}}{Dt} = - \frac{\partial}{\partial z}\left(- K_e \frac{\partial \bar{e}}{\partial z} \right) + K_e \left\{ \left(\frac{\partial \bar{u}}{\partial z} \right)^2 + \left(\frac{\partial \bar{v}}{\partial z} \right)^2 \right\} - \beta K_e \frac{\partial \bar{\theta}}{\partial z} - \varepsilon \quad , \tag{14}$$

where D/Dt is the material derivative following the mean motion and ε the viscous dissipation rate.

The use of such a parameterization for the velocity scale allows for the description of elevated dissipative layers associated with intensive shear in the jet-stream region (Miyakoda and Sirutis, 1977).

It still remains to determine the length scale L_k appearing in (12), as well as the possibly different length-scale L_ε which is usually used to compute ε from the Kolmogorov relation

$$\varepsilon = \frac{\bar{e}^{3/2}}{L_\varepsilon} \quad . \tag{15}$$

Whenever the turbulence is isotropic, like for neutral and unstable stratification, L_k and L_ε are equal to each other, and correspond to the size of the energetic eddies. On the contrary, stably-stratified turbulence corresponds to horizontally elongated eddies, so that L_k may be much less than L_ε . A possible way to deal with such situations is to assume

$$\frac{L_k}{L_\varepsilon} = \frac{\overline{w'^2}}{\bar{e}} \tag{16}$$

which is a simplified form of the more firmly justified parameterization due to Therry and Lacarrere (1983). It would then remain to use two rate equations, one for $\bar{e}$, which is Eq.(14), and one for the vertical component of eddy kinetic energy $\overline{w'^2}$. This has already been done for the quite similar case of the oceanic surface layer (Garwood, 1977).

5. Atmospheric stratified turbulence

Turning first to the unstably stratified case, i.e. to convective regimes, one has to account for the influence of medium-scale, coherent, vertically two-dimensional eddies which can be observed even at very high Rayleigh number. Given the existence of such energetic eddies (see e.g. the laboratory experiment by Willis and Deardorff, 1974), there appears to be no reason why the heat flux due to these big eddies extending throughout the convective zone should relate in a unambiguous manner to the local temperature gradient, as in (11). It is in fact well-known that upward heat fluxes ($\overline{w'\theta'} \geqslant 0$) co-exist with vanishing or even slightly positive temperature gradients (Deardorff, 1966). This amounts once again to counter-gradient heat transfer of the form

$$\overline{w'\theta'} = - K_e \left(\frac{\partial \bar{\theta}}{\partial z} - \Gamma \right) . \tag{17}$$

It should however be emphasized that this counter-gradient property of convective turbulence is due to the existence of non-local transfer in physical space, while the counter-gradient behaviour described in Section 3 is due to averaging only.

This convective counter-gradient behaviour is usually taken into account through more or less sophisticated variants of the idea originally proposed by Deardorff (1966). One can indeed consider a simplified form of the rate equation for the turbulent heat flux

$$\frac{\partial}{\partial t}\overline{w'\theta'} = -\overline{w'^2}\frac{\partial\bar{\theta}}{\partial z} + \beta\overline{\theta'^2} - \tau^{-1}\overline{w'\theta'} + \text{E.T.} \quad , \qquad (18)$$

where τ^{-1} is the inverse relaxation-time of molecular and pressure effects, and E.T. represents extra-terms related to advection following the mean flow and higher-order diffusion. By neglecting transient effects and spatial inhomogeneity, (18) leads indeed to (17) with

$$K_e = \tau\,\overline{w'^2} \quad ; \quad \Gamma = \beta\,\frac{\overline{\theta'^2}}{\overline{w'^2}} \qquad (19)$$

It should however be strongly emphasized that such a parameterization is based only on local processes. The counter-gradient Γ is due to the fact that small-scale, temperature inhomogeneity $\overline{\theta'^2}$ locally induces a positive (upward) heat flux, allowing for warmer parcels ($\theta' > 0$) to rise ($w' > 0$) and colder ones ($\theta' < 0$) to sink ($w' < 0$) whatever the value of the local mean gradient $\partial\bar{\theta}/\partial z$. Convective models based on (17) and (19), or on variant of these equations, may sometime achieve reasonable agreement with experiment, but, if this is the case, it is for unphysical reasons.

The case of stably stratified flows is even more complicated to deal with. It is indeed known that in such a case turbulence co-exists with internal gravity waves, which may considerably modify the turbulence structure. Recent work by Riley et al. (1981) and Lilly (1983) indicates that for sufficiently small Froude number F, i.e. when the effects due to stratification are larger than the inertial ones, the velocity field can be seen as the superposition of two very different flows. One can indeed look at it with a small time scale $T_s = (\beta\,\partial\bar{\theta}/\partial z)^{-1/2}$,in which case the equations read

$$\frac{\partial \underline{u}_H}{\partial t} + F\left(\underline{u_H \nabla_H \cdot \underline{u}_H} + w\frac{\partial \underline{u}_H}{\partial z}\right) = -\underline{\nabla}_H (p/\rho_0)$$

$$\frac{\partial w}{\partial t} + F\left(\underline{u_H \nabla_H}\, w + w\frac{\partial w}{\partial z}\right) = -\frac{\partial}{\partial z}\left(p/\rho_0\right) + \beta\theta$$

$$\frac{\partial \theta}{\partial t} + F\left(\underline{u_H \nabla_H}\, \theta + w\frac{\partial \theta}{\partial z}\right) + w\frac{\partial \bar{\theta}}{\partial z} = 0 \qquad (20)$$

$$\underline{\nabla}_H \cdot \underline{u}_H + \frac{\partial w}{\partial z} = 0$$

Eqs. (20) describve for $F \to 0$ a system of internal gravity waves oscillating at the Brunt-Vaïsala frequency $N = (\beta \partial\bar{\theta}/\partial z)^{1/2}$. At the same time one can also look at the same velocity field with a larger time scale $T_e = V/L$, in which case the equation read

$$\frac{\partial \underline{u}_H}{\partial t} + \underline{u_H \nabla_H \cdot \underline{u}_H} + F^2 w\frac{\partial \underline{u}_H}{\partial z} = -\underline{\nabla}_H (p/\rho_0)$$

$$F^2\left(\frac{\partial w}{\partial t} + \underline{u_H \nabla_H}\, w + F^2 w\frac{\partial w}{\partial z}\right) = -\frac{\partial}{\partial z}\left(p/\rho_0\right) + \beta\theta \qquad (21)$$

$$\frac{\partial \theta}{\partial t} + \underline{u_H \nabla_H}\, \theta + F^2 w\frac{\partial \theta}{\partial z} + w\frac{\partial \bar{\theta}}{\partial z} = 0$$

$$\underline{\nabla}_H \cdot \underline{u}_H + F^2\frac{\partial w}{\partial z} = 0$$

Eqs. (21) describe for $F \to 0$ a two-dimensional, possibly turbulent, flow.

The above analysis indicates that the final stage of stably stratified turbulence corresponds to an horizontally two-dimensional flow where the characteristic length of the eddies keeps growing, as explained by the well-known phenomenology of two-dimensional turbulence, while the kinetic energy of the vertical component is radiated away by a spectrum of internal gravity waves. If such a mechanism takes place in the atmosphere, and due to the growth of induced two-dimensional horizontal structures, it may well be that no spectral gap exists to separate small-scale turbulence from large-scale flow.This would lead to possibly drastic revisions of parameterization schemes for the effects of small-scale turbulence on large-scale flow, as those effects may then not be simply reduced to an eddy viscosity.

6. Perspectives

From the above examples of various features of atmospheric small-scale turbulence, it appears that much work remains to be done to derive adequate schemes to deal with large-scale average properties. The question of predicting the characteristic length scale of small-scale turbulence remains also an open question, specially for the case of stably-stratified flows.

References

André, J.C., 1978, in "Mécanique de l'Atmosphère et Energétique Industrielle", CEA-EDF eds., 31 pp.

André, J.C., 1983, in "Mesoscale Meteorology. Theories, Observations and Models", D.K. Lilly and T. Gal-Chen eds., Reidel, 651-669.

Businger, J.A., J.C. Wyngaard, Y. Izumi and E.F. Bradley, 1971, J. Atmos. Sci., 28, 181-189.

Charney, J.G., 1973, in "Dynamic meteorology", P. Morel ed.,Reidel, 97-351.

Deardorff, J.W., 1966, J. Atmos. Sci., 23, 503-506.

Garwood, R.W.Jr., 1977, J. Phys. Ocean.,7, 455-468.

Hollingsworth, A., 1977, in "The Parameterization of the Physical Processes in the Free Atmosphere", E.C.M.W.F., 230-272.

Kraichnan, R.H., 1967, Phys. Fluids, 10, 1417-1427.

Lilly, D.K., 1983, J. Atmos. Sci.,40, 749-761.

Mahrt, L., C. Berthou, P. Marquet and J.C. André, 1985, to be published.

Miyakoda, K., and J. Sirutis, 1977, Beit. Phys. Atmos.,50, 445-487.

Monin, A.S., and A.M. Obukhov, 1954, Trudy Geofiz. Inst. Acad. Nauk USSR, 24, 163-187.

Monin, A.S., and A.M.Yaglom, 1971, "Statistical Fluid Mechanics. Mechanics of Turbulence", Vol. 1, M.I.T. Press.

Pouquet, A., U. Frisch and J.P. Chollet, 1983, Phys. Fluids, 26, 877-880.

Riley, J.J., R.W. Metcalfe and M.A. Weissman, 1981, in Proc. AIP Conf. Nonlinear Properties of Internal Waves, B.J. West ed., 79-112.

Therry, G., and P. Lacarrère, 1983, Bound.-Layer Meteor., 25, 63-88.

Vinnichenko, N.K., 1970, Tellus, 22, 158-166.

Willis, G.E., and J.W. Deardorff, 1974, J. Atmos. Sci.,31, 1297-1307.

SELF-TURBULIZING FLAME FRONTS

Paul CLAVIN et Geoff SEARBY
Laboratoire de Recherche en Combustion
Centre St Jérôme - 13397 Marseille cedex 13, France

It is well known that the cells appearing on unstable fronts of flames propagating in a laminar flow of premixed gases can exhibit a chaotic motion[1,2]. Two model equations have been proposed by G. Sivashinsky[3,4,5].

The first one, called the "Kuramoto-Sivashinsky" equation, was derived in the context of a diffusion-reaction model corresponding to the crude approximation of a zero gas expansion limit in which hydrodynamic effects disappear. Written in an adimensional form, it takes the following form :

$$\frac{\partial}{\partial t}\alpha = \mathbb{L}^{ks}(\alpha) - \frac{1}{2}|\nabla\alpha|^2 \qquad (1)$$

where the linear operator is given by :

$$\mathbb{L}^{ks} = -\nabla^2 - \nabla^4 \qquad (2)$$

and where $x=\alpha(y,z,t)$ is the locus of the flame front. The instability mechanism is produced by the minus sign in front of the term $\nabla^2\alpha$ which describes destabilizing diffusive effects in the linear operator $\mathbb{L}^{ks}$. The short wave lengths are stabilized by $-\nabla^4\alpha$, this term corresponds to the thermal relaxation of the modifications to the flame temperature[6]. The nonlinear term corresponds to a purely geometrical effect associated with tilting of the front.

A second model equation has been derived by Sivashinsky[3] in the limit of weak but non zero gas expansion $\gamma=(\rho_u-\rho_b)/\rho_u \ll 1$ where ρ_u and ρ_b are the density in the unburnt and burnt gases respectively :

$$\frac{\partial}{\partial t}\alpha = \mathbb{L}^{s}(\alpha) - \frac{1}{2}|\nabla\alpha|^2 \qquad (3)$$

here the linear operator $\mathbb{L}^s$ is given in Fourier-space

$a=e^{i(k_y y+k_z z)}\tilde{a}_k$, by :

$$\tilde{\mathbb{L}}^s_k = \frac{\gamma}{2}k - k^2 \qquad (4)$$

to be compared to equ.(2):

$$\tilde{\mathbb{L}}^{ks}_k = k^2 - k^4 \qquad (5)$$

where k is the modulus of the wave vector. Eqs.(3) with (4) are valid only at the dominant order in the expansion of powers in γ and when the diffusive mechanisms, contrarily to equs.(1),(2) and (5), have a stabilizing effect represented by the $-k^2$ term in equ.(4). The destabilizing $\frac{\gamma}{2}k$ term in equ.(4) arises from non local hydrodynamical effects associated with the deflection of the streamlines across the flame front produced by gas expansion ($\gamma \neq 0$). This instability mechanism is called the Darrieus-Landau instability and has been known for many years (1939) in flame theory. These two model-equations are now very popular because the corresponding partial differential equations describe a transition to intrinsic stochasticity in deterministic systems[7,8] with many degrees of freedom.

Recently, many improvements have been realised in experimental techniques to dynamically stabilize an unstable flame front in a uniform laminar flow[9]. Moreover laser diagnostic techniques for recording the fluctuations in the wrinkled flame position without disturbing the reactive flow are presently available[10,11]. Thus experimental data concerning the statistical characteristics of the chaotic motion of an unstable flame front will be soon obtained. In this context, it is important to have a model equation that could describe real flame fronts.

For real flames, it is known experimentally[11] and theoretically[12,13] that the strong gas-expansion $\gamma \approx 0.85$ produces dominant effects that cannot be fully described in the limit $\gamma \to 0$. Moreover, the effects of gravity must be retained[13] in order to describe the finite size of the cellular structure observed experimentally at the instability threshold for flames propagating downwards[9]. This last effect can be easily included in the model equations (1) or (3) by adding a negative constant proportional to the inverse of the Froude number F_r. Such an equation has been studied in the context of spherical flames[4]. But the corresponding equation does not retain the nonlinear effects associated with the advective motion of the front produced by the flow perturbation induced by the wrinkling of the front. These effects are not negligible for values of gas expansion occuring in ordinary situations. As suggested by a theoretical study of the dynamical properties of flames in a stagnation-point flow, carried out in a simplified model[14], gradients in the tangential component of the flow are expected to strongly influence the flame dynamics.

We will present now a heuristic derivation[6] of a model equation containing the above mentionned effects. The starting point is the local equation relating the normal flame speed U_n to the characteristics, at the front, of the upstream gas flow modified by the presence of the flame[16] ;

$$(U_n - U_L)/U_L = -(\mathcal{L}/d)\,\{(1/\sigma)(d\sigma/dt)\} \tag{6}$$

Where U_L is the laminar flame speed of the planar front and where σ represents an elementary area of the flame front and $(1/\sigma)(d\sigma/dt)$ is the local stretch experienced by the wrinkled flame front under the joint effects of the curvature of the front and of the flow inhomogeneities[16]. $\mathcal{L}$ is the Markstein length caracterizing the reactive mixture, its order of magnitude is the same as the flame thickness d. Equ. 6 has been obtained in the approximation where the characteristic length scale of the wrinkling is large compared to the flame thickness. For small amplitudes of the wrinkles, equ.6, written in the frame where the averaged planar front is steady, yields :

$$\partial\alpha/\partial t + \underline{\nabla}\cdot(\underline{v}\alpha) - u - U_L|\nabla\alpha|^2/2 = \mathcal{L}U_L\{\nabla^2\alpha + \underline{\nabla}\cdot\underline{v}/U_L\} \quad (7)$$

Where $\underline{v}$ and u are the tangential and normal components (at the front) of the fluctuations of the upstream gas flow induced by the wrinkled front. Nonlinear terms which are not important for the qualitative results presented here have been dropped in the r.h.s. of equ.(7). They correspond to a nonlinear modification of the local flame speed, an effect that is expected to be negligible compared to non linear hydrodynamic effects. The model equation is obtained when the values of u and $\underline{v}$ expressed in terms of the fluctuations is the wrinkled flame position α, is introduced in equ.(7). To be fully consistent with equ.(7), the linear approximation for $\underline{v}$ is sufficient whereas the second order in an amplitude expansion has to be retained for u. In other words, a linear solution of the hydrodynamical problem is sufficient to describe the non linear effect of the gas flow induced in the tangential direction. When the corresponding term $\nabla\cdot(\underline{v\alpha})$ is expected to be the most important of the two non linear convection terms associated with the gas expansion, only a linear analysis of the corresponding hydrodynamic problem is required. The second nonlinear term in u can be obtained by a tedious nonlinear second order analysis but is not expected to be important.

The complete linear analysis of the hydrodynamical problem in which the front is considered as a free boundary separating the fresh and burnt gases, can be carried out by using the jump condtions obtained by an asymptotic analysis of the local structure of the wrinkled flame[12,13,15]. When the conditions are sufficiently close to the instability threshold the unsteady effects in the induced flow field can be neglected and one obtains the following expression, written in Fourier space:

$$(\tilde{u}_k/U_L) = (\gamma/2)\{-F_r^{-1} + k/(1-\gamma) - I k^2\}(\tilde{\alpha}/d) \quad (8\text{-a})$$

$$i\underline{k}\cdot\underline{\tilde{v}}_k = -k\tilde{u}_k \quad (8\text{-b})$$

where k is the modulus of the wave vector reduced by d^{-1}. Equ.(8-b) expresses the condition that the gas flow is incompressible outside the preheated zone of thickness d. $\mathcal{I}$ is a positive non dimensional coefficient which, like $\mathcal{L}/d$, depends on the diffusive and reactive properties of the reactive gas mixture. When equs.(8) are introduced in equ.(7), the evolution equation can be written as:

$$\frac{\partial}{\partial t} a = \mathbb{L}(a) - |\nabla a|^2/2 - \underline{\nabla} \cdot (\underline{v} a) \qquad (9\text{-a})$$

where $a = (\alpha/d)$ and $v = (\underline{v}/U_L)$ is given by equ.(8-b) The linear operator $\mathbb{L}$ is similar to equ.(4) but with an additional negative constant expressing the stabilizing effect of acceleration of gravity :

$$\tilde{\mathbb{L}}_k = \frac{\gamma}{2(1-\gamma)} \left\{ -(1-\gamma) F_r^{-1} + k - k^2/k_* \right\} \qquad (9\text{-b})$$

k_*, like the Markstein number $(\mathcal{L}/d)$ characterizes the reactive mixture. By construction $(1-\gamma)^{-1} k_*^{-1}$ is larger than $\mathcal{I}$ appearing in equ.(8-a). Using a different scaling k_* can be set equal to unity as in equ.(4). When $\gamma \to 0$ and when $F_r^{-1} \to 0$ equs.(9) reduce to the Sivashinsky equation(3). Equs.(9) contain an additional non linear term $\underline{\nabla} \cdot (\underline{v} a)$ which has a hydrodynamical origin and which cannot be neglected compared to $|\nabla a|^2$ whenever the gas expansion γ is not negligible.

REFERENCES

(1) G. Markstein, "Nonsteady flame propagation" Pergamon Press (1964)

(2) F. Sabathier, L. Boyer, and P. Clavin, Progress in Astronautics and Aeronautics vol.76 (1981) 246.

(3) G.I. Sivashinsky, Acta Astronautica, 4 (1977) 1177.

(4) G.I. Sivashinsky, Acta Astronautica, 6 (1979) 569.

(5) G.I. Sivashinsky, Am. Rev. Fluid Mech. 15 (1983) 179.

(6) P. Clavin, Prog. Energ. Combust. Sci. (1985) to appear

(7) P. Manneville, These Proceedings.

(8) Y. Pomeau, A. Pumir and P. Pelcé, J. Stat. Phys. 37 (1984) 39.

(9) J. Quinard, G. Searby and L. Boyer, Lectures Notes in Physics, 210 (1984) 331.

(10) L.Boyer, Combust. Flame 39 (1980) 321.

(11) G.Searby, F. Sabathier, P. Clavin and L. Boyer, Phys. Rev. Letters 51 (1983) 1450.

(12) P. Clavin, F.A. Williams, J. Fluid Mech. 116 (1982) 251

(13) P. Pelcé, P. Clavin, J. Fluid Mech. 124 (1982) 219

(14) G.I. Sivashinsky, C.K. Law, G. Joulin, Combust. Sci. Tech. 66 (1982) 777

(15) G. Searby, P. Clavin, Combust. Sci. Tech. (1985) to appear

(16) P. Clavin, G. Joulin, J. Physique Lettres 44 (1983) L-1

SIMULATION AS AN AID TO PHENOMENOLOGICAL MODELING

by
Joel H. Ferziger
Department of Mechanical Engineering
Stanford University
Stanford, CA, USA

I. Introduction

A. The Need for Modeling

The complexity of turbulence makes it necessary to use relatively crude models in the prediction of all but the simplest turbulent flows. Models are derived using a combination of physical insight and approximations to the governing equations. Good models have been developed for relatively narrow ranges of flows; models of moderate accuracy applicable to wide ranges of flows are also available. A single model that can be applied with confidence to a wide range of flows has not been found; indeed, such a model may not exist.

The root of the problem seems to lie in the observation that turbulence is not a single phenomenon. The common element in flows that we call turbulent is chaotic behavior. These flows are non-deterministic in the sense that small changes in the present state lead to large differences in the future state of the flow. However, the mechanism which produces this behavior is not the same in every flow.

Prior to the advent of large computers, models used to represent turbulence had to be relatively simple; models requiring more than a few ordinary differential equations could not be solved. Computers make solution of systems of partial differential equations possible, but direct solution of the Navier-Stokes equations is possible only for simple flows at moderate Reynolds numbers. The most popular current methods are based on solving averaged forms of the Navier-Stokes equations.

Experiments on turbulent flows are difficult; consequently, much of the data needed to do detailed tests of turbulence models is not available. Recent advances, especially in optical diagnostic techniques, may change this. However, at present, data on many quantities that are modeled are not available, so models are tested by comparing their predictions with experimental data. Unfortunately, these flow predictions are prone to errors from sources other than the models. The errors are often difficult to locate; as a result, turbulence models are often blamed for errors due to the numerical approximations.

Full and large eddy simulations of turbulent flows have begun to fill part of this gap. The ability of these methods to compute all the details of turbulent flows makes them valuable for model validation. Indeed, because they cannot be applied directly to technologically interesting flows at present, service as model validators will remain one of the chief functions of these methods for some time. Quite a lot has been accomplished in this area in recent years; this paper will review that work.

B. Reynolds Average Turbulence Models

This paper will concentrate on applications of full and large eddy simulation to the testing of Reynolds average turbulence models. Integral methods are not considered, because they can be validated by experiments. Attention will be concentrated on those aspects of models used in the Reynolds averaged incompressible Navier-Stokes equations which are most difficult to validate experimentally. Reviews of these models were given by Rodi (1980) and Reynolds (1976).

The mean velocity can be defined by ensemble averaging; other kinds of averaging may also be employed. Combined with the notion that the velocity can be regarded as a combination of a mean and turbulent fluctuations, i.e., that $u_i = U_i + u_i'$, where U_i and u_i' represent the mean and fluctuating velocities, these averaging procedures lead to the Reynolds averaged Navier-Stokes equations:

$$\frac{\partial U_i}{\partial t} + \frac{\partial}{\partial x_j} U_i U_j = -\frac{1}{\rho}\frac{\partial p}{\partial x_i} + \nu \frac{\partial^2 u_i}{\partial x_j \partial x_j} - \frac{\partial}{\partial x_j} \overline{u_i' u_j'} \quad (1.1)$$

The terms $\overline{u_i' u_j'}$ in Eq. (1.1) are called the Reynolds stresses and represent the effect of the turbulence on the mean field. Equation (1.1) is not closed; achieving closure is the central problem in turbulent flow simulation.

The most common assumption of turbulence models is that the effect of turbulence on the mean flow can be represented as an increased viscosity:

$$\overline{u_i' u_j'} = -\nu_T \left(\frac{\partial U_i}{\partial x_j} + \frac{\partial U_j}{\partial x_i} \right) \equiv -2\nu_T S_{ij} \quad (1.2)$$

where ν_T is the eddy viscosity and S_{ij} is the strain rate of the mean flow. To complete the model, an expression for ν_T is needed.

The simplest description of turbulence characterizes is by its kinetic energy per unit mass, $k = q^2/2$, and an average length scale, L, in terms of which the eddy viscosity can be written:

$$\nu_T = C_\mu q L \quad (1.3)$$

Models differ in the methods they use to determine the length and velocity scales. In the following two chapters, we shall look at mixing length and two-equation models. Reynolds stress models, which do not require the assumption (1.2), will be considered in Chapter IV. Little attention will be given to algebraic stress models in this paper, because, when applied to the homogeneous flows, they are essentially identical to Reynolds stress models.

II. Mixing Length Models

In mixing length models, q and L are determined directly from the mean velocity field. These models are designed for two-dimensional shear flows for which the choice of length scale is relatively straightforward. For free shear flows, the width of the shear layer is the natural length scale. Near a solid boundary, the distance to the surface becomes the natural length scale.

The model is completed with an expression for the turbulent velocity scale, q. A commonly used approximation,

$$q = SL \quad (2.1)$$

where $S = (S_{ij}S_{ij})^{1/2}$, can be derived in a number of ways (cf. Tennekes and Lumley, 1972). Equation (2.1) is probably the weakest part of the mixing length model.

Mixing length models work very well in two-dimensional shear flows. They can be modified to account for extra effects such as pressure gradients, curvature, and transpiration. Kays and Crawford (1978) have built a model which does an excellent job for a wide range of boundary layers. The major disadvantage of mixing length models is the difficulty they have with complex flows. For example, mixing length models for separated and reattaching flows do not exist.

Because mixing length models are rather specialized, simulations of homogeneous turbulence have but little to contribute to them. However, there is one result worth noting. Although this model is not normally used for this flow, one can define a mixing length for homogeneous shear flow by:

$$\ell = (\overline{u_1'u_2'})^{1/2}/(\partial U_1/\partial x_2) \tag{2.2}$$

and compare it to other length scales of the flow; in this flow $S = \frac{1}{2}(\partial U_1/\partial x_2)$. In particular, the time development of the mixing length and two other length scales, namely q^3/ε, the length scale used in two-equation models (see the following chapter), and $L_{11,2}$, an integral scale, are shown in Fig. 1. We see that the mixing length behaves very differently from the physical length scales.

The reason why the mixing length does not correlate well with the physical scales can be seen by writing Eq. (2.2) as:

$$\frac{\ell}{L} = \frac{(\overline{u_1'u_2'})^{1/2}}{q} \left(\frac{q}{SL}\right) \tag{2.3}$$

In nearly all shear flows, $\overline{u_1'u_2'}/q^2$ is constant; the value is approximately 0.15. In inhomogeneous shear flows, only a small range of $SL/q \approx Sq^2/\varepsilon$ is encountered, so use of mixing length models is reasonable; this parameter, which is nearly the ratio of production to dissipation, is close to unity in these flows. Homogeneous flows cover a broad range of Sq^2/ε, so mixing length models do not apply to them. We conclude that mixing length models apply only to 'equilibrium' flows, i.e., flows in which production and dissipation are approximately equal.

III. Two Equation (k-ε) Models

As Eq. (2.1) is regarded as one of the weak points of mixing length models, an obvious improvement is a model in which the length scale is prescribed but the turbulence kinetic energy is determined by solving a partial differential equation. Unfortunately, these one-equation models are not much of an improvement over mixing length models. We therefore procede directly to two-equation models.

Two-equation models employ the Boussinesq eddy viscosity relation (1.2), but the evolution of the velocity and length scales are determined by a pair of partial differential equations. A PDE for the turbulence kinetic energy determines the velocity scale. This equation can be derived from the Navier-Stokes equations by taking the trace of the Reynolds stress equations given in the next chapter:

$$\frac{\partial k}{\partial t} = \frac{\partial}{\partial x_j} u_j k \neq \text{Production} - \text{Dissipation} + \text{Diffusion} \tag{3.1}$$

The significance of the convective term is clear; the production is the creation of new turbulent energy from the mean velocity field and is given by:

$$P = \overline{u_i' u_j'} \frac{\partial U_i}{\partial x_j} \tag{3.2}$$

Dissipation is the viscous destruction of turbulence energy; diffusion transports energy through the flow and is of no importance in homogeneous flows. The convection term is exact, the production term can be computed from the model itself, the treatment of the dissipation is explained below, and the diffusion is modeled.

We need expressions for the the length scale, L, and the dissipation, ε. It is a well-accepted notion of turbulence theory that these are related by:

$$L = q^3/\varepsilon \tag{3.3}$$

which leads naturally to the idea of using a partial differential equation to define the dissipation and, via Eq. (3.3), the length scale.

An exact equation for the dissipation can be derived from the Navier-Stokes equations by an extension of the method used to derive the energy equation. However, this equation involves correlations of the derivatives of the velocity for which no data are available. Consequently, the modeling of these terms is uncertain. For homogeneous flows, the most commonly used model of the dissipation equation reduces to (Rodi, 1980):

$$\frac{d\varepsilon}{dt} = - c_2 \frac{\varepsilon^2}{k} + c_1 \frac{\varepsilon P}{k} \tag{3.4}$$

where P is the production of turbulent kinetic energy given by Eq. (3.2). The constant c_2 in Eq. (3.4) is obtained by requiring the model to predict the decay of isotropic turbulence; its value is usually taken to be 1.92. The constant c_1 is obtained by requiring the proper prediction of the logarithmic region of wall-bounded flows and is usually taken as 1.44.

This model has had success in predicting a number of flows. However, caution is needed. Although never badly in error, the two-equation model is not always accurate enough for high technology applications. One needs to be especially careful in applying it to flows different from those for which the model was calibrated. For this reason, tests of this model using full-simulation-generated data for homogeneous flows have been made in recent years. Extensions to inhomogeneous flows will be made in the near future. We now review some of these results.

Bardina et al. (1983) showed that this model is incapable of predicting the change in the decay rate of isotropic turbulence caused by system rotation. They found that an additional term must be added to Eq. (3.4) to bring the model and simulation data into agreement; this term is $0.15\ \Omega\varepsilon$, where Ω is the rotation rate. Aupoix and Cousteix (1983) applied large eddy simulation to this flow and suggested an alternative, but similar, modification of the dissipation equation.

A strict test of the k-ε model for any flow can be made when Eq. (3.4) is rearranged to read:

$$\frac{1}{\varepsilon P}\frac{d\varepsilon}{dt} = -C_2\frac{\varepsilon}{P} + C_1 \qquad (3.5)$$

If the model is valid, a plot of $(k/\varepsilon P)\,d\varepsilon/dt$ vs. ε/P should be a straight line. These quantities can be determined from full simulations; $d\varepsilon/dt$ is obtained by numerical differentiation. A plot of this kind (taken from Ferziger and McMillan, 1985), for homogeneous shear flow, is shown in Fig. 2. Although there is a great deal of scatter, much of which can be attributed to uncertainty in the simulation-derived data, the figure shows that the relationship suggested by the model is reasonable. There is no evidence of nonlinear behavior. Figure 2 also shows a least-squares fit and the prediction of the standard model for which the constants were given above. For homogeneous shear flow, the standard model is fairly close to the least-squares fit.

Similar plots for homogeneous strain flows also show the model relationship to be reasonable. However, the 'constants' must be allowed to be time dependent if the model is to be accurate. More than likely, the modeling of these flows would be improved if the three-equation model of Wu et al., which is described below, were used. This has not yet been tested.

Wu et al. (1985) studied compressed turbulence, the flow generated when fluid containing initially isotropic turbulence undergoes compression. The compression rate is comparable to the time scales of the turbulence but slow enough that the Mach number is small; this is is the parameter range encountered in internal combustion engines.

In agreement with Jones and Launder (1972) and Reynolds (1980), they found that a term proportional to the dilatation needs to be added to the dissipation equation. More importantly, they found that the simulation data could not be fit with constant parameters; if the usual form of the model is retained, the parameters need to be allowed to be functions of time. The reason can be traced to the fact that, in this flow (and, probably, many others), the relation (3.3) between the dissipation and the length scales is incorrect. In this flow, the disequilibrium is due to a rapid decrease of the dissipation caused by the density increase.

To decouple the dissipation and the length and time scales, Wu et al. modified the dissipation equation:

$$\frac{d\varepsilon}{dt} = -\frac{\varepsilon}{\tau} + C_2 \frac{P\varepsilon}{k} + C_3\, \varepsilon \frac{\partial u_i}{\partial x_i} \tag{3.6}$$

and introduced a new time scale equation:

$$\frac{d\tau}{dt} = -\frac{S}{11} + C_4 \left(\frac{\varepsilon\tau}{k} - \frac{6}{11}\right) + C_5 S\tau \tag{3.7}$$

The first term is an 'equilibrium' term which is necessary to predict the decay of isotropic turbulence. The second is a 'return to equilibrium' term required to bring the turbulence back to the isotropy after perturbation from it. The last terms account for the effects of external strain. It is particularly encouraging that the effects of various strains appear to be additive; Wu et al. were able to predict four separate flows with a wide range of strain rates with this model.

To sum up this chapter, we have shown that data generated by full simulations of homogeneous flows can be used to investigate the validity of two-equation models, determine their parameters, and suggest improvements in the models. It has also been shown that the commonly used model needs to be modified to account for extra strains.

IV. Reynolds Stress Models

Two-equation models have a number of deficiencies; some were illustrated in the preceding chapter. Others are consequences of the eddy viscosity relationship (1.2) for the Reynolds stresses. Such models are unable to predict counter-gradient fluxes. More significantly, these models produce abrupt and unrealistic changes in the Reynolds stresses when a strain is newly applied to the mean flow.

Reynolds stress models were developed as a way of avoiding the eddy viscosity assumption. They use partial differential equations for the Reynolds stresses themselves; the latter are derived from the Navier-Stokes equations in the following manner. One subtracts the mean equations from the instantaneous equations to obtain equations for the fluctuating velocities. Cross multiplying and adding the resulting equations and averaging the result, we obtain a set of equations:

$$\frac{\partial \overline{u_i' u_j'}}{\partial t} = \text{Convection} + \text{Production} - \text{Dissipation} + \text{Redistribution} \qquad (4.1)$$

Most of the terms in these equations are analogs of terms in the turbulent kinetic energy equation. The new elements are the redistribution terms which represent energy transfer among the components of the Reynolds stress tensor. The new terms, together with the tensor nature of the equations and the increased difficulty of obtaining experimental data, make the constants in Reynolds stress models more difficult to determine than those in the models described earlier. These models also contain more constants, providing hope of greater universality.

Many of the terms that are modeled in Reynolds stress models cannot be measured experimentally. These include the dissipation (which is now a tensor), the diffusion, and the redistribution terms, of which the most important is the pressure-strain correlation:

$$\phi_{ij} = \overline{p'\left(\frac{\partial u_i'}{\partial x_j} - \frac{\partial u_j'}{\partial x_i}\right)} \qquad (4.2)$$

For homogeneous flows, diffusion is of no consequence. The assumption is commonly made that the dissipation is isotropic; it is modeled as in two-equation models. The pressure-strain correlation is usually broken into two terms, one arising from the mean pressure and the second due to the fluctuating pressure. Each term is modeled separately; the models will be given later.

Models of the pressure-strain correlation have been investigated by Feiereisen et al. (1981) and Shirani et al. (1981). Both of these works used simulations of homogeneous shear flow. The former included the effects of compressibility; the latter included a passive scalar, for which results will be presented later.

These simulations showed the dissipation to be anisotropic; in fact, it is nearly as anisotropic as the Reynolds stresses themselves.

This was thought to be a consequence of the low Reynolds numbers, but no dependence of the anisotropy on the Reynolds number was found; the Reynolds number was varied over an order of magnitude. More recent simulations by Lee and Reynolds (1985) confirmed the observation that there is little or no tendency for turbulence to become more isotropic as the eddy size decreases. Indeed, they find the vorticity to be more anisotropic than the Reynolds stresses.

The commonly used model for the slow pressure-strain correlation (the part arising from the fluctuating pressure) is due to Rotta (1951):

$$\phi_{ij}^{(1)} = \frac{C\varepsilon}{k} \left(\overline{u_i' u_j'} - \frac{1}{3} q^2 \delta_{ij} \right) \tag{4.3}$$

The constant must be at least 2.0 if the turbulence is to return to isotropy. The model does not correlate well with the simulation data for the slow part of the pressure-strain correlation, and the constant is considerably smaller than 2. However, if the anisotropic component of the dissipation is added to the slow pressure-strain term, the model represents the combination well; the constant was found to be approximately 2.6, in good agreement with the accepted value. Thus, although the assumptions on which the model is based appear to be faulty, the model is a reasonable one.

For the rapid pressure-strain correlation, the most commonly used model is:

$$\phi_{ij}^{(2)} = \frac{2}{5} (1+A_1) S_{ij} q^2 - \frac{3}{5} A_1 \left[\overline{u_i' u_k'} S_{kj} + \overline{u_j' u_k'} S_{ki} + \frac{2}{3} P S_{ij} \right] - \frac{4}{5} \left(\frac{5}{3} + \frac{7}{12} A_1 \right) \left[\overline{u_i' u_k'} \Omega_{kj} + \overline{u_j' u_k'} \Omega_{ki} \right] \tag{4.4}$$

where $\Omega_{ij} = (\partial u_i/\partial x_j - \partial u_j/\partial x_i)$. Investigation of this model using full simulation results showed that it does not represent the fast pressure-strain correlation accurately. The constants determined by curve fitting are given in Table 1. They are independent of the Reynolds number and other dimensionless parameters but depend on the tensor index. No way of representing these results in a tensorially correct manner has been found, but they can be used as an ad hoc model until a better one is constructed.

Reynolds stress models nearly double the cost of computing a given flow. They rectify some of the shortcomings of simpler models, but in the 1980-81 Stanford meeting (Kline et al., 1981), they were found to be no more accurate, on the average, than the simpler models.

Algebraic stress models reduce the Reynolds stress model equations to algebraic equations by making approximations to the terms which contain spatial derivatives--the convection and diffusion terms. Equations for the turbulence kinetic energy and dissipation are retained. The resulting model is only slightly more expensive to apply than the two-equation model, but, in its current versions, algebraic stress models appear to offer few advantages over two-equation models.

Let us close with some remarks about turbulence models in general. It appears that models can be used with confidence only to simulate flows which are not too different from those used to calibrate them; caution is required when new flows are computed--the models are postdictive rather than predictive. All turbulence models in use today are based on the notion that high-order moments of the fluctuating velocity are unique functionals of the low-order moments. Unfortunately, experimental evidence does not support this argument. In all flows for which there are data, the low-order moments reach their final values before the high-order ones do. In the author's opinion, turbulence models ought to be regarded as engineering correlations, rather than physical laws. This change in viewpoint allows one to search for models that are not universal in character. Giving up the requirement that a single model represent a broad range of physical phenomena allows us to employ different models in physically different parts of a flow. This permits a degree of flexibility that may be valuable in constructing models.

V. Passive Scalars

Turbulent flows containing more than more than one chemical species or temperature variations are of great importance. Accurate prediction of mixing of chemical species and thermal energy is essential to the design of many devices. The simplest situation is one in which the contaminant is a passive scalar (a quantity whose concentration does not affect the fluid properties). This is the obvious and ideal place to start.

Models for predicting of the concentration of a passive scalar in turbulent flows mimic those used for the flows themselves. Thus, mixing length, two-equation, and Reynolds stress-like models have been applied to the prediction of heat and mass transfer. In this chapter, we shall briefly review some results relating to the modeling of passive scalars. The emphasis will be on items in which passive scalar flows differ from flows of simple fluids. Nearly all of the results reported here were obtained by Shirani, Ferziger, and Reynolds (1981), hereafter denoted SFR, or by Antinopoulos-Domis (1981).

SFR considered homogeneous shear flows in which the mean passive scalar field depends linearly on the coordinates. They showed that, in these flows, the passive scalar field rapidly becomes independent of its initial conditions and depends only on the nature of the turbulence and the mean passive scalar gradient. They also computed the turbulent or eddy diffusivity tensor, defined by:

$$D_{ij_T} = \overline{u_i'\theta'}/(d\Theta/dx_j) \tag{5.1}$$

where θ is the concentration of the passive scalar and Θ is its mean. The results obtained, shown in Table 2, are in good agreement with the experiments of Tavoularis et al. (1977).

In the simplest models, including mixing length and two-equation models, the passive scalar is modeled in nearly the same manner as the velocity; the new feature is the inclusion of a turbulent Prandtl (or Schmidt) number which is the ratio of the turbulent diffusivities (assumed scalars) of momentum and species:

$$Pr_T = \nu_T/D_T \tag{5.2}$$

A number of correlations which give the turbulent Prandtl number as a function of other nondimensional quantities are available. SFR investigated several of these and found that none of them correlated the data well; they proposed a new model which yields better correlation.

SFR also investigated Reynolds stress models for the scalar field. The results are very similar to the findings for models applied to the turbulence field and will not be given here.

Antinopoulos-Domis investigated subgrid scale models for passive scalars and found that, for the passive scalar field, the constant in the Smagorinsky subgrid model, which is used in nearly all large eddy simulations, is twice as large as the analogous constant for the velocity field.

VI. Conclusions

It has been shown that full and large eddy simulations of turbulent flows can be used to investigate the validity of turbulence models. Although these methods are limited in the kinds of flows and range of Reynolds numbers they can simulate, they provide detailed data that are not available from laboratory flows and permit new kinds of investigations.

Some particulars provided by the works reviewed in this paper include:

- New insights into the limitations of mixing length models.
- Limits of validity of two-equation models, including the need for a separate time scale equation in strain flows.
- New insight into the approximations required to derive Rotta's pressure-strain correlation model.
- Lack of validity of the model of the rapid pressure-strain correlation.
- Lack of validity of correlations for the turbulent Prandtl number and the construction of a new model.

In sum, it has been shown that a great deal about modeling can be learned from simulations of turbulent flows. Unfortunately, few of these results have been applied in model flow calculations, so it is difficult to assess the importance of the results. However, there is reason to believe that these results will lead to improvements in the quality of predictions of turbulent flows. As larger computers become available, opportunities to do this kind of model validation will increase, and we anticipate growth in the amount of work of this kind that is done.

Acknowledgments

The author wishes to thank NASA-Ames Research Center for sponsoring this work for many years. He also wants to thank his colleague, Prof. W. C. Reynolds, who has taught him much about turbulence and how to model it.

References

Antinopoulos-Domis, M., "Large Eddy Simulation of a Passive Scalar in Isotropic Turbulence," Report QMC P 6037, Queen Mary College, 1979.

Aupoix, B., and Cousteix, J., "Large Eddy Simulation of Turbulence Undergoing Rotation," Proc. Fourth Symp. Turb. Shear Flows, Karlsruhe, 1983.

Bardina, J., Ferziger, J. H., and Reynolds, W. C., "Improved Subgrid Models for Large Eddy Simulation," AIAA paper 80-1357, 1980.

Bardina, J., Ferziger, J. H., and Reynolds, W. C., "Contributions to Large Eddy Simulation of Turbulent Flows," Report TF-19, Dept. of Mech. Engrg., Stanford U., 1983.

Cain, A. B., Reynolds, W. C., and Ferziger, J. H., "Simulation of the Transition and Early Turbulent Regions of a Free Shear Flow," Report TF-14, Dept. of Mech. Engrg., Stanford U., 1981.

Feiereisen, W. J., Reynolds, W. C., and Ferziger, J. H., "Computation of Compressible Homogeneous Turbulent Flows," Report TF-13, Dept. of Mech. Engr., Stanford U., 1981.

Ferziger, J. H. and McMillan, O. J., "Studies of Structure and Modeling in Turbulent Shear Flows, Nielsen Engineering and Research, Mt. View, Ca., 1985 (in press).

Jones, W. P., and Launder, B. E., "The Prediction of Laminarization with a Two-equation Model of Turbulence," Intl. J. Heat Mass Transfer, Vol. 15, 301, 1972.

Kays, W. M., and Crawford, M. E., Convective Heat and Mass Transfer, Second ed., McGraw-Hill, New York, 1978.

Kline, S. J., Lilley, G. M., and Cantwell, B. J., Proceedings of the 1980-81 AFOSR-HTTM-Stanford Conference on Complex Turbulent Flows, Dept. of Mech. Engrg., Stanford University, Stanford, CA, 1981.

Kline, S. J., Ferziger, J. H., and Johnston, J. P., "Calculation of Turbulent Shear Flows: Status and Ten-Year Outlook," ASME J. Fluids Engrg., Vol. 100, 3, 1978.

Launder, B. E., and Spalding, D. B., "The Numerical Computation of Turbulent Flow," Comp. Meth. in Appl. Mech. and Engrg., C3D, 269, 1974.

Lee, M. J. and Reynolds, W. C., "Numerical Experiments of the Structure of Homogeneous Turbulence," Report TF-22, Dept. of Mech. Engrg., Stanford Univ., 1985 (in preparation).

Moser, R. D., and Moin, P., "Direct Simulation of Turbulent Flow in a Curved Channel," Rept. TF-20, Dept. of Mech. Engr., Stanford U., 1984.

Reynolds, W. C., "Phenomenological Turbulence Modeling," in Ann. Revs. Fluid Mech., Vol. 8, p.183, 1976.

Reynolds, W. C., "Modeling of Fluid Motions in Engines--An Introductory Overview," Proc. Symp. Combustion Modeling in Reciprocating Engines, Plenum, p. 41, 1980.

Rodi, W., Turbulence Models and Their Applications in Hydraulics, Intl. Assn. for Hydraulics Research, Delft, 1980.

Rogallo, R. J., "Experiments in Homogeneous Turbulence," Report TM-81315, NASA-Ames Research Center, 1981.

Rotta, J. C., "Statische Theorie Nichthomogener Turbulenz," Z. Physik, C129D, S547, 1951.

Shirani, E., Ferziger, J. H., and Reynolds, W. C., "Simulation of Homogeneous Turbulent Flows Including a Passive Scalar," Report TF-15, Dept. of Mech. Engrg., Stanford U., 1981.

Smagorinsky, J., "General Circulation Experiments with the Primitive Equations. I. The Basic Experiment," Mon. Wea. Rev., C91D, 99, 1963.

Tavoularis, S., "Experiments in Turbulent Transport and Mixing," dissertation, Johns Hopkins U., 1978.

Tennekes, H., and Lumley, J. L., A First Course in Turbulence, MIT Press, 1972.

Tzuoo, K. L., Ferziger, J. H., and Kline, S. J., "Zonal Models for Homogeneous Flows," internal report, Dept. of Mech. Engrg., Stanford Univ., 1984.

Wu, C. T. and Ferziger, J. H., "Simulation and Modeling of Compressed Turbulence," Report TF-21, Dept. of Mech. Engrg., Stanford U., 1985.

WEAK LIMITS OF SEMILINEAR HYPERBOLIC SYSTEMS WITH OSCILLATING DATA

D. McLaughlin,* University of Arizona

G. Papanicolaou,* Courant Institute, New York University

L. Tartar, C.E.N., Limeil, France

Abstract

We consider several examples of nonlinear evolution equations with initial data that are rapidly oscillating functions of the space variable. We obtain an effective system of nonlinear evolution equations for the various moments of the solution by a multiple scale method. We also show how in one case (the Carleman model) compensated compactness gives a very general way of obtaining the effective equations without the use of multiple scales.

1. Introduction

We are interested in the behavior of solutions of deterministic, nonlinear evolution equations or systems when the initial data are rapidly oscillating functions of the space variable. Such problems cannot be solved numerically in a direct way. They are frequently analyzed by obtaining a hierarchy of equations for the various moments of the solution. This hierarchy is infinite due to the nonlinearity of the equations and it is frequently rendered finite by various ad hoc closure procedures.

We shall examine closely two examples of equations with rapidly oscillating data in the following sections: the Carleman equations and the Broadwell equations. These examples are analyzed by the usual asymptotic methods of multiple scales and modulation theory. The form of the effective equations or of the hierarchy is then discussed and compared with the hierarchy obtained directly. In the case of the

---

*Research supported by the Air Force Office of Scientific Research, Grant No. AFOSR-80-0228.

Carleman equations one can also use the theory of compensated compactness [1]. We show how this is done and how more complete the resulting analysis is when that theory is applicable.

The main conclusions of this paper are that

(i) nonlinear equations with rapidly oscillating data give rise to effective equations that are substantially different from the original ones;

(ii) the correct form of the hierarchy of moment equations is simpler than the full hierarchy obtained directly but is still an infinite one.

In the cases where the compensated compactness method does not apply or gives incomplete information the correct result is obtained by the asymptotic analysis but we have to introduce many more hypotheses about the nature of the oscillating data than otherwise.

There are of course many other interesting physical problems to which an analysis such as the one presented here could be applied. For example the Euler equations [2] and Burger's equation. The analysis of the Euler equations in three dimensions is considerably more involved and has been carried out only at a formal computational level.

2. Carleman's equations

This is a system of equations for the function u(t,x) and v(t,x), $t > 0$, $x \in \mathbb{R}^1$,

(2.1)
$$\begin{aligned} u_t + u_x + u^2 - v^2 &= 0 \\ v_t - v_x + v^2 - u^2 &= 0 \end{aligned}$$

with given initial conditions. It is a very simple model for a Boltzmann equation where u(t,x) and v(t,x) represent the average densities of particles at time t at position x moving with velocity one and minus one, respectively. The "collisions" or interactions in (2.1) are represented by the nonlinear terms and are not physically realistic. For example, two left going particles collide to produce a right going particle per unit time, and the right going particles collide to produce a left going one.

We consider (2.1) as a nonlinear system and let

$$(2.2) \qquad \begin{aligned} u(0,x) &= \tilde{U}(x, \frac{x}{\varepsilon}) \\ v(0,x) &= \tilde{V}(x, \frac{x}{\varepsilon}) \end{aligned}$$

where $\tilde{U}(x,y)$ and $\tilde{V}(x,y)$ are given nonnegative functions, periodic in y and smooth as functions of x. The parameter $\varepsilon > 0$ denotes the scale of rapid variation of the initial data. Problem (2.1), (2.2) is not taken from the context of some particular physical problem or associated model. We want to understand in as simple a setting as possible the manner in which the rapid oscillations of the initial data will affect the evolution. Note that problem (2.1), (2.2) is well posed for $\varepsilon > 0$ fixed when $\tilde{U}$ and $\tilde{V}$ are positive, bounded functions. Instead of $\tilde{U}$ and $\tilde{V}$ being periodic in y we may assume that they are stationary random processes. To unify the analysis we shall assume that the data are random even in the periodic case and write

$$(2.2') \qquad \begin{aligned} u(0,x) &= U(x, \frac{x}{\varepsilon}, \omega) \\ v(0,x) &= V(x, \frac{x}{\varepsilon}, \omega) \end{aligned}$$

where $U(x,y,\omega) = \tilde{U}(x,y+\omega)$, $V(x,y,\omega) = \tilde{V}(x,y+\omega)$ and ω is a random variable distributed uniformly over the period of $\tilde{U}$ and $\tilde{V}$. In general we shall take (2.2′) to hold with $\omega \in \Omega$, where $(\Omega,\underline{F},\underline{P})$ is a probability space. Thus, $u = u(t,x,\omega)$, $v = v(t,x,\omega)$ and the solution of (2.1) will be stationary random processes. We shall denote expectation by $\langle\ \rangle$ i.e.

$$(2.3) \qquad \langle u\rangle = \langle u(t,x)\rangle = \int_\Omega u(t,x,\omega)\, P(d\omega)$$

and we shall frequently omit the argument ω to simplify the notation. The solution also depends on ε so we write $u = u^{\varepsilon}$, $v = v^{\varepsilon}$ when neded for emphasis.

Taking expectations in (2.1) we obtain

$$(2.4)\qquad \begin{aligned} &(\partial_t + \partial_x)\langle u^{\varepsilon}\rangle + \langle (u^{\varepsilon})^2\rangle - \langle (v^{\varepsilon})^2\rangle = 0 \\ &(\partial_t - \partial_x)\langle v^{\varepsilon}\rangle + \langle (v^{\varepsilon})^2\rangle - \langle (u^{\varepsilon})^2\rangle = 0 \end{aligned}$$

with initial data

$$(2.4')\qquad \begin{aligned} \langle u^{\varepsilon}(0,x)\rangle &= \overline{U}(x) = \langle U(x,y)\rangle \\ \langle v^{\varepsilon}(0,x)\rangle &= \overline{V}(x) = \langle V(x,y)\rangle \end{aligned}$$

Note that we have used the stationarity of $U(x,y,\omega)$ and $V(x,y,\omega)$ for each x fixed. Of course (2.4) is not a closed system since $\langle (u^{\varepsilon})^2\rangle$ and $\langle (v^{\varepsilon})^2\rangle$ are involved. Equations for these quantities can be obtained and eventually the whole hierarchy of moment equations is generated this way. The problem is to find the asymptotic form of this hierarchy; does it for example simplify as $\varepsilon \to 0$?

We can get the correct form of the asymptotic limit by multiple scales as follows. Let

$$(2.5)\qquad \tau = t/\varepsilon \quad \text{and} \quad y = x/\varepsilon$$

and assume that u^{ε} and v^{ε} have an expansion of the form

$$(2.6)\qquad \begin{aligned} u^{\varepsilon}(t,x,\omega) &= u_0(\tau,y,t,x,\omega) + \varepsilon u_1(\tau,y,t,x,\omega) + \dots \\ v^{\varepsilon}(t,x,\omega) &= v_0(\tau,y,t,x,\omega) + \varepsilon v_1(\tau,y,t,x,\omega) + \dots \end{aligned}$$

Inserting this into (2.1) and equating coefficients of powers of ε yields

$$(2.7)\qquad \begin{aligned} &(\partial_\tau + \partial_y)u_0 = 0 \\ &(\partial_\tau - \partial_y)v_0 = 0\ , \end{aligned}$$

$$(2.8)\qquad \begin{aligned}(\partial_\tau+\partial_y)u_1 + (\partial_t+\partial_x)u_0 + u_0^2 - v_0^2 &= 0\\ (\partial_\tau-\partial_y)v_1 + (\partial_t-\partial_x)v_0 + v_0^2 - u_0^2 &= 0\ ,\end{aligned}$$

etc.

From (2.7) we conclude that

$$(2.9)\qquad \begin{aligned}u_0 &= u_0(y-\tau\ , \ t,x,\omega)\\ v_0 &= v_0(y+\tau\ , \ t,x,\omega)\end{aligned}$$

Let

$$(2.10)\qquad \eta = y + \tau\ , \qquad \xi = y - \tau$$

Then, in view of (2.1) we may write (2.9) in the form

$$(2.11)\qquad \begin{aligned}2\,\partial_\eta u_1 + (\partial_t + \partial_x)u_0 + (u_0^2 - v_0^2) &= 0\\ 2\,\partial_\xi v_1 - (\partial_t - \partial_x)v_0 + (v_0^2 - u_0^2) &= 0\end{aligned}$$

where

$$(2.9') \qquad u_0 = u_0(\xi,t,x,\omega)\ , \qquad v_0 = v_0(\eta,t,x,\omega)$$

are as yet undetermined stationary random functions for each t and x.

From (2.11) we see that u_1 and v_1 will not be stationary processes in general. However, if $u_0(\xi,t,x,\omega)$ and $v_0(\eta,t,x,\omega)$ are chosen to satisfy

$$(2.12)\qquad \begin{aligned}(\partial_t + \partial_x)u_0 + u_0^2 - \langle v_0^2\rangle &= 0\\ (\partial_t - \partial_x)v_0 + v_0^2 - \langle u_0^2\rangle &= 0\end{aligned}$$

with initial conditions

$$(2.12')\qquad \begin{aligned}u_0(\xi,0,x,\omega) &= U(x,\xi,\omega)\\ v_0(u,0,x,\omega) &= V(x,\eta,\omega)\ ,\end{aligned}$$

then (2.11) will have a solution that will grow slower than linearly in ξ and η as

$|\xi| \to \infty$ and $|y| \to \infty$. This is simply a consequence of the ergodic theorem for the stationary processes

$$(\partial_t + \partial_x)u_0 + u_0^2 - v_0^2 ,$$

$$(\partial_t - \partial_x)v_0 + (v_0^2 - u_0^2) ,$$

as functions of (η,ω) and (ξ,ω) respectively since both of these quantities are stationary and have mean zero.

Note that the form of the equations (2.12) is dictated by the suppression of secular terms (linearly growing terms in ξ and η) in (2.11). We also have

$$\text{(2.13)} \quad \begin{aligned} u_1(\xi,\eta,t,x,\omega) &= \frac{1}{2}\int_0^{\eta} (v_0^2(s,t,x,\omega) - \langle v_0^2(t,x)\rangle)\, ds \\ v_1(\xi,\eta,t,x,\omega) &= -\frac{1}{2}\int_0^{\xi} (u_0^2(s,t,x,\omega) - \langle u_0^2(t,x)\rangle)\, ds \end{aligned}$$

Returning now to (2.6) we see that, formally, we have

$$\text{(2.14)} \quad \begin{aligned} u^{\varepsilon}(t,x,\omega) &= u_0\Big(\frac{x-t}{\varepsilon},t,x,\omega\Big) + o(1) \\ v^{\varepsilon}(t,x,\omega) &= v_0\Big(\frac{x-t}{\varepsilon},t,x,\omega\Big) + o(1) \end{aligned} \qquad \varepsilon \to 0$$

One can verify that for stationary, second order processes for any $0 \le T < \infty$

$$\frac{1}{T}\int_0^T \langle u_0(\frac{x-t}{\varepsilon},t,x)\; v_0(\frac{x+t}{\varepsilon},t,x)\rangle\, dt \to \langle u_0(t,x)\rangle \langle v_0(t,x)\rangle , \qquad \varepsilon \to 0$$

and hence

$$\text{(2.15)} \quad \frac{1}{T}\int_0^T \langle u^{\varepsilon}(t,x)\; v^{\varepsilon}(t,x)\rangle\, dt \to \langle u_0(t,x)\rangle \langle v_0(t,x)\rangle$$

We conclude that the hierarchy of moment equations beginning with (2.4), which is quite involved, simplifies considerably and asymptotically has the form associated with (2.12) i.e.

$$(2.16)\quad \begin{aligned} &(\partial_t+\partial_x)\langle u_0^m\rangle + m\langle u_0^{m+1}\rangle - m\langle v_0^2\rangle\langle u_0^{m-1}\rangle = 0 \\ &(\partial_t-\partial_x)\langle v_0^m\rangle + m\langle v_0^{m+1}\rangle - m\langle u_0^2\rangle\langle v_0^{m-1}\rangle = 0\ , \\ &\langle u_0^m(0,x)\rangle = \langle U^m(x)\rangle \\ &\langle v_0^m(0,x)\rangle = \langle V^m(x)\rangle \qquad\qquad m = 1,2,\ldots \end{aligned}$$

Solving the hierarchy of moment equation (2.16) is equivalent to solving the nonlinear system (2.12) with random initial data (2.12′).

3. Carleman′s equation by compensated compactness

The equation (2.1) with initial data $u^\varepsilon(0,x)$, $v^\varepsilon(0,x)$ that converge as $\varepsilon \to 0$ weakly to some functions were analyzed in [3,4] using the ideas of compensated compactness [1]. We shall review briefly this analysis.

The main fact we need from [1] is as follows. Let $u^\varepsilon(x)$ and $v^\varepsilon(x)$ be functions from $\underset{\sim}{R}^N \to \underset{\sim}{R}^N$ that converge as $\varepsilon \to 0$ in $(L^\infty(\underset{\sim}{R}^N))^N$ weak * . That is for every function $\phi(x)$ in $(L^1(\underset{\sim}{R}^N))^N$

$$(3.1)\quad \begin{aligned} &\int_{\underset{\sim}{R}^N} u^\varepsilon(x)\cdot\phi(x)\,dx \to \int_{\underset{\sim}{R}^N} u(x)\cdot\phi(x)\,dx \\ &\int_{\underset{\sim}{R}^N} v^\varepsilon(x)\cdot\phi(x)\,dx \to \int_{\underset{\sim}{R}^N} v(x)\cdot\phi(x)\,dx \end{aligned}$$

where u(x) and v(x) are the limit functions. Clearly $u^\varepsilon(x)\cdot v^\varepsilon(x)$ will not converge to $u(x)\cdot v(x)$ in general. However, if

$$(3.2)\quad \text{div}\ u^\varepsilon(x)\ \text{ is bounded in }\ L^\infty(\underset{\sim}{R}^N)\ ,\ \text{uniformly in}\ \varepsilon$$

and

$$(3.3)\quad \text{curl}\ v^\varepsilon(x) = \left(\frac{\partial v_i^\varepsilon}{\partial x_j} - \frac{\partial v_j^\varepsilon}{\partial x_i}\right)\ \text{ is uniformly bounded}$$

$$\text{in } (L^\infty(\underset{\sim}{R}^N)^{N^2}\ ,\ \text{ uniformly in } \varepsilon$$

Then $u^\varepsilon(x)\cdot v^\varepsilon(x)$ will converge to $u(x)\cdot v(x)$ weakly, that is for every function $\psi(x)$ in $C_0^\infty(\underset{\sim}{R}^N)$

$$\int_{\underset{\sim}{R}^N} u^\varepsilon(x)\cdot v^\varepsilon(x)\ \psi(x)\ dx \to \int_{\underset{\sim}{R}^N} u(x)\cdot v(x)\ \psi(x)\ dx$$

To apply this result to (2.1) we note that if the initial data

$$u^\varepsilon(0,x) = U^\varepsilon(x)\ , \qquad v^\varepsilon(0,x) = V^\varepsilon(x)$$

are bounded in $L^\infty(\underset{\sim}{R})$ uniformly in ∞ then any solution of (2.1) will be bounded in $L^\infty(\underset{\sim}{R} \times R)$ (space-time) uniformly in ε. There will exist therefore a weak * convergent subsequence which we will denote still by $u^\varepsilon(t,x)$, $v^\varepsilon(t,x)$. From the equation (2.1) we see that div $u^\varepsilon = (\partial_t + \partial_x)u^\varepsilon$ and curl $v^\varepsilon = (\partial_t - \partial_x)v^\varepsilon$ are in $L^\infty(\underset{\sim}{R} \times R)$ uniformly in ε. Thus, by the above quoted theorem $u^\varepsilon(t,x)v^\varepsilon(t,x)$ converges weakly (in space-time) to $u(t,x)v(t,x)$ where $u(t,x)$ and $v(t,x)$ are the weak * limits of $u^\varepsilon(t,x)$ and $v^\varepsilon(t,x)$, respectively.

This is not quite what we want since it corresponds to only finding how first moments behave, that is

$$(\partial_t + \partial_x)u(t,x) + u_2(t,x) - v(t,x) = 0$$

$$(\partial_t - \partial_x)v(t,x) + v_2(t,x) - u(t,x) = 0$$

where $u_2(t,x)$ and $v_2(t,x)$ are the weak * limits of $(u^\varepsilon(t,x))^2$ and $(v^\varepsilon(t,x))^2$, respectively. To get the full hierarchy we note that from (2.1) we obtain

$$(3.4)\quad \begin{aligned} &(\partial_t + \partial_x)(u^\varepsilon(t,x))^m + m(u^\varepsilon(t,x))^{m-1}((u^\varepsilon(t,x))^2 - (v^\varepsilon(t,x))^2) = 0 \\ &(\partial_t - \partial_x)(v^\varepsilon(t,x))^m + m(v^\varepsilon(t,x))^{m-1}((v^\varepsilon(t,x))^2 - (u^\varepsilon(t,x))^2) = 0 \end{aligned}$$

$$(3.5)\quad (u^\varepsilon(t,x))^m = (U^\varepsilon(x))^m\ , \quad (v^\varepsilon(t,x))^m = (V^\varepsilon(x))^m\ , \quad m = 1,2,\ldots\ .$$

Let $u_m(t,x)$, $v_m(t,x)$, $U_m(x)$, $V_m(x)$ denote the weak * limits of $(u^\varepsilon(t,x))^m$, etc. Since the initial data are bounded in $L^\infty(\underset{\sim}{R})$ uniformly in ε so will any solution of

(3.4) be bounded in $L^\infty(\underset{\sim}{R} \times \underset{\sim}{R})$ for $m = 1,2,\ldots$. But now the theorem quoted above helps substantially in passing to the limit in (3.4), (3.5). In fact we get

$$(3.6) \qquad \begin{aligned} &(\partial_t + \partial_x)u_m + m\, u_{m+1} - m\, u_{m-1}v_2 = 0 \\ &(\partial_t - \partial_x)v_m + m\, v_{m+1} - m\, v_{m-1}u_2 = 0 \end{aligned}$$

$$u_m(0,x) = U_m \ , \qquad v_m(0,x) = V_m \quad , \qquad m = 1,2,\ldots$$

and this is precisely system (2.16), except for notation.

There are several points that need be made in comparing the analysis of this section with that of the previous one. The most important one is that the initial data $U^\varepsilon(x)$ and $V^\varepsilon(x)$ had completely general dependence on ε. They did not have to have some particular form like $U(x, \frac{x}{\varepsilon})$, $V(x, \frac{x}{\varepsilon})$ as is necessary when using multiple scale techniques. In addition, existence of solutions for (2.1) implies existence of solutions for (3.6) since the limit functions necessarily satisfy (3.6). This is not the case in the asymptotic analysis; one must prove existence separately for (2.1) and (2.16). Uniqueness is an issue when either method is used, however. Finally, regularity in x required from the initial data is minimal with the compensated compactness method.

4. Broadwell's equation

This is another system of semilinear equations that models the Boltzmann equation [5]

$$(4.1) \qquad \begin{aligned} &u_t + u_x + uv - w^2 = 0 \\ &v_t - v_x + uv - w^2 = 0 \\ &w_t + \frac{1}{2}(w^2 - uv) = 0 \end{aligned}$$

The initial data is asumed to have the form

$$(4.2)\qquad \begin{aligned} u^{\varepsilon}(0,x) &= U(x, \tfrac{x}{\varepsilon}, \omega) \\ v^{\varepsilon}(0,x) &= V(x, \tfrac{x}{\varepsilon}, \omega) \\ w^{\varepsilon}(0,x) &= W(x, \tfrac{x}{\varepsilon}, \omega) \end{aligned}$$

with $U(x,y,\omega)$, $V(x,y,\omega)$, $W(x,y,\omega)$ given stationary random functions for such $x \in \underset{\sim}{R}$, with $\omega \in \Omega$, $(\Omega,\underline{F},P)$ a probability space.

To find the behavior of the solution when ε goes to zero we let τ and y be defined by (2.5) and expand

$$(4.3)\qquad \begin{aligned} u^{\varepsilon}(t,x) &= u_0(\tfrac{t}{\varepsilon}, \tfrac{x}{\varepsilon}, t,x,\omega) + \varepsilon u_1(\tfrac{t}{\varepsilon}, \tfrac{x}{\varepsilon}, t,x,\omega) + \dots \\ v^{\varepsilon}(t,x) &= v_0(\tfrac{t}{\varepsilon}, \tfrac{x}{\varepsilon}, t,x,\omega) + \varepsilon v_1(\tfrac{t}{\varepsilon}, \tfrac{x}{\varepsilon}, t,x,\omega) + \dots \\ w^{\varepsilon}(t,x) &= w_0(\tfrac{t}{\varepsilon}, \tfrac{x}{\varepsilon}, t,x,\omega) + \varepsilon w_1(\tfrac{t}{\varepsilon}, \tfrac{x}{\varepsilon}, t,x,\omega) + \dots \end{aligned}$$

Inserting this into (4.1) and equating coefficients of powers of ε we obtain

$$(4.4)\qquad \begin{aligned} (\partial_\tau + \partial_y)u_0 &= 0 \\ (\partial_\tau - \partial_y)v_0 &= 0 \\ \partial_\tau w_0 &= 0 \end{aligned}$$

$$(4.5)\qquad \begin{aligned} (\partial_\tau+\partial_y)u_1 + (\partial_t+\partial_x)u_0 + u_0v_0 - w_0^2 &= 0 \\ (\partial_\tau-\partial_y)v_1 + (\partial_t-\partial_x)v_0 + u_0v_0 - w_0^2 &= 0 \\ \partial_\tau w_1 + \partial_t w_0 - \tfrac{1}{2}(u_0v_0-w_0^2) &= 0 \end{aligned}$$

From (4.4) we conclude that

$$(4.6)\qquad \begin{aligned} u_0 &= u_0(\xi,t,x,\omega) \\ v_0 &= v_0(\eta,t,x,\omega) \\ w_0 &= w_0(y,t,x,\omega) \end{aligned}$$

where ξ and η are defined by (2.10). Therefore (4.5) becomes

$$(4.7)\qquad \begin{aligned} 2\partial_\eta u_1 + (\partial_t+\partial_x)u_0 + u_0v_0 - w_0^2 &= 0 \\ 2\partial_\xi v_1 + (\partial_t-\partial_x)v_0 + u_0v_0 - w_0^2 &= 0 \\ \partial_\tau w_1 + \partial_t w_0 - \tfrac{1}{2}(u_0v_0-w_0^2) &= 0 \end{aligned}$$

The condition that u_1 , v_1 , w_1 have no linearly growing terms in ξ, η or τ, y gives the defining system for u_0, v_0, w_0.

$$\text{(4.8)}\qquad \begin{aligned} &(\partial_t+\partial_x)u_0 + u_0\langle v_0\rangle - \langle w_0^2\rangle = 0\\ &(\partial_t-\partial_x)v_0 + v_0\langle u_0\rangle - \langle w_0^2\rangle = 0\\ &\partial_t w_0 + \frac{1}{2}(w_0^2 - \langle u_0v_0\rangle_\tau) = 0 \end{aligned}$$

Here we have defined

$$\text{(4.9)}\quad \langle u_0v_0\rangle_\tau = \lim_{T\uparrow\infty} \frac{1}{T}\int_0^T u_0(y-\tau, t,x,\omega)\, v_0(y+\tau, t,x,\omega)\, d\tau$$

This limit exists almost everywhere and is a stationary process (in y, ω) for each x and t. This can be seen from the spectral representation for the processes u_0 and v_0 . For stationary processes that are periodic or almost periodic (hence have discrete spectrum) the function $\langle u_0v_0\rangle_\tau$ defined by (4.9)is a nontrivial random function of y as can be seen by using the Fourier series for u_0 , v_0 . For processes that have mixing properties the limit (4.9) will simplify considerably and we will have

$$\text{(4.9')}\qquad \langle u_0v_0\rangle_\tau = \langle u_0\rangle\,\langle v_0\rangle$$

For example, this will be the case when $u_0 = v_0 =$ the stationary Ornstein-Uhlenbeck process. Of course u_0 and v_0 are themselves the stationary random solutions of (4.8) with initial conditions

$$\text{(4.10)}\qquad \begin{aligned} &u_0(\xi,0,x,\omega) = U(x,\xi,\omega)\ , \qquad v_0(\eta,0,x,\omega) = V(x,\eta,\omega)\\ &w_0(y,0,x,\omega) = W(x,y,\omega) \end{aligned}$$

Thus, it is not obvious under what conditions on U, V and W the limit $\langle u_0v_0\rangle_\tau$ at t

> 0 will simplify to the form (4.9′). When the initial data are almost periodic or periodic then it will not simplify; it will be a nontrivial random function of y.

The hierarchy of equations for the moments of u_0, v_0 and w_0 is easy to write down if (4.9′) holds but in general when $\langle u_0 v_0 \rangle_\tau$ is a function of y it is complicated and not particularly illuminating. Notice that when (4.9′) holds product moments of u_0, v_0 and w_0 factor as they did for u_0 and v_0 in the Carleman equation in section 2. Of course here this is not due to compensated compactness since we do not have enough information and control on derivatives of u_0, v_0, w_0. The simplification (4.9′) is entirely due to the propagation of the mixing properties of the initial data, when that occurs. In the periodic case the hierarchy of moments is quite involved and could not be obtained by compensated compactness.

With u_0, v_0, w_0 determined by (4.8) and (4.10) the solution of (4.7) takes the form

(4.11)
$$
\begin{aligned}
u_1 &= \frac{1}{2}\int_0^\eta [w_0^2(x-\tau,t,x,\omega) - \langle w_0^2\rangle \\
&\quad - u_0(\xi,t,x,\omega)\,(v_0(u,t,x,\omega) - \langle v_0\rangle)]\,d\eta \\
v_1 &= \frac{1}{2}\int_0^\xi [w_0^2(\xi+\tau,t,x,\omega) - \langle w_0^2\rangle \\
&\quad - v_0(\eta,t,x,\omega)\,(u_0(\xi,t,x,\omega) - \langle u_0\rangle]\,d\xi \\
w_1 &= \frac{1}{2}\int_0^\tau [u_0(+y-\tau,t,x,\omega)\,v_0(y+\tau,t,x,\omega) - \langle u_0 v_0\rangle_\tau]\,d\tau
\end{aligned}
$$

The analysis in this section may be summarized by using that the solution u^ε, v^ε, w^ε of (4.1), (4.2), where the initial data is random and rapidly varying, has the form

(4.12)
$$
\begin{aligned}
u^\varepsilon(t,x,\omega) &= u_0\Big(\frac{x-t}{\varepsilon},t,x,\omega\Big) + o(1) \\
v^\varepsilon(t,x,\omega) &= v_0\Big(\frac{x+t}{\varepsilon},t,x,\omega\Big) + o(1) \\
w^\varepsilon(t,x,\omega) &= w_0\Big(\frac{x}{\varepsilon},t,x,\omega\Big) + o(1)
\end{aligned}
$$

where u_0 , v_0 and w_0 are stationary random functions of the fast argument $(x-t)/\varepsilon$, $(x+t)/\varepsilon$ or x/ε and are defined by (4.8), (4.10).

5. Concluding remarks

The analysis of sections 2 and 4 can be repeated for several other nonlinear systems such as the 3-wave equations and the equations of selfinduced transparencies which are coupled systems of wave equations, not Boltzman like equations. Of course these problems are exercises in understanding how oscillations propagate. Physical problems such as the one in [2] are very difficult because the nonlinearity involves the convective term $u\cdot\nabla u$ which has derivatives. Even the Korteweg de Vries equation with rapidly oscillating data cannot be fully analyzed at present although formal asymptotics can go quite far.

Compensated compactness is a powerful tool when the oscillations have simple structure, like in the Broadwell model, or in other nonlinear systems where one wants to show that oscillations do not develop at $t > 0$ if they are not there initially. The scope of this method is not yet fully understood.

References

1. L. Tartar, Compensated compactness and applications to partial differential equations, in Nonlinear Analysis and Mechanics: Heriot-Watt Symposium vol. IV, Research Notes in Math. vol. 39 Pitman, London, 1979,pp. 136-212.
2. D. W. McLaughlin, G. Papanicolaou and O. Pironneau, Convection of microstructure and related problems, SIAM J. Appl. Math., 1985, to appear.
3. L. Tartar, Solutions oscillantes des équations de Carleman, Séminaire Goulaouic-Meyer-Schwartz, 1980-1981, n° XII.
4. L. Tartar,Etude des oscillations dans les equations aux derivées partielles non lineaires, Springer Lecture Notes in Physics #195, Trends in applications of pure mathematics to mechanics, P. Ciarlet and M. Roseau, editors, 1984.
5. J. E. Broadwell, Shock structure in a simple discrete velocity gas, Phys. Fluids 7, 1964, 1243-1247.

LARGE SCALE OSCILLATORY INSTABILITY FOR SYSTEMS WITH TRANSLATIONAL AND GALILEAN INVARIANCES

P. Coullet
Laboratoire de Physique de la Matière Condensée
Parc Valrose, 06034 Nice
and
Observatoire de Nice, BP 139, 06003 Nice

S. Fauve
Groupe de Physique des Solides de l'Ecole Normale Supérieure
24 rue Lhommond, 75005 Paris

1. Introduction

In the past few years a considerable amount of attention has been paid to the study of cellular patterns that arise in non equilibrium systems. The concept of phasedynamics plays a central role in the stability analysis of these patterns. Actually part of these ideas trace back to wave theory, where it has been shown that, under some conditions, the nonlinear propagation of a wave could be described only through the dynamics of it slowly varying phase [1]. More recently used in the context of pattern stability theory [2][3], this concept allows to describe various large scale instabilities, as for exemple the Eckhaus and the zig-zag instabilities. In these cases the phase modes are diffusive in character and the instability is associated with a diffusion coefficient that becomes negative as a control parameter is varied.

Up to now phase theory was not able to take into account a very important large scale instability frequently observed in the convective structures ; the so called *oscillatory instability* [4]. The physical description of this instability requires a new field, the vertical vorticity, generated by a torsion of the rolls, which in turn acts as a restoring force [5]. Indeed, the very presence of this velocity field is a direct consequence, at least with stress-free vertical boundary conditions, of the

Galilean invariance of the convective system. Our main goal is to show that, for a wide class of hydrodynamical systems, *the coupling between Galilean and translational invariances* can lead to a *large scale oscillatory instability* which is described by a *propagative phase.*

2. Symmetry considerations

we consider a one-dimensional hydrodynamical system described by a scalar field $A(x,t)$ and a velocity field $B(x,t)$ that obey the following equations

$$\begin{aligned} \partial_t A &= f(A,B;\partial_x) \\ \partial_t B &= g(A,B;\partial_x) \end{aligned} \tag{1}$$

where ∂_x denotes the partial derivative with respect to x. The physical system described by (1) is assumed to be invariant under

* Space translations $x \to x + \phi$, which implies that f and g do not explicitely depend on x

$$\frac{d}{dx} f(A,B;\partial_x) = \frac{Df}{DA}\partial_x A + \frac{Df}{DB}\partial_x B \tag{2.a}$$

$$\frac{d}{dx} g(A,B;\partial_x) = \frac{Dg}{DA}\partial_x A + \frac{Dg}{DB}\partial_x B \tag{2.b}$$

where D/DA stands for $\partial/\partial A + \partial/\partial A_x \partial_x + \partial/\partial A_{xx}\partial_x^2 + ...$ where the subscripts mean partial derivatives, and $A, A_x, A_{xx}, ...$ are considered as independent variables.

* Space inversion $x \to -x$, $B \to -B$ which implies that

$$f(A,-B;-\partial_x) = f(A,B;\partial_x) \tag{3.a}$$

$$g(A,-B;-\partial_x) = -g(A,B;\partial_x) \tag{3.b}$$

* Galilean transformations $x \to x + \psi t$, $B \to B - \psi$ which implies that

$$f(A,B-\psi;\partial_x) - \psi\partial_x A = f(A,B;\partial_x) \tag{4.a}$$

$$g(A,B-\psi;\partial_x) - \psi\partial_x B = g(A,B;\partial_x) \tag{4.b}$$

We assume that equation (1) has a stationary solution $A_0(x)$, $B_0 = 0$ describing a periodic pattern. This requirement is in itself a new restriction since $A_0(x,t)$ has to obey simultaneously to the differential equations :

$$f(A_0,0;\partial_x) = 0 \tag{5.a}$$

$$g(A_0,0;\partial_x)=0 \tag{5.b}$$

Such a solution breaks the *translational and Galilean invariances*. We will show in the next sections that two phase variables are associated with these broken symmetries.

3. Translational invariance and diffusive phasedynamics.

The stability of the periodic pattern is investigated through the variational equation

$$\partial_t \begin{pmatrix} a \\ b \end{pmatrix} = L(x;\partial_x)\begin{pmatrix} a \\ b \end{pmatrix} + N(x;\partial_x;A,B) \tag{6}$$

where

$$a(x,t)=A(x,t)-A_0(x)$$
$$b(x,t)=B(x,t)$$

and where the jacobian operator $L(x;\partial_x)$ depends periodically on x

$$L(x;\partial_x)=\begin{pmatrix} Df/DA|_{A_0,0} & Df/DB|_{A_0,0} \\ DG/DA|_{A_0,0} & DG/DB|_{A_0,0} \end{pmatrix} \tag{7}$$

and N is a nonlinear operator.

We first pay attention to the linear part of equation (6) in order to study the stability of the periodic pattern. Taking the derivative with respect to x of (5.a) and using (2.a) one easily shows that the vector $\Phi(x)=\begin{pmatrix} \partial_x A_0 \\ 0 \end{pmatrix}$ is a marginal mode for equation (6) i.e.

$$L(x,\partial_x)\Phi(x)=0 \tag{8}$$

A perturbation of the form $A(x,t)=A_0(x)+\phi\partial_x A_0 \simeq A_0(x+\phi)$ is precisely the phase shift of the periodic pattern. More generally the phase branch is obtained by allowing a slow dependence for ϕ both in space and time. Technically, one look for a perturbation in the form

$$a(x,t)=\phi(X,T)\partial_x A_0+\mathcal{A}(x,\partial_X;\phi) \tag{9}$$

where $\mathcal{A}$ depends linearly on $\phi,\phi_X,\phi_{XX},...$, and X,T represent respectively slow space and time variables. The compatibility between (9) and the linear part of (6) is then got through a linear equation for ϕ

$$\partial_X=\mathcal{F}(\phi,\partial_X) \tag{10}$$

Taking into account the invariance properties (2) and (3) yields at the lowest orders

$$\partial_T\phi=\alpha\phi_{XX}+\beta\phi_{XXXX}+.... \tag{11}$$

The so-called phase branch is defined as

$$s_k = -\alpha k^2 + \beta k^4 + ... \tag{12}$$

For $k = 0$ one recovers a constant phase shift and (11) becomes

$$\partial_T \phi = 0 \tag{13}$$

If all the other eigenvalues of the operator L have negative real parts, the full non linear equations for a and b can be reduced to a nonlinear phase equation by allowing $\mathcal{A}$ and $\mathcal{F}$ to depend nonlinearly on $\phi, \phi_X, \phi_{XX}, ...$ Actually this is not the case in our problem.

4. Galilean invariance and propagative phasedynamics.

The *Galilean invariance* implies that the zero eigenvalue corresponding to the phase shift is *doubly degenerate.* Using the infinetesimal version of (4) one can easily show that the vector $\mathbf{\Psi} = \begin{pmatrix} 0 \\ -1 \end{pmatrix}$ also belongs to the kernel of L^2 since

$$L\mathbf{\Psi} = \mathbf{\Phi}(\mathbf{x}) \tag{14}$$

The physical meaning of this degeneracy is the following : a perturbation of the form $\phi\mathbf{\Phi}(\mathbf{x}) + \psi\mathbf{\Psi}$ amounts to look for a solution

$$\begin{aligned} A &= A_0(x) + \phi\partial_x A_0 \approx A_0(x+\phi) \\ B &= -\psi \end{aligned}$$

which, thanks to the Galilean invariance, is solution of (1) only if

$$\phi = \psi T \tag{16}$$

or equivalently

$$\begin{aligned} \partial_T \phi &= \psi \\ \partial_T \psi &= 0 \end{aligned}$$

Let us remark that this can be simply recovered from (8) and (14) since the restriction of L to the basis $(\mathbf{\Phi}, \mathbf{\Psi})$, is nothing else than the Jordan block

$$\begin{pmatrix} 0 & 1 \\ 0 & 0 \end{pmatrix}$$

Hence the *translational and Galilean invariances are coupled.* A linear change in time of the spatial phase induces a constant velocity and reciprocally the addition of a constant flow produces a linear variation for the phase of the periodic pattern. This is the typical resonant behavior of an oscillator at "zero frequency" [6][7]. We next show that this singular behavior disappears as soon as one considers more general perturbations for which ϕ and ψ are now allowed to vary slowly in space and time. In order to reduce (6) to a phase equation, we look for a solution in the form

$$\begin{aligned} a(x,t) &= \phi(X,T)\partial_x A_0 + \mathcal{A}(x,\partial_X,\phi,\psi) \\ b(x,t) &= -\psi(X,T) + \mathcal{B}(x,\partial_X,\phi,\psi) \end{aligned} \tag{18}$$

The compatibility between (19) and (6) will be insured by the dynamical equations for ϕ and ψ

$$\begin{aligned} \partial_T\phi &= \psi + \mathcal{F}(\phi,\psi,\partial_X) \\ \partial_T\psi &= \mathcal{G}(\phi,\psi,\partial_X) \end{aligned} \tag{19}$$

In order to determine $\mathcal{A}$,$\mathcal{B}$,$\mathcal{F}$ and $\mathcal{G}$, these quantities are expanded in powers of ϕ, ϕ_X, ..., ψ, ψ_X, ..., considered as independent variables ; at a given order (i,j) one has to solve two singular linear non homogeneous equations

$$\mathcal{L}\begin{pmatrix} \mathcal{A}_{(i,j)} \\ \mathcal{B}_{(i,j)} \end{pmatrix} = \begin{pmatrix} I_{(i,j)} \\ J_{(i,j)} \end{pmatrix} - \mathcal{F}_{(i,j)}\Phi - \mathcal{G}_{(i,j)}\Psi \tag{20}$$

where $\mathcal{R}_{(i,j)}$ represents the coefficient of $\phi^{i_0}\phi_X^{i_1}...\psi^{j_0}\psi_X^{j_1}...$ in the formal expansion of $\mathcal{R}$ which stands for $\mathcal{A}$, $\mathcal{B}$, $\mathcal{F}$ and $\mathcal{G}$; $\mathcal{L}$ is the linear operator

$$\mathcal{L} = \psi\frac{\partial}{\partial\phi} + \psi_X\frac{\partial}{\partial\phi_X} + ... - L(x;\partial_x) \tag{21}$$

and $I_{(i,j)}$ and $J_{(i,j)}$ are known quantities at this order.

As usual in singular perturbation methods, one first determines $\mathcal{F}_{(i,j)}$ and $\mathcal{G}_{(i,j)}$ with the help of solvability conditions, and then solves the equations for $\mathcal{A}_{(i,j)}$ and $\mathcal{B}_{(i,j)}$. The technique sketched here is reminiscent of the one used for the Hopf bifurcation "at zero frequency" [8]. Analitycal calculations are not easy since $L(x;\partial_x)$ is, in general, a complicated periodic linear operator. As far as the general form of the phase equation only matters, the knowledge of the structure of the kernel of $\mathcal{L}$ and the symmetry properties (2), (3) and (4) allow us to simplify (19) to a great extent.

* Thanks to the structure of $\mathcal{L}$ all the terms in $\mathcal{F}$ are "non resonant", and can be choosen equal to zero. For the same reason one can suppress the terms ψ^2, $\psi\psi_X$, ψ_X^2, $\psi\psi_{XX}$, and some other higher order terms in $\mathcal{G}$.
* Thanks to the translational invariance $\mathcal{F}$ and $\mathcal{G}$ cannot depend explicitely on ϕ.
* Thanks to the Galilean invariance $\mathcal{F}$ and $\mathcal{G}$ cannot depend explicitely on ψ.
* Thanks to the space reflection invariance

$$\begin{aligned} \mathcal{F}(-\phi,-\psi;-\partial_X) &= -\mathcal{F}(\phi,\psi;\partial_X) \\ \mathcal{G}(-\phi,-\psi;-\partial_X) &= -\mathcal{G}(\phi,\psi;\partial_X) \end{aligned} \tag{22}$$

Using these arguments, (19) yields at the lowest orders

$$\frac{\partial^2}{\partial T^2}\phi = \alpha\phi_{XX} + \beta\frac{\partial}{\partial T}\phi_{XX} + \gamma\phi_{XXXX} + \delta\frac{\partial}{\partial T}\phi_{XXXX} + g\phi_X\phi_{XX} + g'(\phi_X\frac{\partial}{\partial T}\phi_X)_X \tag{23}$$

the divergent form of the right hand side of (23) being clearly connected with the conservation of the spatial average of ψ.

Equation (23) governs the long wavelength perturbations of the periodic pattern. If $\alpha > 0$ and $\beta > 0$, we get a pair of damped propagative modes with a propagation speed $\sqrt{\alpha}$. A stationary instability exists for $\alpha = 0$, and a Hopf bifurcation occurs for $\beta = 0$. The effect of the nonlinear terms is such that the stationary instability is subcritical whereas the oscillatory instability can be supercritical if $gg' < 0$.

5. Conclusion.

We have shown that the coupling between *translational and Galilean invariances* could lead to a *large scale oscillatory instability* which is universally described, for one-dimensional systems, by a nonlinear second order in time phase equation. The same type of conclusion is true if one considers two-dimensional perturbations of unidimensional periodic patterns, although the phase equation is somewhat different; in that case two kinds of oscillatory instability can occur; one, longitudinal to the periodic pattern, corresponds to the instability discussed here; the other, tranverse, induces an oscillatory torsion of the pattern; because of the incompressibility condition only the second type of instability occurs in Rayleigh-Benard convection; it is the well known oscillatory instability of the roll structure.

Acknowledgements : this work has been partly supported by the CNRS (ATP "Dynamique des fluides géophysiques et astrophysiques") and the CPAI. We have benefitted from discussions with , M.E. Brachet and L. Tuckermann.

References

1. L.N. Howard and N. Kopell, Studies in Applied Math. 56, 95-145 (1977).

2. Y. Pomeau and P. Manneville, J. Phys. Lettres 40, 609-612 (1979).

3. Y. Kuramoto, Prog. Theor. Phys. 71, 1182-1196 (1984).

4. F.H. Busse, J. Fluid Mech. 52, 97-112 (1972).

5. E.D. Siggia and A. Zippelius, Phys. Rev. Lett. 47, 835-838 (1981).

6. V.I. Arnold, Geometrical Methods in the Theory of Ordinary Differential Equations, Springer Verlag (1977).

7. J. Guckenheimer and P. Holmes, Non Linear Oscillations, Dynamical Systems, and Bifurcations of Vector Fields, Springer Verlag (1984).

8. P. Coullet and E.A. Spiegel, SIAM J. Appl. Math. 43, 775-821 (1983).

Note added in proof: An example of elastic oscillatory behavior for the case of Kuramoto-Shivashinski equation may be found in the paper by Z.S. SHE, U. FRISCH, O. THUAL: These proceedings, p.1.

THE KURAMOTO-SIVASHINSKY EQUATION : A CARICATURE OF HYDRODYNAMIC TURBULENCE ?**

by

Y. Pomeau*+, S. Zaleski+

*Service de Physique Théorique
CEN-Saclay
91191 Gif-sur Yvette Cedex, France

+Laboratoire de Physique de l'ENS, GPS
24 rue Lhomond, Paris 75005
France

One of the central problems in theoretical studies of hydrodynamic turbulence is the difficulty of computer simulations of the Navier-Stokes equations in "realistic" situations with a large Reynolds number and, say, rigid boundary conditions. So it is still worth pursuing other ways of research, at least until the (possible ?) advent of computational capabilities allowing "easy" simulations of the real equations.

This way has been followed, in particular in order to gain a better understanding of the dynamical process at work in extended turbulent systems. We tried, for instance, to check numerically the idea of cascading transfer of energy and to enumerate the number of degrees of freedom with respect to the volume of the system. It turns out to be possible to do many very detailed numerical studies for the Kuramoto-Sivashinsky equation (KSE) in one space dimension.

This KSE is truly remarkable, because it may have a spontaneously turbulent behavior with many degrees of freedom, and has some of the qualitative properties of the Navier-Stokes equations. In what follows, we shall first present some "elementary" properties of this KSE, as well as its physical meaning. Then we shall recall some facts discovered in numerical studies, some of them being rather puzzling and unexplained as well. Finally, we shall look at the long term fluctuations of the conserved quantities, with "rigid" boundary conditions.

One possible version of the KSE is the following one :

$$\varphi_t + \varphi\varphi_x + \varphi_{xx} + \varphi_{xxxx} = 0 \tag{1}$$

$(\varphi_t = \frac{\partial\varphi}{\partial t}, \varphi_x = \frac{\partial\varphi}{\partial x}, \ldots)$.

It is formally the 1d version and x-derivative ($\varphi = \psi_x$) of

---

** This summarizes two talks presented at the workshop "Modélisation macroscopique des écoulements turbulents", Nice, Décembre 10-14 (1984).

$$\psi_t + \frac{1}{2}(\vec{\nabla}\psi)^2 + \Delta\psi + \Delta^2\psi = 0 \qquad (1')$$

that is -actually- the dimensionless form of the equations derived by Kuramoto[1] and Sivashinsky[2].

It is now well known that this equation was derived by Kuramoto for describing the non-linear stage of evolution of the phase instability of a string of weakly connected identical self oscillating systems distributed homogeneously. The function ψ is the local time phase of the oscillators. Sivashinsky proposed equation (1') for describing in 2d the non linear evolution of the Landau-Darrieus instability of the flame fronts. The boundary condition for (1) can be either the periodicity in space : $\varphi(x) = \varphi(x+L)$, L fixed or the Neumann and Dirichlet (N+D) conditions $\varphi = \varphi_x = 0\big|_{x=0,L}$.

A simple property of (1) is the linear instability of long wavelength perturbations around the steady solution $\varphi = 0$. For an infinite length L, the time dependence of those linear periodic perturbations is $\delta\varphi_q(x,t) = a_q \sin(qx + \alpha_o)\, e^{(q^2-q^4)t}$, so it is unstable for any wavenumber q such that $\sigma(q) \equiv q^2 - q^4$ is positive, that is for $q > 1$. Note also that the fluctuations with very large wavenumbers are linearly stable*.

By integrating (1) over x, one derives the following conservation relation :

$$Q_t^{[1,2]} + j(x_1,t) - j(x_2,t) = 0 \qquad (2)$$

where

$$j(x_i,t) = \frac{\varphi^2}{2} + \varphi_x + \varphi_{xxx}\Big|_{x=x_i}$$

and

$$Q^{[1,2]}(t) = \int_{x_1}^{x_2} dx\, \varphi(x,t) \quad .$$

With the N+D b.c., the boundary value of j is

$$j(x,t)\big|_\partial = \varphi_{xxx}\big|_\partial$$

---

*The connection between the growth of short wavelength perturbations and well posedness and/or smooth evolution may lead to rather subtle mathematical questions, especially if one wants to include the effect of non linearities. The famous conjecture by Birkhoff for the Kelvin Helmholtz instability claims that, if $\sigma(q)$ grows like q at large q, then analytical initial data yield a smooth evolution over a finite time interval only. This has been proved to be true[3] for a Mullins-Sekerka like instability when nonlinearities are present, by a deep analysis. However, at least from the point of view of the "linear-like" analysis, C^∞ but not C^ω initial data may yield a catastrophy after a finite time for any $\sigma(q)$ growing at large q, so that it appears reasonable to impose the damping of fluctuations with a large wavenumber, as in the KSE, to get well posed problems. As emphasized below, the mathematical well posedness of the KSE has been studied in depth[5,6], important and non trivial results have been obtained for periodic b.c.

Below, we shall denote as Q (without supercript) the invariant Q defined by taking the total length, i.e. $x_1 = 0$ and $x_2 = L$. From eq. (2), Q can be considered as a conserved quantity in the usual sense : its variation depends on boundary fluxes only. It may be thought as similar to the energy of a fluid layer in the Rayleigh-Bénard instability or as the fluid angular momentum in the Taylor-Couette instability. It seems likely that Q is the only non trivial conserved quantity made of algebraic combination of φ and its derivatives, Gervois proved this recently[4] for the 1d case, by a purely algebraic method.

A number of mathematical properties of (1)/(1') have been shown recently. This started with the work of Aimar and Pesnel[5], and has culminated with the article by Scheurer et al[6]. From the point de view of the "physicist" (that would be probably considered as an over simplification by a mathematician !), one can say that, at least for periodic b.c., the behavior of the KSE is reasonably well understood : it is known to have smooth and bounded solutions and there exists too a L-dependent upper bound for its number of degrees of freedom, this number being defined by counting the largest Lyapunov exponents. However, the mathematical results for the *rigid* (N+D) b.c. are much less abundant.

Indeed, the KSE is interesting because it has the rare property of having solutions with a spontaneously turbulent behavior, as soon as the length L becomes large enough[1,2]. For sometime, it was believed that this was quite independent of the b.c. (periodic or N+D), but more recent "experiments" show that for periodic b.c., the chaotic behavior is only transient : a final periodic steady state is reached. This kind of steady solution was already found by Aimar and Pesnel[5]. For rigid b.c. on the contrary, the chaotic behavior seems to be permanent at least on time scales needed for leaving the chaotic state with periodic b.c. . This striking sensitivity of the long term behavior to the details of the b.c. was discovered by Paul Manneville[10]. It shows -if there were any need for this- how far we are from understanding even simple properties of those chaotic systems.

We have investigated some simple statistical properties of KSE particularly in the large L limit. As said before, it seems that, for N+D b.c., the chaotic behavior is self sustained, and appears for "almost all" initial data. However, one must be very careful with this sort of claim, owing in particular to the lack of well defined and physically meaningful measure in the (huge) space of all possible initial data. Once a permanent chaotic state has settled, one may try to analyze the fluctuations of $\varphi(x,t)$ as, say, the velocity fluctuations of a real turbulent flow. The most obvious quantities to measure are the Fourier spectra of those fluctuations, both in space and time. The space spectrum is flat near $q = 0$ (q = wavenumber in the Fourier space), it has a sharp maximum near the most "unstable" wavenumber, i.e. the one such that the logarithmic increment $\sigma(q) = q^2 - q^4$ is maximum ($q \approx 1/\sqrt{2}$), at wavenumbers slightly larger than this maximum the spectrum decays approximately as k^{-4}, and finally as an exponential for still higher wavenumbers (see communication by Paul Manneville at this conference). It is difficult to draw any conclusion from those broad features, concerning the connection of the KSE with the general problem of fully developed turbulence. Certainly, an important difference with the situation of real turbulence is the existence of an intrinsic length in the KSE. This length is obtained by comparing the second and fourth order space derivatives in (1) or (1'). It has been set to 1 in our writing of the KSE. On the contrary, for fully developed turbulence, the length scale is much less obvious : the Kolmogoroff length scale depends in a highly non trivial way on the macroscopic scale and on the Reynolds number. This Kolmogoroff length scale is connected to highly *non linear* phenomena, contrary to what happens to the internal length scale in the KSE. This existence of an intrinsic length scale is likely what explains one of the simplest results concerning the KSE in the chaotic regime : the number of degrees of freedom increases linearly with L. In short, they are as many degrees of freedom as one may insert non overlapping segments of width of the order of the internal length scale. To enumerate those degrees of freedom, we have followed a rather straightforward method : we simply counted the number of positive Lyapunov exponents. Using rather standard algorithm developed for statistical mechanics on stripes, it is possible to get up to a few hundreds of such exponents. And it turns out that, at large L, this number grows as L with a good accuracy. Furthermore, the repartition of the positive exponents tends to an apparently smooth and well defined distribution, without any outstanding feature.(See Paul Manneville's lecture in this conference for more details on this).

Thus, the main qualitative conclusion of this is the fact that the chaotic dynamics of the KSE is dominated by "events" with a well defined space scale. This has various consequences : those "events" are responsible of the energy transfer from the unstable to the stable modes. And they are indications[7] that this transfer does not occur through a cascade, where each step is an interaction between structures of similar size. It seems that, for the KSE, the energy transfer occurs in some sense in "one step" only, although it is not obvious to devise an accurate and convincing way of testing the idea of Onsager-Kolmogoroff cascade in numerical computations. This absence of energy cascade in the usual sense can be seen in the absence of short scale intermittency : Alain Pumir[8] by looking at the statistics of the higher space derivatives of $\varphi(x,t)$ has not seen any evidence of increase of the non gaussianity with the derivation order. Thus, from this point of view, the KSE is certainly quite different from what is observed in real turbulent flows[9].

Nevertheless, not all statistical properties of the KSE are so easy to understand. Motivated by a previous work of Manneville[10] on low frequency noise of the conserved quantity $Q(t)$ defined before, we looked at its statistical properties. Recall that

$$Q(t) = \int_0^L \varphi(x,t)dx \quad ,$$

where $\varphi(x,t)$ is the solution of (1) with the N+D b.c. and that (with these b.c.)

$$\frac{dQ}{dt}(t) = j'(L,t) - j'(0,t)$$

with

$$j'(L/0,t) = -\varphi_{xxx}\Big|_{x=L/0} .$$

Very loosely speaking, one can think of Q(t) as the fluctuation of the number of particles in a box drawn in a Lorentz gaz at thèrmal equilibrium, the walls of this box being either a geometrical fiction or permeable to the gas particles. This fluctuation has two very simple elementary properties :

(1) its mean square value $(\overline{Q^2})^{1/2}$ is of the order of the volume of the system,

(2) its dynamics is governed by a diffusion process (according to the Onsager principle of regression of equilibrium fluctuations). This yields an autocorrelation function

$$\overline{Q(0)\ Q(t)} \simeq \overline{Q^2}\ e^{-\tilde{D}At} ,$$

where A is the lowest eigenvalue of the Laplacian in the box (with the Dirichlet b.c.) and D is the diffusion coefficient. Indeed $A \sim L^{-2}$, where L is a typical box size. Whence one expects from this analogy and after a few simple transformations that

$$\chi(\tau) \equiv \overline{[Q(t+\tau) - Q(t)]^2} \sim \overline{Q^2}\ \tilde{D}A\tau \tag{3}$$

for $\tau << (\tilde{D}A)^{-1}$, (where the average is a gliding average over t), and that $\chi(\tau) \underset{\tau\to\infty}{\to} 2\overline{Q^2}$ (Q(t+τ) and Q(t) become independent variables at large time separation). The law given in (3) can be seen as a simple consequence of the fact that, for short times, the changes in Q are due to the random addition or subtraction of contributions arising from the boundary current, that the mean square derivation of Q increases as τ , because of the law of large number and the (presumed) absence of correlation in the fluctuations of the boundary current for long time differences.

We investigated numerically the KSE by looking for the statistical properties of Q, with N+D b.c. In what follows, an overbar will denote a time average (obtained numerically), assumed to be also a probabilistic average with a well defined invariant measure. As explained before, we looked at the time dependent variance $\chi(\tau)$ defined in (3)

$$\chi(\tau) = \overline{(Q(t+\tau) - Q(t))^2} ,$$

The above mentioned ergodic assumption implies that χ is a function of τ (and not of t).

The standard expectation for $\chi(\tau)$ would be

$$\chi(\tau) \underset{1 \ll \tau \ll L^{-2}}{\simeq} 2D\tau$$

with

$$D = \int_0^\infty d\tau \, \overline{J(0)\ J(\tau)}$$

and

$$J(t) = j(L,t) - j(0,t)$$

We plotted on Figs. 1 and 2 the records of $\chi(\tau)$ and $\overline{J(0)J(\tau)}$. Both figures are drawn from a numerical integration of (1) with N+D b.c. and L = 200. On fig. 1, χ is displayed as a function of τ . For $T_1 < \tau < T_2$, where T_1 and T_2 can be roughly estimated from Fig.1, there is an anomalous scaling of χ, that is :

$$\chi(\tau) \sim C\tau^{\mu} \quad ,$$

with $\mu \simeq 0.55 \pm 0.05$. Above T_2, χ saturates, although it becomes difficult to have accurate data, due to poor statistics for such long times. The saturation level, $\chi(\infty)$, has been estimated to be roughly of order L^2, so that T_2 would be of order L^4. Those results, if confirmed, would point to an anomalous diffusion process in real space.

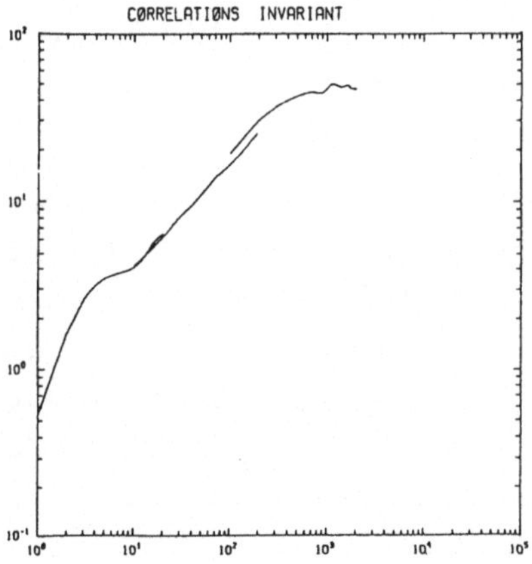

Fig.1. The function $(Q(t+\tau) - Q(t))^2$ is represented as a function of the time τ. The three branches correspond to averaging on time windows of various scales, the larger being of size $T=1.1\ 10^5$. The discrepancy between the three branches in the overlapping regions gives a rough estimate of the averaging error. The length L of the space interval was 200. The saturation effect at large times τ is more clearly visible for experiments with smaller values of L.

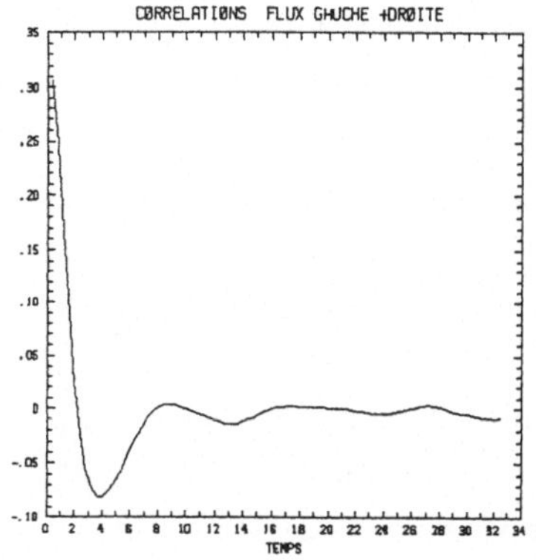

Fig.2. Correlations of the total flux, i.e. $\overline{J(0)J(\tau)}$ are represented as a function of τ. The size of the window was $1.5\ 10^3$. The remaining fluctuations on the right of the curve are due to averaging errors.

The short time scale T_1 can be understood from the analysis of $\overline{J(0)J(t)}$ as a function of time. It oscillates on a short time scale of order 4. This time is independent of L. The damping of $\overline{J(0)J(t)}$ occurs for $t \simeq T_1$, although it is again difficult to extract its mean value out of the noise. If, as the computation suggests, $\int_0^t \overline{J(0)J(\tau)}\, d\tau$ is near zero for $t \gtrsim T_1$, the anomalous behavior would naturally start for time scales above T_1. To check the statistical independency of the two boundaries we examined independently $j(0,\tau)$ and $j(L,\tau)$. Actually, one can distinguish a fluctuating part $\tilde{j}$ and the mean part $\bar{j}$

$$j(x,\tau) = \tilde{j}(x,t) + \bar{j}(x) \quad .$$

We found

$$\bar{j}(0) = \bar{j}(L) \simeq 0.4$$

and

$$\overline{\tilde{j}(0,t)\ \tilde{j}(L,t)} \simeq 0$$

There is a continuous "current" from one side to the other, a fact to be connected to the advection term $\varphi\varphi_x$ of (1). But the fluctuations on each side are not correlated. Further, $\overline{\tilde{j}(0,t)\ \tilde{j}(0,0)}$ and $\overline{J(0)\ J(t)}$ have a similar shape.

It is difficult to give any theoretical interpretation of these results without more computational data on similar systems and related quantities.[11] However, a diffusion process in the real space would point to the φ_{xxxx} term of (1). Another tentatively related phenomena would be some intermittency of the motion on the attractor perhaps related to the problems of random walk on fractals.

The integrations were made using a standard Adams-Bashforth Crank-Nicholson scheme. At the boundaries special care was taken to insure a representation of (1) with N+D b.c. as accurate as in the bulk, i.e. with $\mathcal{O}(h^2)$ errors only. The results were checked with reduction of time and space steps. Computations were realized on the CRAY-1 of the CCVR, Palaiseau, France, and on a CRAY of the Los Alamos National Laboratory. 80 minutes of CPU were necessary to get the result of figure 1. Part of this work was done while the authors were visiting the Center for Non Linear Studies at Los Alamos, which we acknowledge for their support.

REFERENCES

[1] Y. Kuramoto, "*Chemical oscillations, waves and turbulence*" Springer Verlag Berlin, (1984).

[2] G.I. Sivashinsky, Act. Astronautica 4, 1177 (1977) ; 6, 659 (1979).

[3] B. Shraiman, D. Bensimon ; Phys. Rev. A30, 2840 (1984).

[4] A. Gervois, private communication

[5] M.T. Aimar, Thèse 3ème Cycle, Univ. Provence, Marseille, (1982).

[6] B. Nikolaenko, B. Scheurer, R. Teman; to appear in Physica D.

[7] A. Pumir, Y. Pomeau, P. Pelcé ; J. of Stat. Phys. 37, 39 (1984).

[8] A. Pumir, to appear in Phys. Rev. A.

[9] See, for instance, A.J. Monin, A.M. Yaglom *"Statistical Fluid Mechanics"* M.I.T. Press (1972).

[10] Communication at this Conference.

[11] If the diffusion coefficient $D = \int_0^\infty \overline{J(0)J(t)}dt$ turns out to vanish, it may happens that the mean fourth power $\overline{[Q(t)]^4}$ becomes of order D't at large times. Usually this fourth power is dominated by a term as $3D^2t^2$. The coefficient D' is a triple time integral of the four time correlation of J(.). In that case one might thus expect a similar growth of $\overline{Q^2}$ as $t^{1/2}$ (instead of the more usual t), which would agree with our exponent $0.55 \pm .05$.

Computation of a dimension for a model of fully developed turbulence

R.Grappin[1],A.Pouquet[2],J.Leorat[1]
(1) Observatoire de Meudon F-92190 Meudon
(2) Observatoire de Nice BP252 F-06007 Nice

Abstract

One computes the Lyapounov dimension of a model of fully developed turbulence studied in ref.(1,2). Non-linear interactions act between nearest neighbours in a discretized wavenumbers space and conservation properties are verified. Equations are of the form:

$$dX_n/dt = k_n \Sigma A_{ij} X_i X_j - \nu k_n^2 X_n + \delta_{n1}$$

where A_{ij} are coupling constants and i,j =n-1,n, or n+1, and n goes from 1 to N. The wavenumbers k_n are discretized in a geometric way: k_{n+1}/k_n=const; X_n are scalar velocity or magnetic field amplitudes, ν is the diffusivity (kinetic or magnetic) and δ_{n1} is a forcing term acting only on the first kinetic mode. This model exhibits time fluctuations at all scales, and time-averaged power-law spectra, as well as intermittency at small (dissipative) scales. We have studied here the case where the maximum over minimum wavenumber is 256, varying the dissipation. We found that the maximal Lyapounov exponent is scaled by the inverse of the turn-over time of the smallest scale in the inertial range. The Lyapounov dimension is approximatively given by the total number of modes which lie in the inertial range. (We found in particular no indication of saturation of dimension with the Reynolds number).
Figure 1 below (left) shows the kinetic and magnetic energy spectra averaged over about $5\ 10^5$ time steps (T=1024). One sees the growth of the inertial range when viscosity is reduced. Figure 2 (right) gives the Lyapounov exponents; Table below gives resulting Lyapounov dimensions.

viscosity	10^{-2}	10^{-3}	10^{-4}
number of modes in inertial range	7	10	16
Lyapounov dimension	5.9	10.7	15.5
Maximal Lyapounov exponent	0.3	1.1	3.4

We conjecture that these results hold also for real fully developed 3-dimensional Navier-Stokes turbulence:
1) the maximum Lyapounov number scales as $Re^{1/2}$;
2) the Lyapounov dimension scales as $Re^{9/4}$.

Note that, at a given resolution, there exists a state of maximal Reynolds (minimum viscosity) compatible with the formation of a power-law energy spectrum: its Lyapounov dimension will be nearly equal to the total number of degree of freedom of the system. There exists also a (statistical) equilibrium state with Lyapounov dimension exactly equal to the number of degrees of freedom, namely the Gibbs thermal equilibrium obtained by taking a zero visosity (leading to equipartition of energy between all modes). Thus the concept of dimension, which has proved to be an interesting tool in the study of transition to turbulence, does not seem to be so useful for developped turbulence, being unable to distinguish between so fundamentally different states as fully developped turbulence and Gibbs thermal equilibria.

References

1. C.Gloaguen Thèse de 3è cycle Université Paris7, 1983
2. C.Gloaguen,J.Léorat,A.Pouquet,R.Grappin, A scalar model for MHD turbulence, submitted to PhysicaD

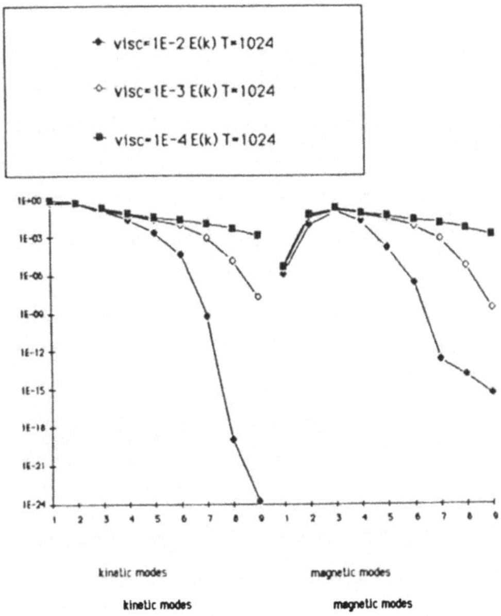

Figure1:
Kinetic and Magnetic spectra
abcissa: $\log_2(k)$, where k is the wavenumber.

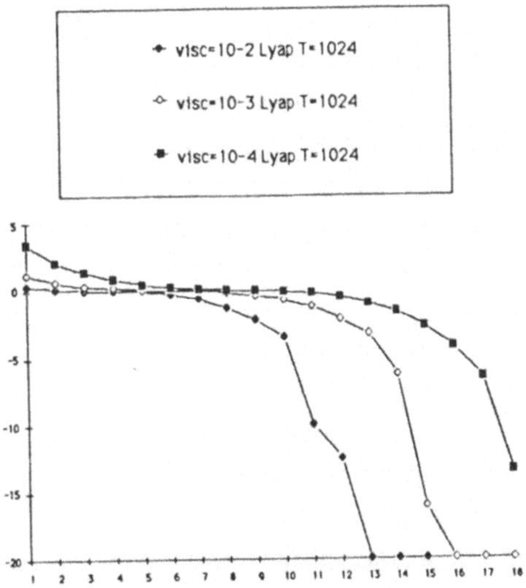

Figure 2:
Lyapounov exponents
Note that the three last exponents for the two highest viscosities are actually lower than -20.

Pattern Formation by Particles Settling in Viscous Flows

Leonard A. Smith
Goddard Institute for Space Studies, NASA, New York 10027 and
Physics Department, Columbia University, New York 10027

E. A. Spiegel
Astronomy Department, Columbia University, New York 10027

1. Introduction

The discovery of an increasing number of fractal objects in the natural environment prompts the present study of the settling of particles in a moving fluid. We describe here simulations of the motion of small objects in two-dimensional cellular flows. One of our aims is to see how an initial distribution of such particles deforms in time. We assume throughout that the particles do not interact with each other or affect the fluid, whose motion is two-dimensional. Even without sedimentation effects, and without many other complications that real fluids provide, intricate structures in the swarming dust arise through their chaotic motions.

It is not hard to construct incompressible flows in three dimensions with chaotic Lagrangian orbits (Arter, 1983). In two-dimensional time-dependent flows with open steamlines, the fluid particles can have chaotic orbits, as in Aref's (1984) explorations. Even when the fluid streamlines are closed, the motion of particles moving in the fluid may be chaotic. Our preliminary results indicate that chaotic motion of swarms of noninteracting particles in cellular flows may produce fractal structures whose dimension seems insensitive to the control parameters of the flow.

We consider small bodies immersed in a fluid with stream function $\psi(x,y,t)$. With the neglect of hydrodynamic mass, the nondimensional equations are

$$\ddot{x} = -\mu(\dot{x} - \psi_y) \qquad (1.1a)$$

$$\ddot{y} = -\gamma - \mu(\dot{y} + \psi_x) \qquad (1.1b)$$

where μ and γ are constants. The structures seen in this problem are interesting, as a preprint of Maxey and Corrsin shows. We here consider the reduced problem of very viscous flows and neglect inertial terms. This limit has geological interest (Huppert, 1984).

If particle acceleration is negligible,

$$\dot{x} = \phi_y \;; \qquad \dot{y} = -\phi_x \tag{1.2a,b}$$

where the stream function for particle motion is

$$\phi = \gamma x + \psi. \tag{1.3}$$

These reduced equations are equivalent to a Hamiltonian system with one degree of freedom and a time dependent Hamiltonian equal to ϕ.

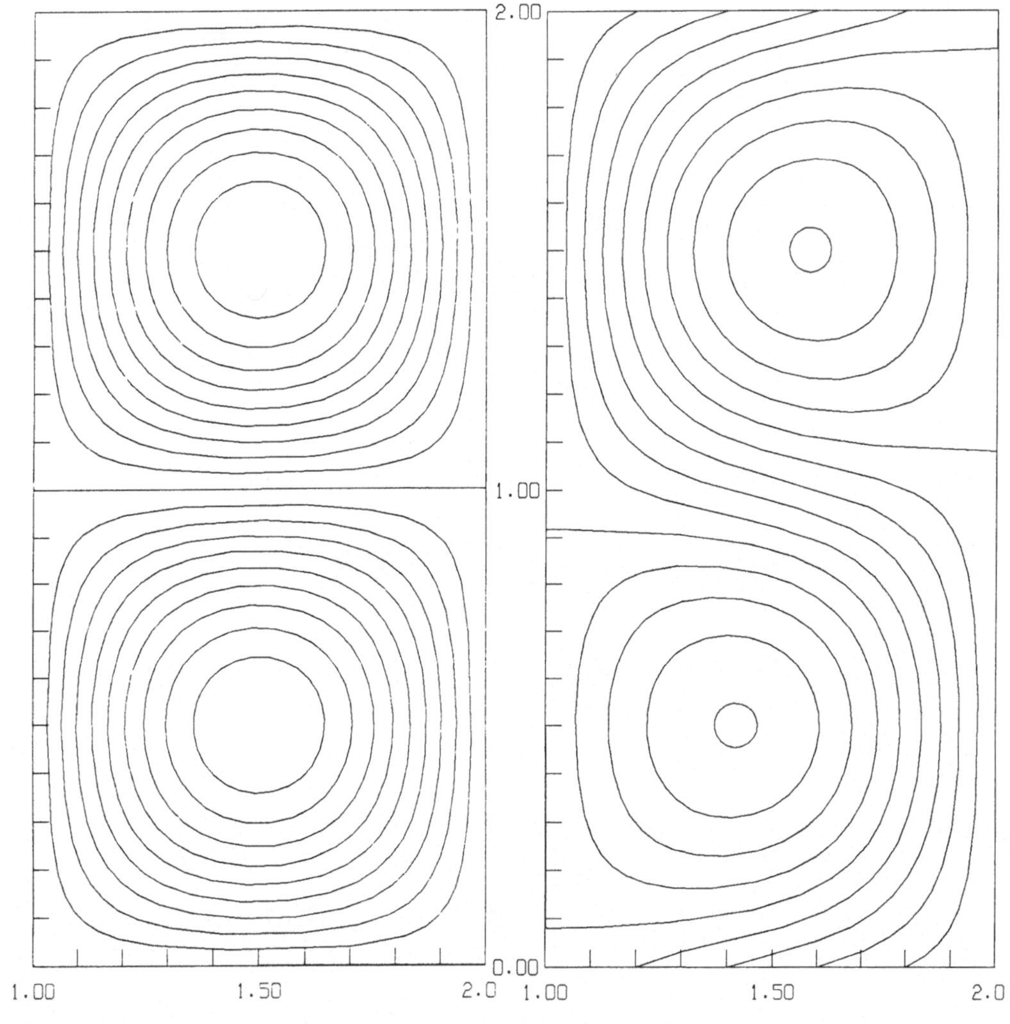

Fig. 2.1. (a) Fluid streamlines (ψ) according to (2.1). (b) Particle streamlines (ϕ) according to (1.3) and (2.1). For $A=1$ and $\gamma=0.25$.

2. Steady Flow

Stommel (1949) considered the motion of particles for steady ψ with closed streamlines. His results are illustrated in Fig. 2.1 which shows the situation for the case

$$\psi = \frac{A}{\pi} \sin \pi x \sin \pi y. \tag{2.1}$$

where A is a constant. In Fig. 2.1a we show the streamlines of ψ and in Fig. 2.1b the streamlines of ϕ. Experiments suggested by Stommel's results have been reported by Toobey, Wick and Isaacs (1977).

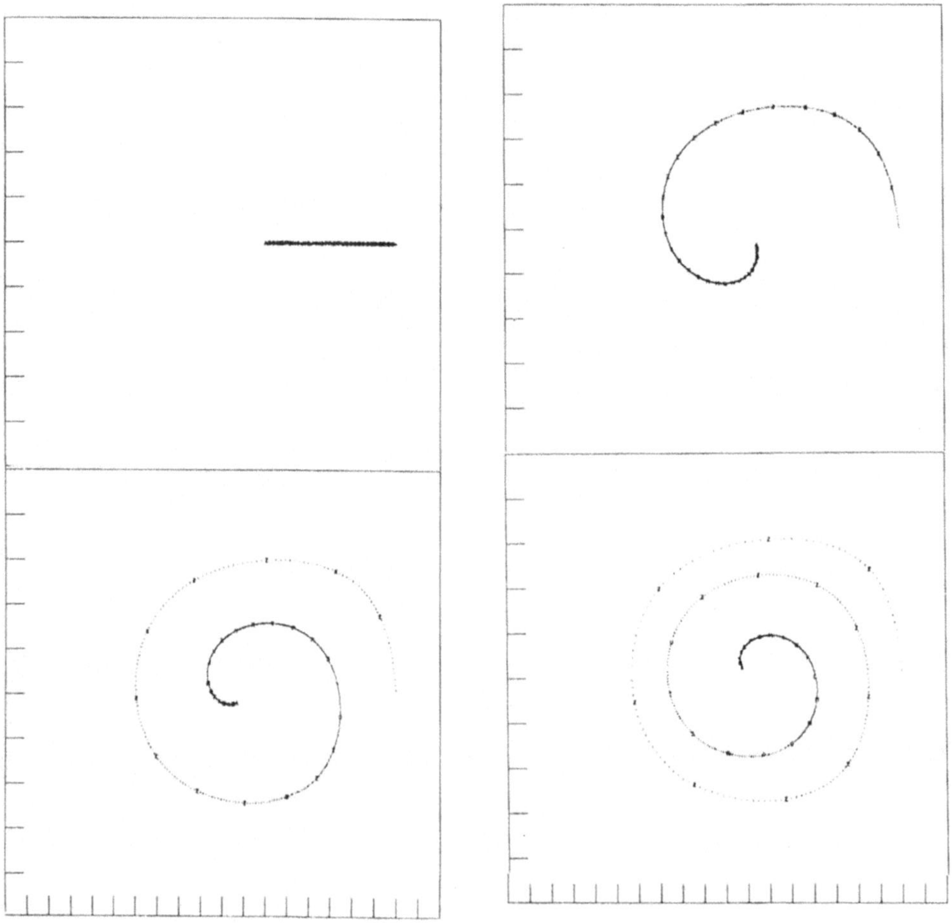

Fig. 2.2. The evolution of a line of particles with the stream function of Fig. 2.1. The distributions are for t = (a) 0 (b) 18 (c) 35 (d) 54 and the unit of length marked off is 0.1 in the vertical.

The problem with steady ψ has interesting consequences which may be taken over from a stellar dynamical study by Quinn (1984). A sheet of particles introduced into a fluid layer one cell deep is rolled up by the flow (2.1) as in Fig. 2.2. The density of the particles, when projected onto a horizontal plane, will evolve as in the upper panels of Fig. 2.3. If the particles are visible in the fluid, they may look like the lower panels of this figure. The well-defined structures in the projected density of the dust is a familiar phenomenon to those who have watched vortex rollup in suitably dyed fluids. Quinn calls the process phase wrapping (since y is momentum in his case). Stommel motivated his original study with a discussion of patterns formed in the sea. The results of Quinn may bear on such questions as dune formation in shallow water along beaches, though the inertial effects may be signifigant in such cases.

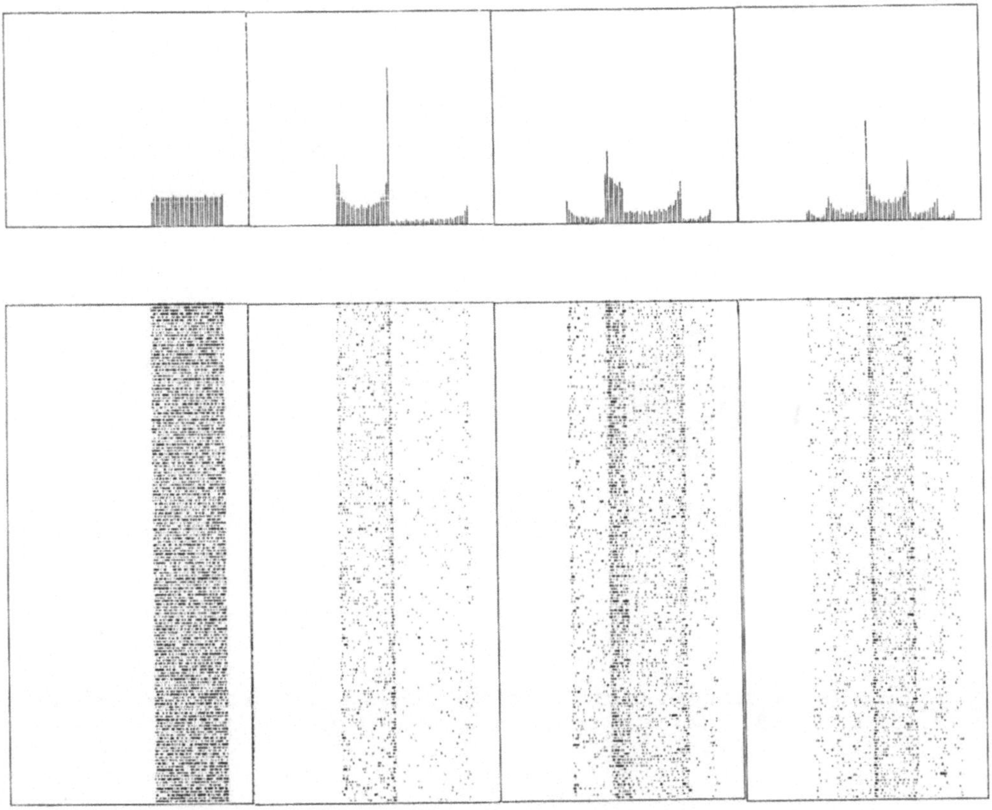

Fig. 2.3. The particle densities integrated (in y) for the four states of Fig. 2.2. The upper panels show the particle number explicitly and the lower panels are 'dust plots' simulating the appearance of the convected dust.

3. Dusty Chaos

Stommel (1949) studied particle trajectories in the steady cellular flow of Fig. 2.1a. He found regions of retention where particles are trapped indefinitely. Particles are trapped when the stream function of the fluid has an oscillatory dependence on time as well. Let

$$A = 1 + \varepsilon\cos(\omega t). \qquad (3.1)$$

We find that particles remain suspended in the fluid even when $\varepsilon \sim 10$, for $\omega \sim 1$. Results of this kind are easy to obtain numerically and surfaces of section of spatial orbits are easily drawn.

In Fig. 3.1 we illustrate the motion of a particle for the parameters indicated in the caption. In the first panel we show an orbit in the x-y plane. In 3.1b we show a stroboscopic view of the same orbit, that is, the x and y coordinates of the particle at the succession of times $t=0,P,2P,\ldots$, where $P=2\pi/\omega$.

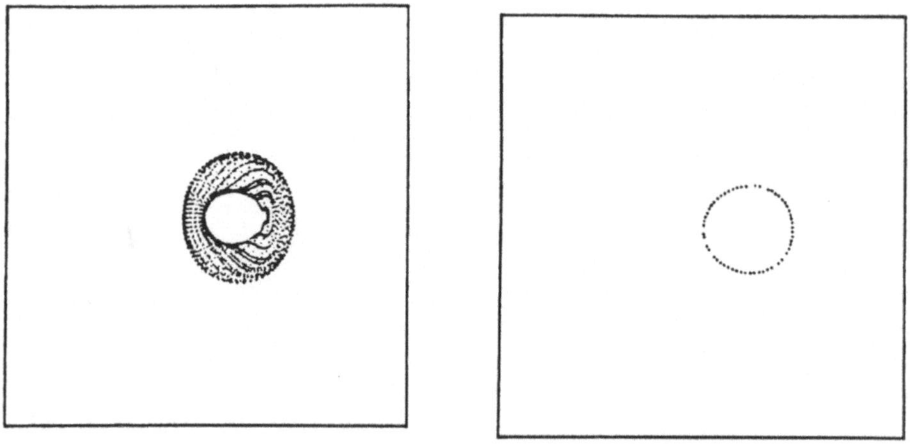

Fig. 3.1. (a) A trapped orbit in the time-dependent case for $\varepsilon=0.5$, $\gamma=0.25$, $\omega=\pi/2.2$. (b) The orbit seen stroboscopically.

Fig. 3.2 gives a corresponding pair of plots, for different values of the parameters, showing islands. The system displays the textbook behavior of Hamiltonian chaos (Lichtenberg and Lieberman, 1983) but we are more interested in the behavior of swarms of particles.

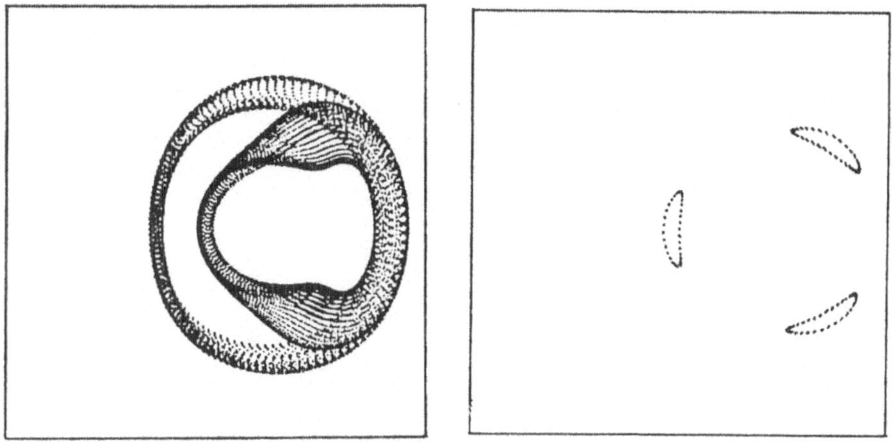

Fig. 3.2. Same as 3.1 but with $\omega = \pi/2.25$.

Fig. 3.3a shows stroboscopic views of the orbits of 32 particles that were uniformly distributed on the line $y = 3/2$ at $t = 0$. The largest of the concentric closed tori in the upper right outlines a region of particle retention. The particles shown outside this region fall to the bottom boundary of the cell where they are dealt with by the periodic boundary conditions. In 3.3a, the blank region surrounding the region of retention corresponds to chaotic motion. The region of chaotic motion is shown in Fig. 3.3b for a long run following four particles. The chaotically falling particles spend some time entrained in individual cells. The large lacuna on the lower left in 3.3a is another region of particle trapping, into which no particles can enter from the outside. Fig. 3.4 is similar to 3.3 but with an increase in ε to show the development of the region of chaotic fallout; the intial conditions have been changed to emphasize the islands.

The two large islands embedded in the chaotic region of Fig. 3.4a contain particles that remain in the initial cell. Their orbits resonate with the fluid oscillations. The fluid velocity is maximum whenever they are at their extreme y-values. A higher order resonance produces the island chain forming within these large islands.

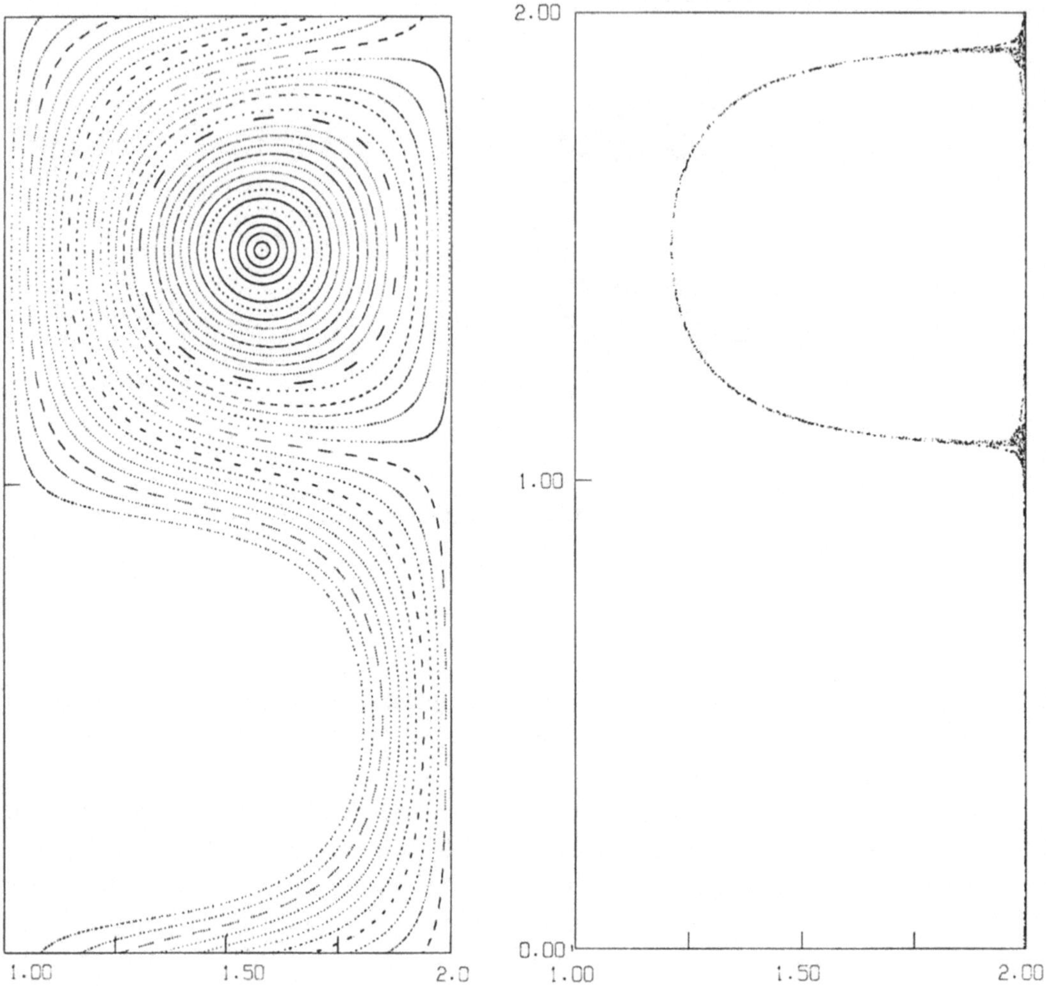

Fig. 3.3. Orbits with falling particles subjected to periodic boundary conditions and shown stroboscopically for $\varepsilon = 0.01$, $\gamma = 0.25$, $\omega = \pi/2.25$. (a) For thirty-two particles starting out uniformly on $y = 3/2$ and followed for 300 oscillation periods of the fluid flow. (b) For four particles starting out on $y = 2$, so as to be in the chaotic region, and followed for 500 periods.

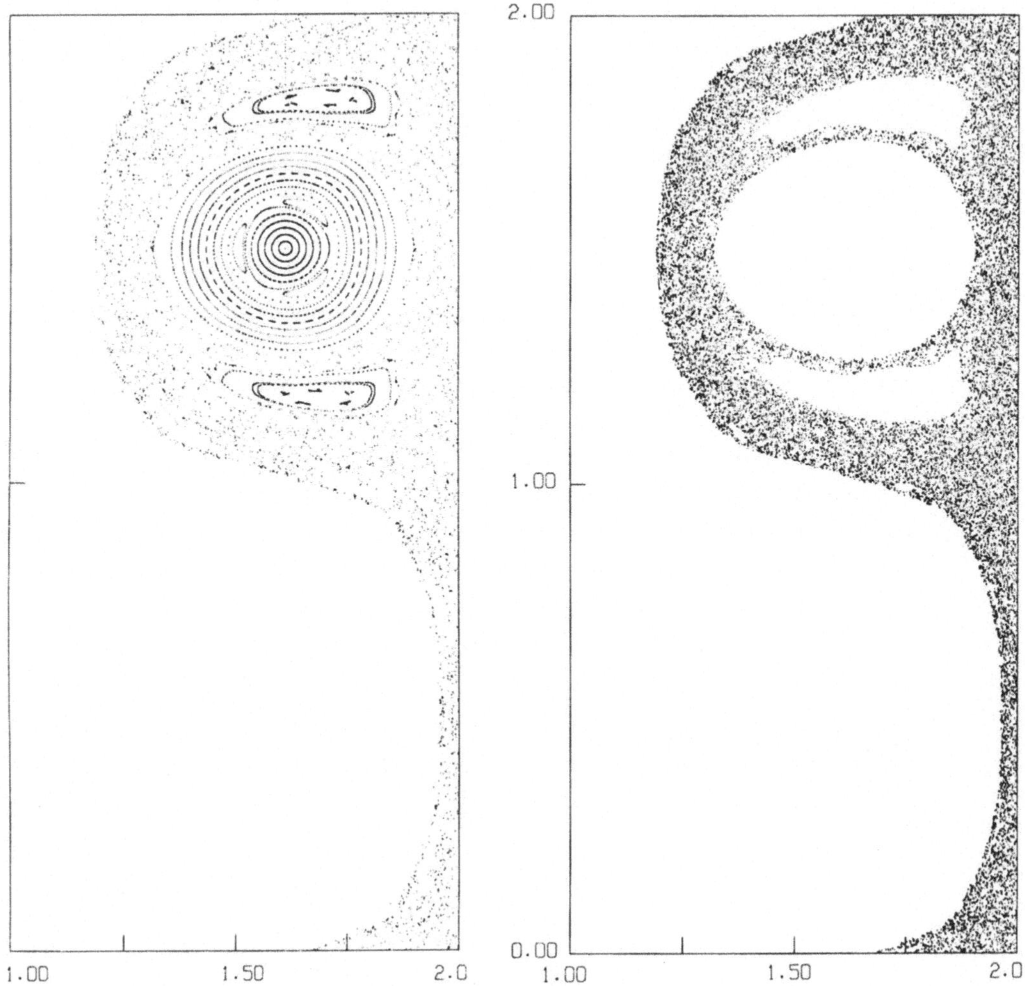

Fig. 3.4. Same as in Fig. 3.3 but with $\varepsilon = 0.05$. This time, the orbits in (a) have been followed for 400 oscillation periods and those in (b) for 3000. To the left edge of the chaotic region integrable orbits are found, though they are not shown in the figure.

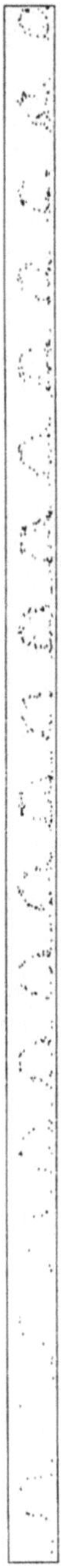

Strung out along the left of this page is Fig. 3.5 showing the locations of 512 particles that were initially spread along the line y = 2.0 in the x interval [1.85,1.90]. Particles from this interval fall through chaotically. The particles spread in vertical extent, but stay within the original cell width. In this case, the abcissa is the value of y (not mod 2). We wish to see whether a fractal structure develops. However, there are not enough points in Fig. 3.5 to do this. In Fig. 3.6 we show the results of a calculation designed to suggest the detail in the loops of Fig. 3.5. In this calculation we applied the periodic boundary counditions in y and Fig. 3.6 shows only the upper half of a cell. There are long intertwined filaments extending into the lower half of the cell where the segment was stretched exponentially. This poses a resolution problem that we circumvented in two ways. First, we projected the points in Fig. 3.5 onto a vertical line and used the Grassberger-Procaccia algorithm (1983a,b) to calculate a correlation exponent of 0.78.

←
Fig. 3.5. The locations after 50 periods of particles that started in the interval $1.85 \leq x \leq 1.90$ on $y = 2.0$. For $\varepsilon = 0.5$, $\gamma = 0.25$, $\omega = \pi/2.25$.

→
Fig. 3.6. A closeup of one of the loop-like features in Fig. 3.5 but calculated as explained in the text.

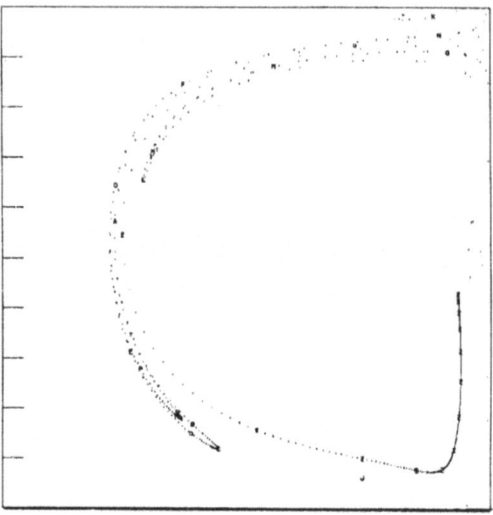

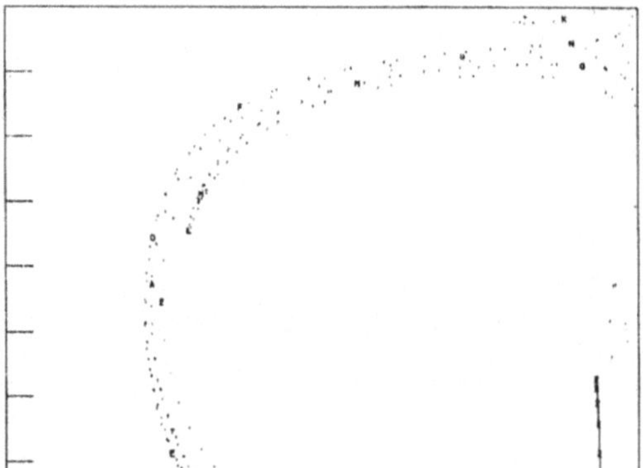

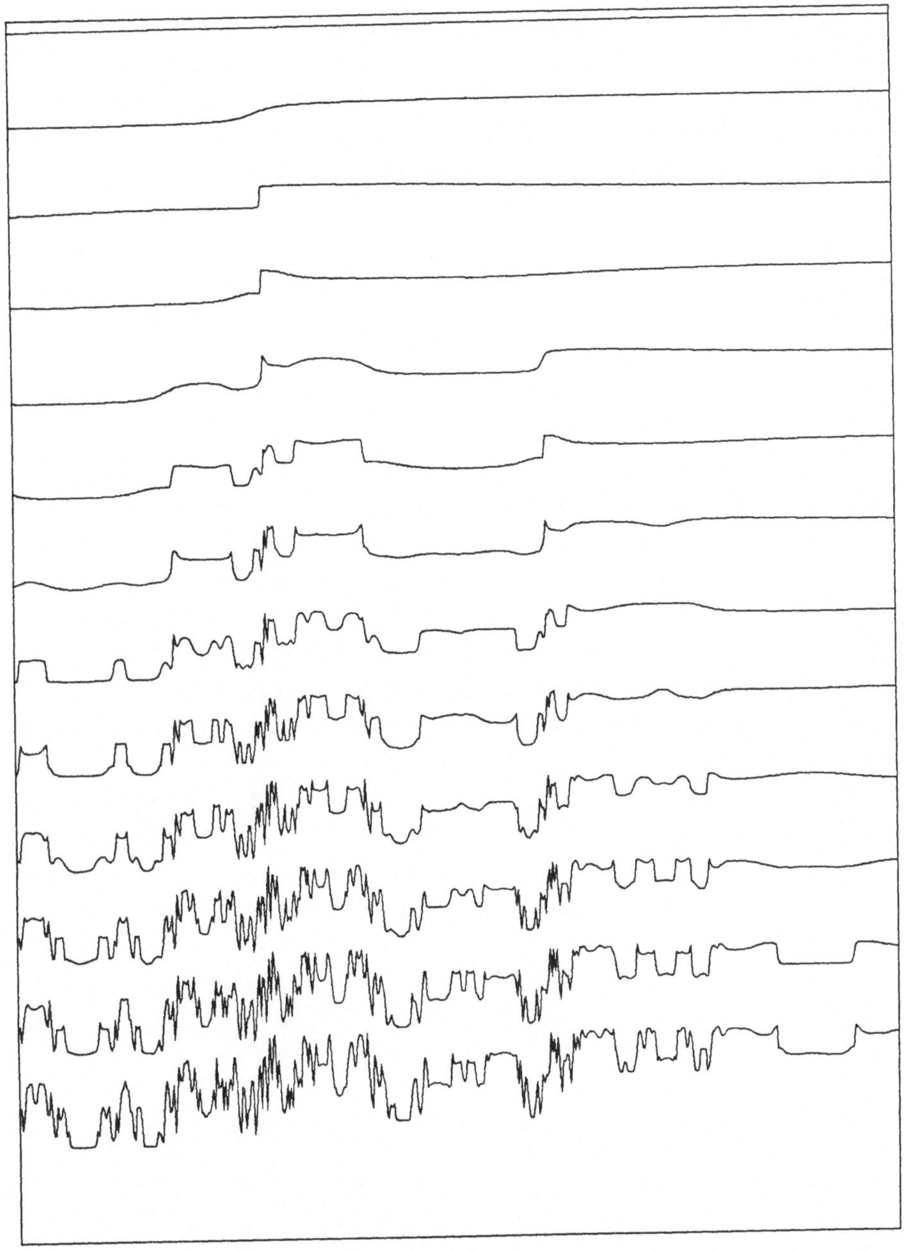

Fig. 3.7. The evolution of an initially simple particle distribution. Shown for a sequence of times separated by one period of oscillation of the fluid. The range in x is 0.0 to 0.1. The height of the box is 110 cell heights. The vertical offset of each curve is linear with time.

The second method focused on the vertical separation of initially nearby particles. In Fig. 3.7 we show the (scaled) y-coordinate of each particle as a function of x_0, the initial x-coordinate, for a sequence of times. Portions of these curves with very large slopes correspond to initial line segments of particles that have been stretched over many cell heights. The stretching and folding in x-y space of the original line of particles produces the self-similar structure seen in Fig. 3.8. We broke the x_0 coordinate into steps of size Δx_0 and found the length, L, of each curve as a function of Δx_0. Then we fit L to the formula

$$L \propto (\Delta x_0)^{-\sigma}. \tag{3.2}$$

As a function of time, σ behaves as shown in Fig. 3.9. The precise form of this evolution depends on how we perform the calculations; the plateau at $\sigma = 0.76$ is characteristic and is independent of the initial particle density and extent. This plateau does not last indefinitely at fixed resolution. As expected, it persists longer if the initial density is higher. We therefore assign the value $\sigma = 0.76$. We do not have a good way to determine the precision of this result but, on the basis of many such calculations, we would estimate that the internal errors are less than 10%.

We found that σ was not sensitive to ε and γ. We have results when these parameters are in the range 0.1 to 0.8. For ε less than about 0.1, the effect of time dependence is so weak, that the particle spreading is very slow, while for large ε the time steps required become prohibitively small.

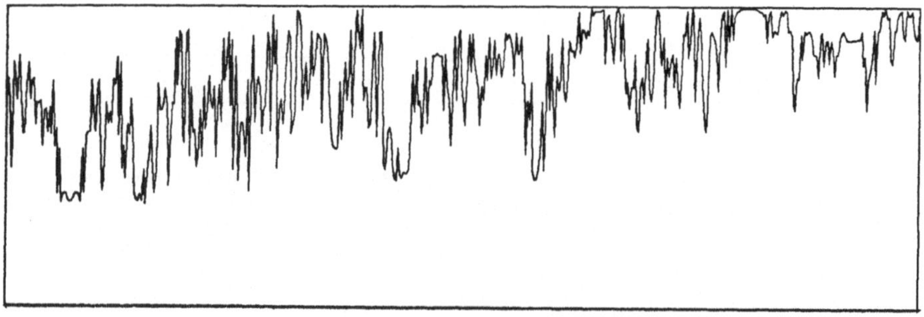

Fig. 3.8. A sequel to Fig. 3.7 twelve periods later.

The increase of L with decreasing Δx_0 suggests that a fractal object is being formed. We expect its dimension to be approximately $1+\sigma$ (Mandelbrot, 1975). The agreement of σ with the Grassberger-Procaccia correlation exponent reinforces this belief. Like others before us, we have learned that it is easier to find a process that makes fractal objects than to understand its dynamics.

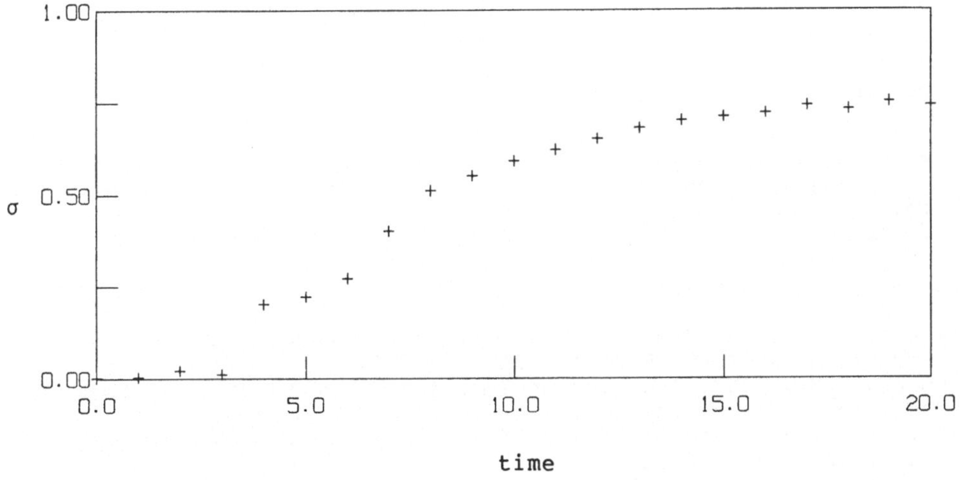

Fig. 3.9. The evolution of σ of (3.2) in time.

4. Conclusion

When studying sedimentation, one often neglects the fluid motion. Yet, as we have seen, the effect of fluid motion on the trajectories of settling particles can organize particle motion. When particle effects on the fluid are allowed, we have a self-consistency problem that may be relevant to the formation of structures in fluid dynamics. While their delicate features are sensitive to noise, diffusion and Brownian motion, mean effects may survive to feed back on the diffusive process. Our calculations already reveal features that hint at such macroscopic implications.

The problem that has most interested us in this study is the formation of fractal swarms. This work raises the problem of finding a dynamical argument to postdict the dimension of the swarms. This question of the dynamics of deformation of a line of passive particles is a first step in the understanding of the stretching of an active vortex line (compare Neuringer, 1968 to Cocke, 1969). We have thought it worth pointing out that simple numerical simulations provide a hint about the former problem.

This study is a direct outcome of work (Smith, 1984) begun as a result of a lecture by Herbert Huppert in the GFD Summer Program at the Woods Hole Oceanographic Institution. We are indebted to Hassan Aref and Walter Robinson for helpful discussions. The work has received support from the NSF under grant PHY 80-2371 to Columbia University and from the NASA Cooperative Agreement NCC 5-29 through GISS.

5. References

Aref, H., 1984, Stirring by chaotic advection, J.Fluid Mech., **143**, 1-21.

Arter, W., 1983, ERGODIC STREAM-LINES IN STEADY CONVECTION, Phys. Lett., **97A**, 171-174.

Cocke, W.J., 1969, Turbulent Hydrodynamic Line Stretching: Consequences of Isotropy, Phys. Fluids, **12**, 2488-2492.

Grassberger, P. and Procaccia, I., (1983a) Characterization of Strange Attractors, Phys. Rev. Lett., **50**, 346-349.

Grassberger, P. and Procaccia, I., (1983b) MEASURING THE STRANGENESS OF STRANGE ATTRACTORS, Physica, **9D**, 189-208.

Huppert, H.E., 1984, Lectures on Geological Fluid Dynamics, Woods Hole Ocenaographic Institution, G.F.D. course, to be expected.

Lichtenberg, A.J. and Lieberman, M.A., 1983, Regular and Stochastic Motion, Springer-Verlag.

Mandelbrot, B.B., 1975, Les objets fractal: forme, hasard et dimension. Paris: Flammarion.

Neuringer, J.L., 1968, GREEN'S FUNCTION FOR AN INSTANTANEOUS LINE PARTICLE SOURCE DIFFUSING IN A GRAVITATIONAL FIELD AND UNDER THE INFLUENCE OF A LINEAR SHEAR WIND, Siam J. Appl. Math., **16**, 834-841.

Quinn, P.J., 1984, ON THE FORMATION DYNAMICS OF SHELLS AROUND ELLIPTICAL GALAXIES, Astrophys. J., **279**, 596-609.

Smith, L.A., 1984, PARTICULATE DISPERSAL IN A TIME-DEPENDENT FLOW, Fellow's lecture in G.F.D. Course of W.H.O.I., with several figures.

Stommel, H., 1949, Trajectories of small bodies sinking slowly through convection cells, J. Mar. Res., **8**, 24-29.

Toobey, P.F., Wick, G.L. and Isaacs, J.D., 1977, The Motion of a Small Sphere in a Rotating Velocity Field: A Possible Mechanism for Suspending Particles in Turbulence, J. Geophys. Res., **82**, 2096-2100.

LIAPOUNOV EXPONENTS FOR THE KURAMOTO-SIVASHINSKY MODEL

Paul MANNEVILLE
IRF/DPhG/SPSRM, CEN Saclay
91191 GIF SUR YVETTE Cedex, France.

1) INTRODUCTION

The transition to stochasticity in deterministic systems has been the subject of considerable attention recently. Many experiments, especially in the field of convection, have confirmed the validity of an approach in terms of dissipative dynamical systems and powerful tools have been developed to measure the amount of "weak turbulence" present in a given system. When confinement effects are strong the number of degrees of freedom is small and chaos is mainly temporal. When they are weak, many degrees of freedom can be excited even close to the instability threshold. Then weak turbulence becomes truly spatio-temporal, with structural defects and "phase variables" playing a crucial role [1]. Within this context, the importance of studies of the Kuramoto-Sivashinsky equation [2] has already been emphasized by previous contributors to this Workshop. Here we shall present results obtained using the 1-dimensional version of this model in the form:

$$\partial_t \varphi + \partial^2_{xx} \varphi + \partial^4_{xxxx} \varphi + 2 \varphi \partial_x \varphi = 0 \qquad (1a)$$

with boundary conditions:

$$\varphi = \partial_x \varphi = 0 \quad \text{at } x = 0 \text{ and } x = L \qquad (1b)$$

where L is the parameter which controls the number of linearly unstable modes and, further, the amount of turbulence.

In the study of weak turbulence the key concept is that of the Liapounov exponent (LE in the following) which measures the divergence of nearby trajectories [3]. The amount of chaos can be evaluated from the Liapounov spectrum in terms of the number of non-negative LEs, the entropy h (sum over the positive LEs [4]), or the Liapounov dimension of the attractor [5].

Liapounov exponents are derived from a set of solutions of the tangent equation, that is, the equation linearized about the trajectory under investigation:

$$\partial_t \psi + \partial^2_{xx} \psi + \partial^4_{xxxx} \psi + 2 \psi \partial_x \varphi + 2 \varphi \partial_x \psi = 0 \qquad (2a)$$

with boundary conditions:

$$\psi = \partial_x \psi = 0 \quad \text{at } x = 0 \text{ and } x = L \qquad (2b)$$

LEs and tangent solutions associated with them are nothing but the generalisations to nonlinear systems of eigenvalues and eigenvectors defined for linear systems with constant coefficients. In practice LEs cannot be calculated separately but rather recursively. Benettin and coworkers have shown that the LEs λ_i could be extracted from the series of growth rates μ_i of parallelepipeds of increasing dimensions in phase space. The growth rate μ_1 of a 1-d parallelepiped (a segment) yields the largest LE λ_1. For a parallelogram one gets: $\mu_2 = \lambda_1+\lambda_2$, in three dimensions: $\mu_3 = \lambda_1+\lambda_2+\lambda_3$, etc. Thus $\lambda_i = \mu_i - \mu_{i-1}$.

The Liapounov dimension is defined as [6]:

$$D_L = N_L + \left(\sum_{j=1}^{N_L} \lambda_j\right)/|\lambda_{N_L+1}|$$

where N_L is the largest integer for which the sum $\lambda_1+\ldots+\lambda_{N_L}$ is non negative. Recalling the relation between the μ's and the sum of the λ's we see that D_L also reads $N_L + (0-\mu_{N_L})/(\mu_{N_L+1}-\mu_{N_L})$ so that we can interpret the Liapounov dimension as that which interpolates linearly between the dimension of parallelepipeds which suffer an average global expansion ($\mu_{N_L}>0$) and those which are contracted ($\mu_{N_L+1}<0$). A graphical example is given in fig.1 for L=50. Being a time average, this dimension takes into account the probablistic features of the dynamics. This makes heuristically plausible the idea that this dimension is the Haussdorff dimension of the invariant measure on the attractor of the dynamical system.

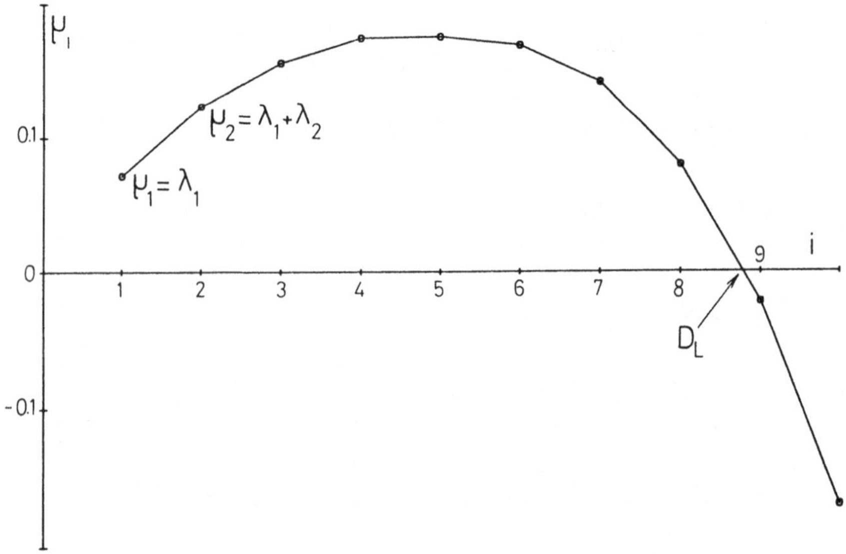

Fig.1: Growth rate μ of parallelepipeds as a function of their dimensions and determination of the Liapounov dimension by linear interpolation. Example given for the Kuramoto-Sivashinsky equation with L=50.

2) RESULTS

We have performed simulations of eqs.1 and 2 using a finite difference second order Crank-Nicolson/Adams-Bashforth scheme. Calculations were performed on the Cray-1S computer of the CCVR for which a completely vectorized code has been written making use of efficient built-in recurrence solvers and scalar products. Though a rather high resolution ($\delta x<0.125$, $\delta t<0.025$) was required to insure convergence, statistically meaningful results were obtained at lower resolution ($\delta x=0.5$, $\delta t=0.1$) as soon as L was made large enough, typically $L > 50$. Getting stable temporal averages and a well ordered Liapounov spectrum required simulations over very long durations ($\Delta T > 5000$). In practice we stopped the simulation when the series of LEs remained monotonically decreasing and when the LE which should be zero (due to the autonomous character of the flow) was much smaller, one order of magnitude if possible, than its immediate neighbours. These criteria were difficult to meet at the largest length considered L=400 which required the computation of as many as 100 LEs.

The Liapounov spectrum turns out to be a discrete decreasing series bounded from above by some maximum value λ_m. This maximum value increases more or less regularly from 0 (limit cycle for L<30) to a saturation value ~ 0.1 reached at about L=150 [7]. Our results are gathered in the table below [8].

L	λ_m	h	$N_{\geq}$	D_L	$h/N_>$
50	0.072	0.18	5	8.8	4.39
100	0.092	0.46	12	20.4	4.22
200	0.097	1.01	25	43.3	4.21
400	0.101	2.10	51	89.4	4.20
					$\times 10^{-2}$

The number of non-negative LEs $N_{\geq}$ increases as:

$$N_{\geq} = 0.131 \times L - 1.30$$

The Liapounov dimension grows as:

$$D_L = 0.230 \times L - 2.70$$

while the entropy follows a similar law:

$$h = 5.48 \times 10^{-3} \times L - 9.0 \times 10^{-2}$$

letting the average entropy per positive LE remain roughly constant.

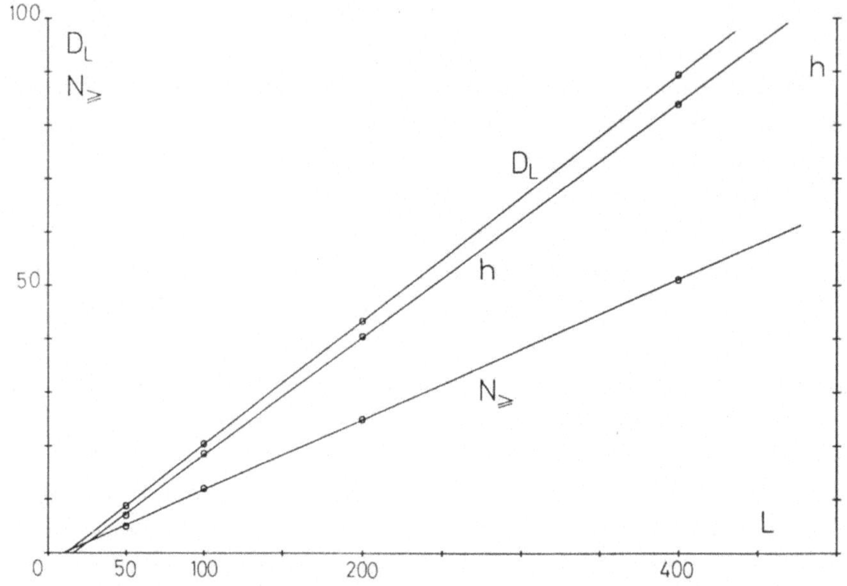

Fig.2: Variation of the number of non negative Liapounov exponents, the Liapounov dimension (left scale) and the entropy (right scale) as functions of the length L.

All these results are presented graphically in fig.2 where the linear behaviour is strongly evidenced (coefficients given above where obtained by least square fits with confidence factors equal to 1 within less than 0.001). The internal consistency of these results is proven by the fact that extrapolation towards small L yields comparable values for the first onset of weak turbulence. Indeed the onset should correspond to $h \longrightarrow 0$ giving L_t = 16.4 and to $D_L \longrightarrow 1$ (chaos after limit cycle period doubling for example) giving L_t = 16.1 while the extrapolation for $N_\geqslant$ --> 1 gives 17.6 again in good agreement.

In addition, the distribution $\mathcal{D}(\lambda)$ of LEs seems to tend towards a limit when L increases. Fig.3 displays the LE indices as a function of their values, that is to say, the integral $\int_{\lambda}^{\lambda_m} \mathcal{D}(\lambda')d\lambda'$ which is much less "noisy" than $\mathcal{D}(\lambda)$ from a numerical point of view. Here indices are scaled by the number of non negative exponents. A justification of this scaling (as well as an equivalent one using the Liapounov dimension) can be found in the fact that one should find the point corresponding to the "extrapolated onset of chaos" ($\lambda \longrightarrow 0$, $N_\geqslant$ or $D_L \longrightarrow 1$) close to the limit function, which is indeed the case. There seems to be an inflexion at $\lambda = 0$. This fact that could be related to Ruelle's conjecture of a singularity at λ=0 for Navier-Stokes fully developed turbulence [9], but here the singularity, if it exists, must be much weaker.

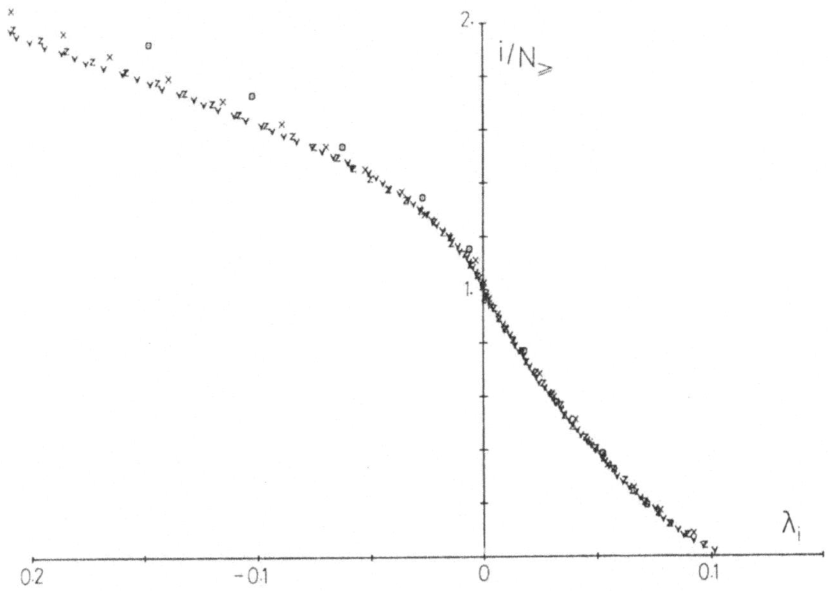

Fig.3: Integral of the distribution of Liapounov exponents for different lengths, ⊙: L = 50, x : L = 100, z: L = 200, v : L = 400.

Indeed there is no evidence of accumulation of LEs around $\lambda=0$ since the slope there is definitely finite. In fact the strongest singularity could be a (small) slope discontunuity corresponding to a (small) jump for $\mathcal{D}(\lambda)$, more probably a curvature jump (a cusp for $\mathcal{D}$) or possibly a higher order singularity (a rounded maximum for $\mathcal{D}$).

Finally the spatio-temporal evolution of basis functions in tangent space is also of interest. Fig.4 displays the evolution of the solution and of three basis functions or "Liapounov vectors" (LV in the following) for L=100 between t=10000 and 10050. The profile of the solution (fig.4a) presents roughly periodic though fluctuating perturbations. This could be anticipated from the power spectrum of spatial fluctuations which presents a broad maximum at the linearly most unstable wave-vector $1/\sqrt{2}$ [10]. On the other hand, Liapounov vectors display only localized perturbations superimposed on a smooth background. The size of the perturbations are again of the order of the inverse of the most unstable wave-vector. The first LV (fig.4b) corresponding to the largest LE presents a very small number of such perturbations. It gives the correction to the solution displayed in fig.4a which would be the most strongly amplified.

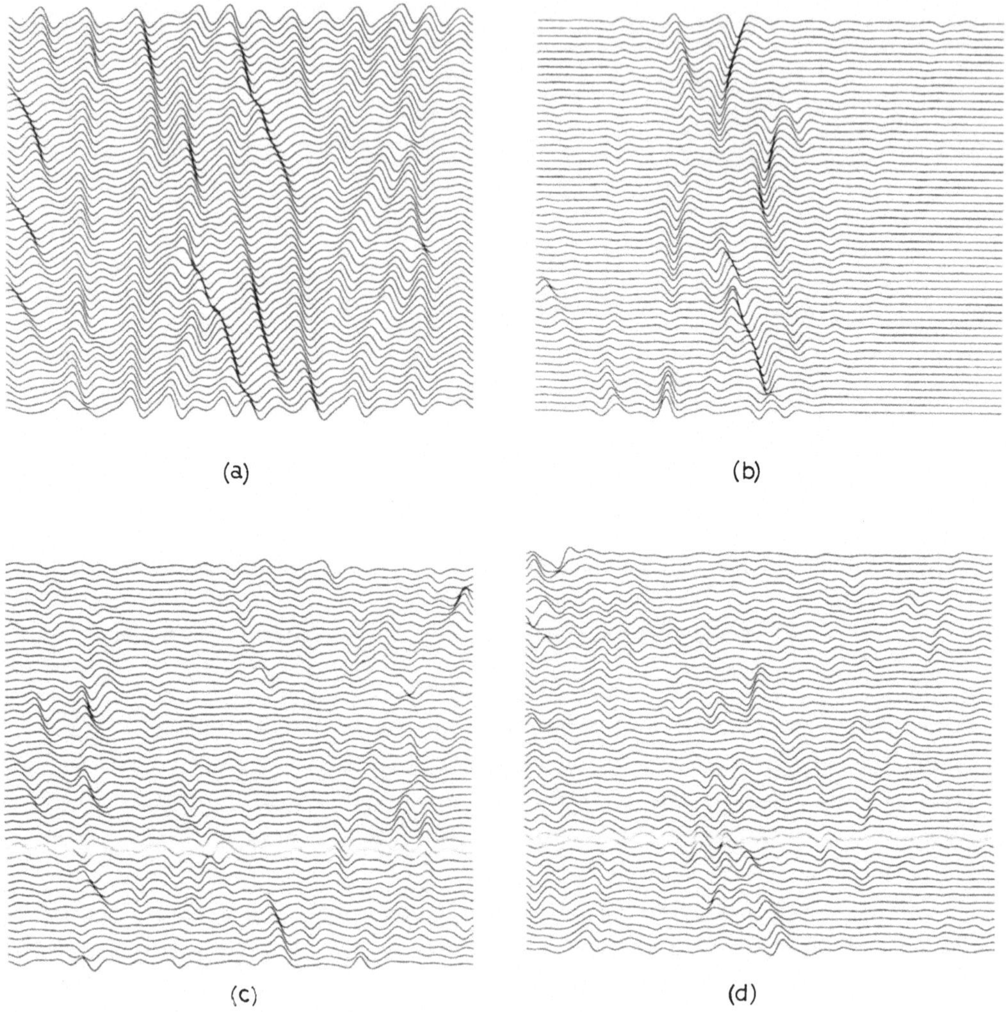

Fig.4: Evolution of (a) the solution, (b) the first Liapounov vector ($\lambda = 0.092$), (c) the 12th LV ($\lambda \cong 0$), and (d) the 25th LV ($\lambda = -0.24$). Curves are drawn every time unit and time run upwards from t=10,000 to 10,050.

Considering the solution itself, we can see that the behaviour is somewhat more regular at the two ends, $x = 0, L$, even nearly periodic on the left. On the other hand the evolution of the central part is much more irregular. Accordingly, the first LV is nearly structureless on both sides and presents strong localized perturbations only in the middle. The complexity of the structure seems to increase with the index of the LV, which explains the increasing power at small wave-vectors in the spectra of ref.11. The evolution of the 12th LV is displayed in fig.4c just below that of the solution itself.

This has been done intentionally since it corresponds to $\lambda = 0$ and thus should resemble the tangent vector in the direction of the flow. This is indeed the case since one can clearly go ahead in time by adding to the solution a small perturbation proportional to this LV (consider again the nearly periodic pattern on the left). Fig.4d displays the pattern corresponding to the last (25th) LV. For L=100 the Liapounov dimension is about 20.4 and thus parallelepipeds with 25 dimensions are already strongly contracted. By superposition of transparencies one can convince oneself that the localized perturbations remain nethertheless correlated to those of the solution itself. This fact gives us confidence in the validity of the extraction of LEs at least up to this index and somewhat beyond.

More refined simulations using a pseudo-spectral method are in progress [12]. A parallel study of the case with periodic boundary conditions stopped short. The reason was the unexpected observation of a steady state obtained after long turbulent transients for L=64, seemingly independent of the resolution used. These transients were very long, of the order of a few thousand time units, which attested to the smallness of the basin of attraction of the steady state. However this dissuaded us from calculating LEs since some doubt was cast on the very existence of sustained turbulence in the periodic case.

3) CONCLUSION

Results presented here clearly demonstrate the linear increase of the amount of turbulence in the Kuramoto-Sivashinsky model. As far as we know, only weaker bounds have been deduced analytically [13] and a detailed theory is still lacking for this system. The behaviour observed in this partial differential equation is somewhat different from that for ordinary differential equations with delay which form another class of infinite dimensional dynamical systems. In this latter case [14] while the Liapounov dimension increased linearly with the delay, the entropy remained roughly constant, the maximum Liapounov exponent decreased, and the structure of the Lyapounov vectors became finer and finer when their indices were increased.

As a partial differential equation with turbulent solutions, the Kuramoto-Sivashinsky model could have been an appealing model of hydrodynamic turbulence. As a matter of fact, a study more in line with the conventional approach to "fully developed turbulence" has been started [11]. Results on the energy spectrum and related time-space correlation functions have been tentatively interpreted in light of the results summarized above but much remains to be done for a rigorous theory. In fact, as far as three-dimensional

Navier-Stokes turbulence is concerned, our results may be partly misleading due to the specificities of the model considered (one-dimensional character, no true inertial range) but an approach to the problem within the framework of dynamical system theory may be rewarding, especially concerning the turbulent evolution of large scale coherent structures [15].

ACKNOWLEDGMENTS

Most of the calculations have been performed thanks to a CPU time allocation on the CRAY-1 computer of the CCVR which the author acknowledges for this support. He also wants to thank Y.Kuramoto, Y.Pomeau, A.Pumir and L.Tuckerman for many discussions related to this work.

REFERENCES

[1] for a review see related communications at: "Common Trends in Particle and Condensed Matter Physics" Les Houches 1983, Physics Reports 103(1984).

[2] Y.Kuramoto: Suppl.Prog.Theor.Phys.64(1978)346.
G.I.Sivashinsky: Acta Astronautica 4(1977)1177.

[3] see: G.Benettin and L.Galgani: in "Intrinsic Stochasticity in Plasmas", G.Laval and D.Gresillon Eds. (Editions de Physique,Orsay,1979).

[4] G.Benettin, L.Galgani and J.M.Strelcyn: Phys.Rev.A14(1976)2338.

[5] J.D.Farmer, E.Ott, J.A.Yorke: Physica 7D(1983)153.

[6] P.Frederickson, J.L.Kaplan, E.D.Yorke, J.A.Yorke: J.Diff.Eq.49(1983)185.

[7] A.Pumir: Thèse 3ème Cycle Univ. Paris VI, Dec.1982.

[8] P.Manneville: presentation of preliminary results at "Interdisciplinary Turbulence" IHES Bures sur Yvette Nov.1983 unpublished.

[9] D.Ruelle: Comm.Math.Phys.87(1982)287.

[10] T.Yamada, Y.Kuramoto: Prog.Theor.Phys.56(1976)681
see also fig.1 of ref.11 below.

[11] Y.Pomeau, A.Pumir and P.Pelce: J.Stat.Phys.37(1984)39.

[12] in collaboration with L.Tuckerman.

[13] B.Nicolaenko, B.Scheurer, R.Temam: C.R.Acad.Sc.Paris298II(1984)23
B.Nicolaenko, B.Scheurer: Physica 12D(1984)391
B.Nicloaenko, B.Scheurer, R.Temeam: "Some global dynamical properties of the Kuramoto-Sivashinsky equations: nonlinear stability and attractors", Physica D to appear.

[14] D.J.Farmer: Physica 4D(1982)366.

[15] Two dimensional turbulence is being studied in this spirit by B.Legras LMD, ENS Paris.

VORTICES AND VORTEX-COUPLES IN TWO-DIMENSIONAL TURBULENCE
or
LONG-LIVED COUPLES ARE BATCHELOR'S COUPLES

C. Basdevant[†], Y. Couder[††] and R. Sadourny[†]

November 1984

† Laboratoire de Météorologie Dynamique, Ecole Normale Supérieure, 24, rue Lhomond, 75231 Paris Cedex 05, France.

†† Groupe de Physique des Solides, Ecole Normale Supérieure, 24, rue Lhomond, 75231 Paris Cedex 05, France.

Several works have been recently devoted to the study of macroscopic coherent structures in two-dimensional incompressible turbulent flows. Existence of coherent structures is observed in large-scale meteorological fields, in oceanic flows as well as in laboratory experiments such as wakes behind an "infinite" body, shear layers or in Magneto-hydro-dynamics when the fluid motion is constrained to be two-dimensional by an imposed magnetic field. Theoretical works also predict the existence of stable macroscopic structures in two-dimensional flows. Many numerical experiments exhibit a tendency for the vorticity of the flow to concentrate into circular eddies ; a recent paper by Mc William (1984) is devoted to the description of this behaviour.

At present many questions remain unanswered about these coherent structures : an important question is whether two-dimensional flows systematically tend to construct coherent structures, and, if true, does this tendency invalidate the statistical theories of two-dimensional turbulence which are intrinsically unable to take into account the dynamical effect of such structures. Another question is how these structures are constructed and what is their role in the phase of transition to turbulence.

In part I of this paper we shall focus our attention on a particular form of coherent structures : vortex couples. We shall demonstrate their existence and important dynamical role, both experimentally and numerically ; we shall also study their structure and generation.

In part II we shall compare the evolution of vorticity with a passive pollutant in a 2-d turbulent flow, demonstrating the inability of the classical phenomenological theory to take into account some important aspects of inertial range dynamics.

I. Solitary vortex couples

Several theoretical works (Thomson 1867, Batchelor 1967, Stern 1975, Larichev and Reznik 1976, Deem and Zabusky 1978 a,b, Flierl and al 1980, Pierrehumbert 1980) predict the existence, in two-dimensional flows, of stable, isolated couples of vortices of opposite sign with a fast translation motion. Such couples have been studied in a meteorological context and called modons (Stern, 1975, Larichev and Reznik, 1976, Flierl and al 1980) : in the β-plane approximation large scale atmospheric motions are modelled by the 2D Navier-Stokes equation together with a linear dispersive term representing the variation of Coriolis forces with latitude ; these modons have been found to be soliton-like solutions of the equation.

In the absence of Coriolis forces, the stability of similar couples was demonstrated by the technique of contour dynamics in the case of the plain two-dimensional Euler equation (Deem and Zabusky 1978 a et b, Pierrehumbert 1980) and then called V-states.

We present here laboratory observations of such vortex-couples in a two-dimensional turbulent wake. Throughout the present report, we will call these objects "vortex couples" leaving the term of "vortex pairs" to groups of two vortices of the same sign.

A. Vortex couples near a Von Karman street

The wake of an infinite cylinder in a fluid is characterized by the Reynolds number $Re = Vd/\nu$ where d is the diameter of the cylinder, V its velocity and ν the kinematic viscosity of the fluid. For $Re > 50$, the wake is a regular Von Karman street of alternate vortices. For $Re > 150$, the alternate shedding of vortices behind the cylinder apparently retains its regularity but the wake becomes turbulent further downstream. Experimentally, a three-dimensional evolution of the vortices is observed where their axis bends and they stretch into contorted shapes. This usually masks the two-dimensional destabilization. However, experiments by Taneda (1959) showed that at a large distance downstream the initial wake could be transformed by vortex merging processes into a new street of periodicity 1.8 to 3.3 times the initial one.

Several numerical simulations (Christiansen and Zabusky 1973, Aref and Siggia 1981) of the two dimensional evolution of a vortex street show this merging process (though generally as a systematic pairing that would lead to a doubling of periodicity). The simulation of Aref and Siggia (1981) was the first to show the possibility of creation of vortex couples (that they call neutral pairs) which can escape out of the wake and form a gas of travelling objects. These couples however were never before observed experimentally.

In order to observe in a laboratory experiment the two dimensional instabilities of a wake, it is necessary to inhibit any possibility of three dimensional evolution. We achieve this by forming a Von Karman street in a thin liquid soap film where the fluid motions are necessarily confined to the film plane. The technique has been already used to observe the decay of two-dimensional grid turbulence (Couder 1984), the reader will find in this previous work the nature of the soap solution, the viscosity of such films and the discussion on the effects of friction on the surrounding air. A detailed description of the experiment is given in Couder, Basdevant and Thomé (1984), we will give here just a brief description. We use a horizontal film stretched on a rectangular frame 46x16 cm. A disk of aluminium paper 20 μm thick is inserted in the film. It can then be towed at a velocity V by means of a thin needle passing through its center. With velocities $V < 100$ cm/s, Reynolds number up to 1500 could be obtained for disks of diameter 0,6 cm.

The figure 1 illustrates the three different regimes that are obtained at different Reynolds numbers :

- for low Reynolds numbers (fig 1a) we did not observe any destabilization of the vortex street.
- for $Re \gtrsim 150$ (fig. 1.b) the wakes become unstable and neighbouring vortices are no longer equal and equidistant downstream : weaker vortices get stretched and sucked by the axial flow between their preceeding neighbours and finally coalesce (at least partially) with them to form larger structures. As in Taneda's (1959) experiment, the ratio of the resulting period to the initial one is not necessarily 2.

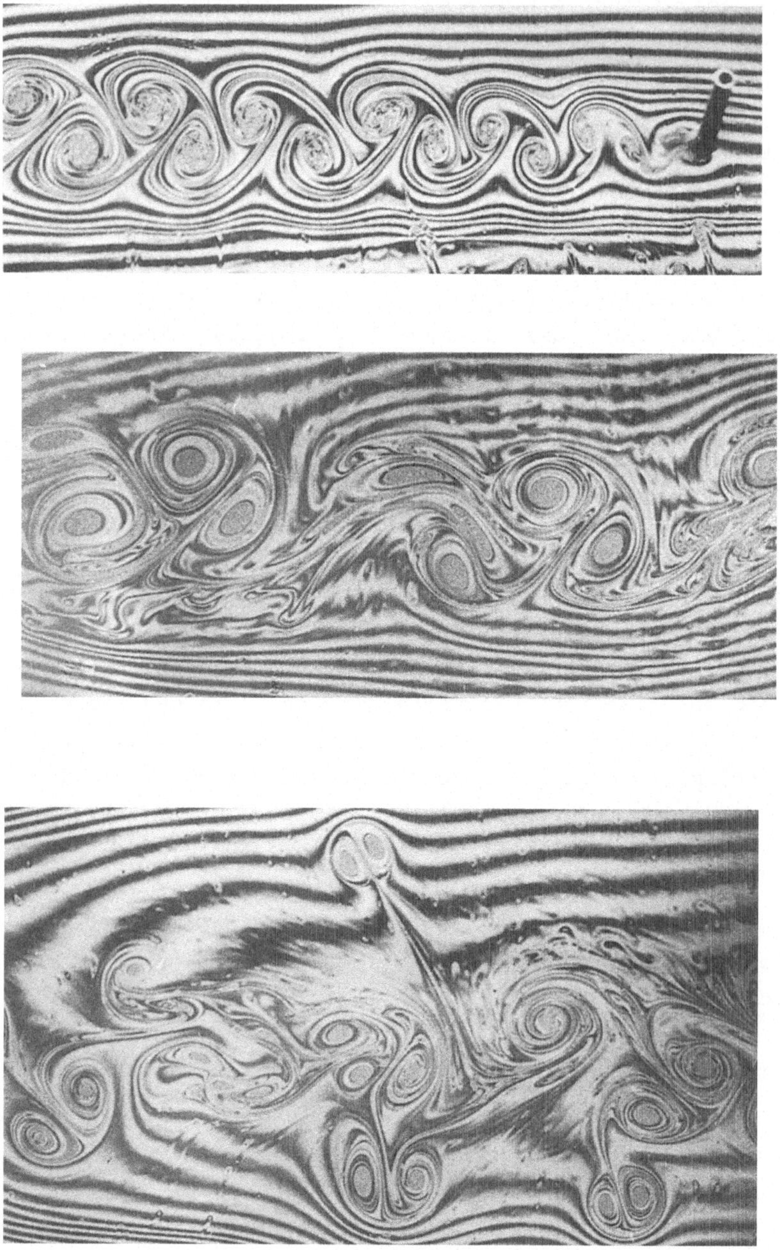

Fig. 1 : The interference fringes of the film behind a cylinder ; (a) a regular Von Karman street ; (b) stretching and absorption of vortices by a couple of opposite vortices located before them ; (c) the escape of vortices couples away from the initial wake.

- for larger values of the Reynolds number (Re $\gtrsim$ 300) a more remarkable phenomenon occurs (Fig. 1c). The wake seems to explode with the formation of couples of vortices of opposite sign having (due to their own velocity field) a fast translation motion which makes them escape out of the initial zone of the wake. The typical aspect of a couple has a circular envelope with two inner opposite circulations.

It could be objected that in a thick film local variations of thickness can be created by the motion. The motion of thick regions similar to drops could then create effects similar to those observed. The numerical simulation will show that the formation of solitary couples does not depend on this effect.

The numerical model is also described in Couder, Basdevant and Thomé (1984) ; let us recall that the two-dimensional Euler equation is integrated in time using a pseudo-spectral (128x128) code on a periodic square of side 10 cm. A super-dissipativity is used to model subgrid-scale effects (Basdevant et Sadourny 1983). The initial state of the flow is a collection of vortices which corresponds to a realistic representation of a Von Karman street. It is built by superposing on the diagonal of the domain 18 alternate vortices, each of them with a gaussian streamfunction (Fig. 2a). The figures 2b-2d show the aspect of the vorticity field at various times.

A complete similarity between the experiment in the soap film and the numerical calculation is observed. In both cases the evolution is dominated by two phenomena : the interaction between vortices leads either to pairing and merging and thus to an increase of the structure size or to the formation of couples of opposite vortices.

In this latter case the couples move away from the initial central zone of the wake and form an expanding cloud around it. Each couple has a circular envelope and moves under the influence of its own velocity field with a velocity much larger than the velocity of the wake. A couple formed of equal strength vortices moves in a straight line ; otherwise its trajectory is curved towards the side of the stronger vortex.

The detailed structure of a couple, once far from the influence of other vortices, will be studied in section I-B. However we may note at this stage that the couples we find are structured objects, in contrast to the numerical results of Aref and Siggia (1981) where couples

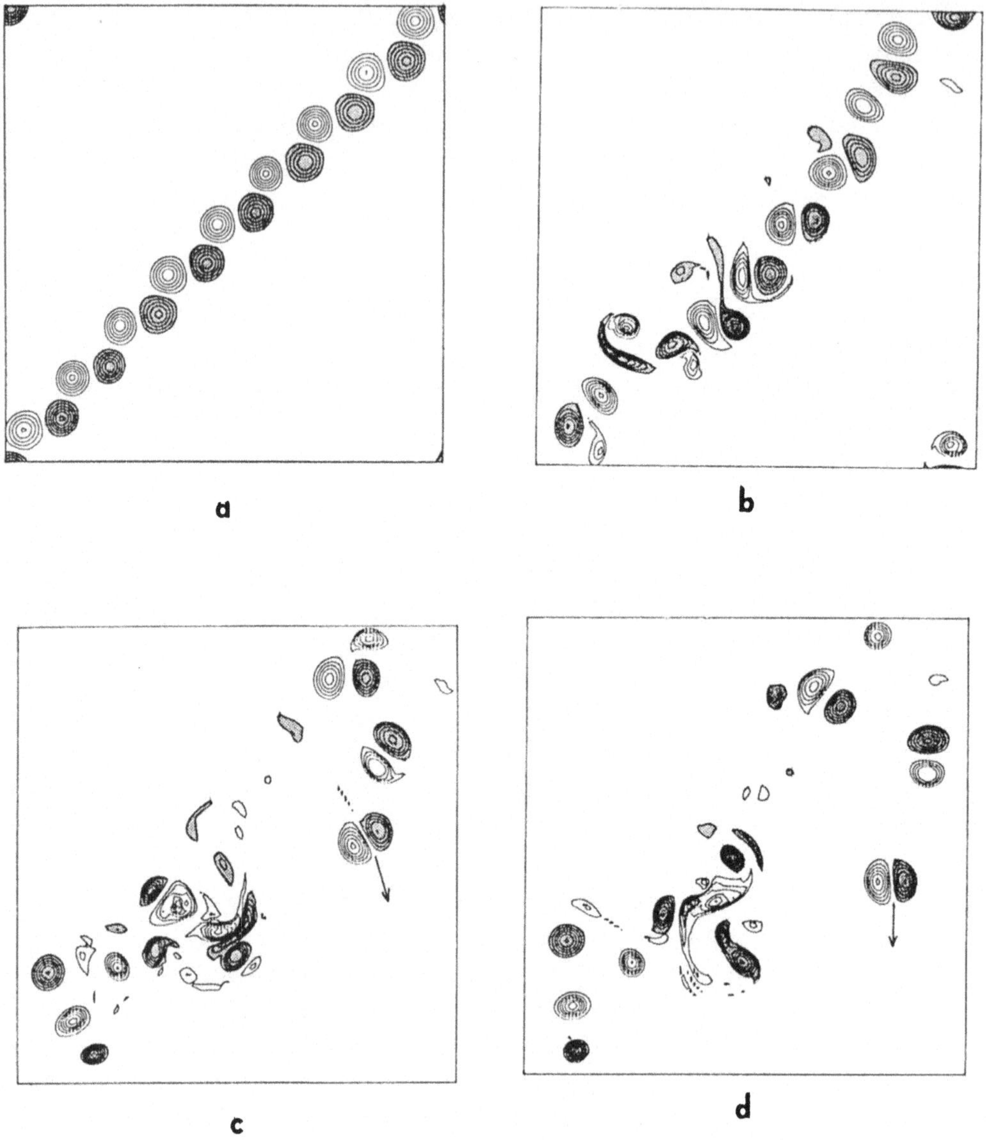

Fig. 2 : Isolines of vorticity at times t = 0, t = 3.5s, t = 5s, and t = 6s. The positive vortices (in dark) are initially below the diagonal, the negative are above, so that all vortices move towards the lower end of the diagonal at velocity 0.5 cm/s. The further evolution shows that the marked couple (c and d) moves at 2.3 cm/s. The side of the square equal to 10 cm gives the scale.

look like simple juxtaposition of two vortices (this may be due to the vortex-in-cell method they used).

These couples are intrinsically stable. Their lifetime is essentially limited by their interaction with other vortices. Collision between couples lead to various behaviours. We observed experimentally and numerically the coalescence of two couples following each other, the deflexion of trajectories in the case of near collision and the exchange of partners in lateral or frontal collisions. Interactions of that type have been predicted by numerical simulation in the case of modons (Mc Williams and Zabusky 1982).

B. Structure and birth of the vortex couples

The soap film experiment provides a good visualisation of the flow and clearly shows (Fig. 1c) the circular envelope of the vortex couples. However it does not lend itself to easy measurement of the velocity field. For that purpose we extracted from one of the Von Karman simulations a vortex couple that had reached a stable shape and was far away from any disturbance. It is shown on Fig. 3, we compared its velocity and vorticity profile to those of the theoretical model given by Batchelor (1967). This model corresponds to a stationary solution of the two dimensional Euler equation. The vorticity is zero outside a circular envelope of radius a and distributed inside the envelope

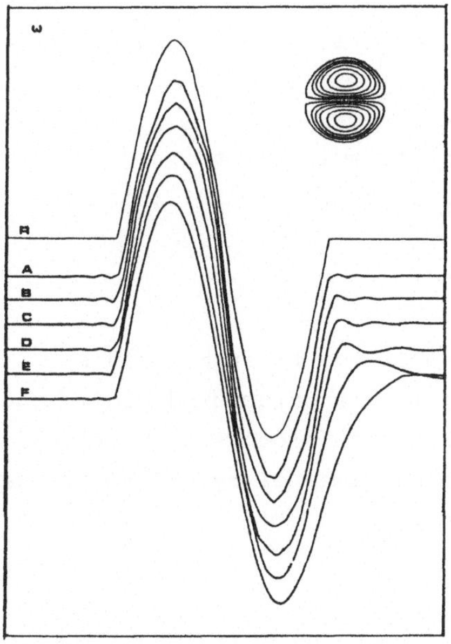

Figure 3 : Vorticity profiles across the couple drawn in the upper right corner, compared to the model given by Batchelor. Curve R displays the Batchelor's reference (formula 1) at $\theta = \pi/2$. Curves A to F display $\omega/\lambda \sin\theta$, where ω is the vorticity, θ the polar axis of the section, λ a correcting factor ;
A : $\theta = \frac{\pi}{2}$, $\lambda = 1$;
B : $\theta = \frac{\pi}{3}$, $\lambda = 1$;
C : $\theta = \frac{\pi}{4}$, $\lambda = 1.05$;
D : $\theta = \frac{\pi}{6}$, $\lambda = 1.23$;
E : $\sin\theta = 0.2, \lambda = 1.49$;
F : $\sin\theta = 0.1, \lambda = 1.52$.

with

$$\omega = \left(\frac{3,82}{a}\right)^2 \psi \qquad (1)$$

The stream function is then

$$\psi = - C\, J_1\left(\frac{3,83r}{a}\right) \sin\theta$$

and the velocity of the couple

$$U = \frac{1}{2}\, C\, \frac{3,83}{a}\, J_o(3,83)$$

where J_o and J_1 are the two first Bessel functions.

Various vorticity sections of the observed couple are compared to the corresponding section of the Batchelor's model on Fig.3. The comparison is strikingly good ; except for differences due to the influence of the rest of the fluid, a small discrepancy with the theoretical model stands in the slight asymmetry between front and back of the numerical couple. Nevertheless the vortex couples obtained numerically are very close to Batchelor's theoretical couples.

In a subsequent series of numerical experiments we studied (in collaboration with H. Fabre and B. Laviron 1984) the processes by which two vortices of opposite sign coalesce into a structured couple. The experiments were done with the numerical model mentioned in section A

In the first series of experiments we tried to form a couple from only two isolated vortices. The initial states were then defined by two synthetic vortices at various distances from each other. The model vortices that we used were of two types (both with distributed vorticity) named respectively G and S.

- a G vortex is defined by a gaussian vorticity field (figure 4a)

$$\omega = \omega_o e^{-\frac{r^2}{a^2}}$$

- a S vortex, is defined by a gaussian streamfunction field ; as a consequence the vortex is surrounded by a screen of vorticity of opposite sign. (Fig. 4b)

$$\omega = \omega_o \left(1 - \frac{r^2}{a^2}\right) e^{-r^2/a^2}$$

The results of these experiments may be summarized as follows : for two isolated vortices to form a structured couple the initial existence of screens of opposite sign around the vortices is essential. Initialization with two vortices of type G never leads

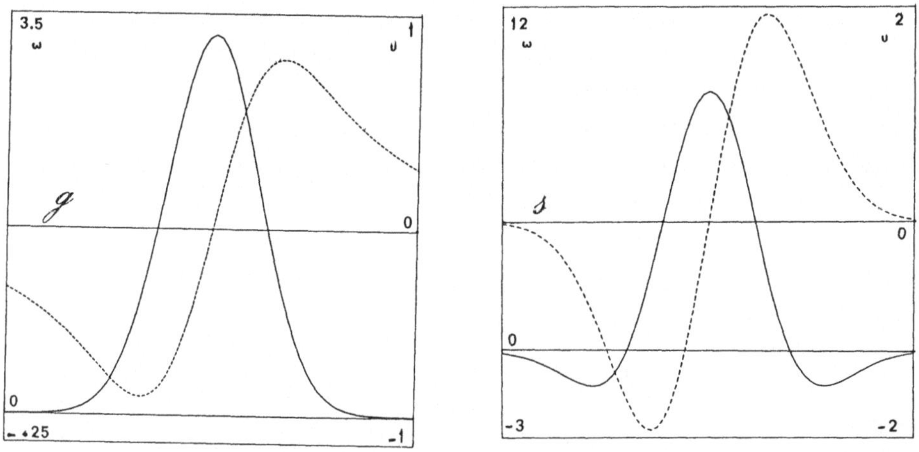

(a) (b)

Fig. 4 : Vorticity (solid line) and velocity (dashed line) profiles ; (a) vortex of type G, (b) vortex of type S.

to a couple similar to Batchelor's couple; they remain (as in Aref and Siggia 1981) two juxtaposed vortices (Fig.5) whose translating velocity

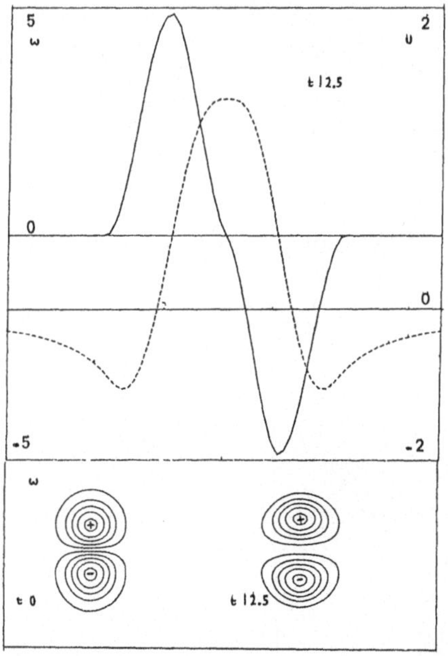

Fig. 5 : Weakly-linked couple initiated with two vortices of type G. Lower part : vorticity contours at time 0 and 12.5. Upper part vorticity (solid line) and velocity (dashed line) profiles at time 12.5 : Note the variation of vorticity gradient between the two vortices.

is less than the velocity of a coherent couple. Conversely, initialization with two vortices of type S (sufficiently close from each other) leads to the birth of a structured couple. This process is described on figure 6. The screen of each vortex appears broken by the interaction with the other vortex. The two strong cores are then located between two weaker vortices that are the remains of the screens. Those weaker vortices act upon the stronger one, slowing them down and compressing them into a couple. Then the structured couple moves on leaving behind the remains of the screens. Afterwards the structured couple moves faster than its unstructured analog. It appears that an initial compression effect is needed to form structured couples. This can be achieved by several configurations when more than two vortices are initially involved.

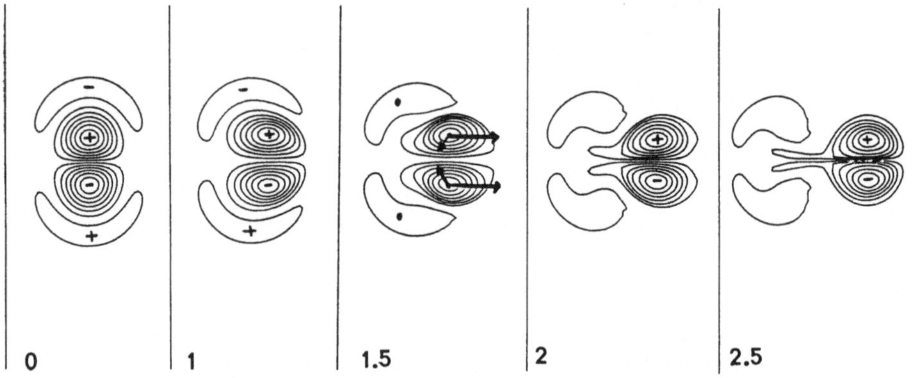

Fig. 6 : Couple initiated with two vortices of type S. Vorticity contours at times 0,1, 1.5, 2 and 2.5. At time 1.5, to illustrate the compression effect we draw arrows indicating the action on a vortex of the other vortex and of its own screen, as if they were point vortices.

It can for instance be the result of the collision of two weakly linked couples. This is illustrated by figure 7. In this experiment two

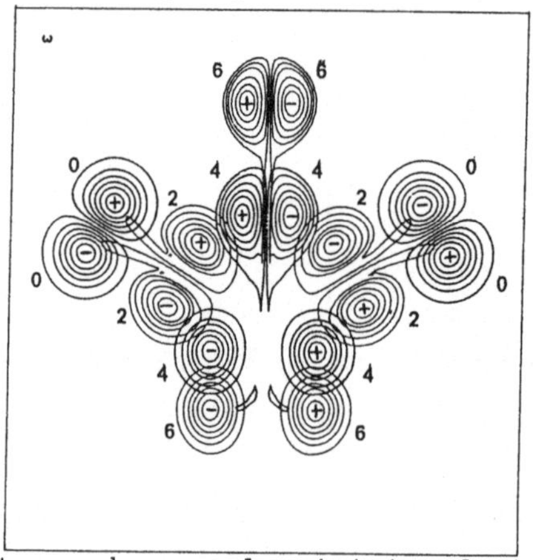

Fig. 7 : Collision of two weak -couples initiated with G-type vortices. They collide at an angle of 120°. Vorticity contours at times 0, 2, 4 and 6 are superimposed on the same figure.

weak-couples were initiated, using vortices of type G, in such a way that their trajectories cross with an angle of 120°. We see, as a result of the collision, that the two inner vortices join to form a structured couple while the two outer vortices are left away. A head on collision of this type would create two structured couples. Simple laboratory experiment confirms this behavior : two thin disks are placed perpendicular to the surface of the film in such a way that they are almost tangent to the film. When rotated impulsively in opposite directions during a short time they create two opposite jets of air tangent to the film. The jets set the film into motion and create couples that move towards each other. They collide at an angle of 100° on fig. 8a and head-on on fig. 8b and c. After collision, one (fig. 8a) or two (fig. 8c) structured couples move away.

The ease with which such couples are formed in a turbulent wake can now be better understood. The initial state of the flow is formed of alternate vortices of the same intensity. If turbulence or modulation of the velocity brings two opposite vortices closer together than average they will start moving relatively to the wake. If this motion is longitudinal in the wake it will lead to pairing and coalescence with other vortices. If it is transverse the couple will escape out of the wake. An extra "pinching" however will still be needed to form a structured couple. It will result of the influence of the two closest vortices which compress the couple into a circular shape. This is confirmed by a very recent experiment where we show that the threshold for the appearance of couples be considerably lowered when vibrations of the cylinder modulate the position of the vortices in the wake.

Two opposite sign vortices of comparable strength always form a travelling dipole solution of the Euler equation. Amongst those the couples described by Batchelor appear to be singular by their compact shape and by the impulse needed to form them.

The compression effect needed to create these couples of circular envelope would seem to make them relatively rare. The fact that they are the most frequently observed both experimentally and numerically near a turbulent wake might be related to two factors : they escape faster out of the wake and their range of interaction with outer structures is shorter so that they are not easily destroyed.

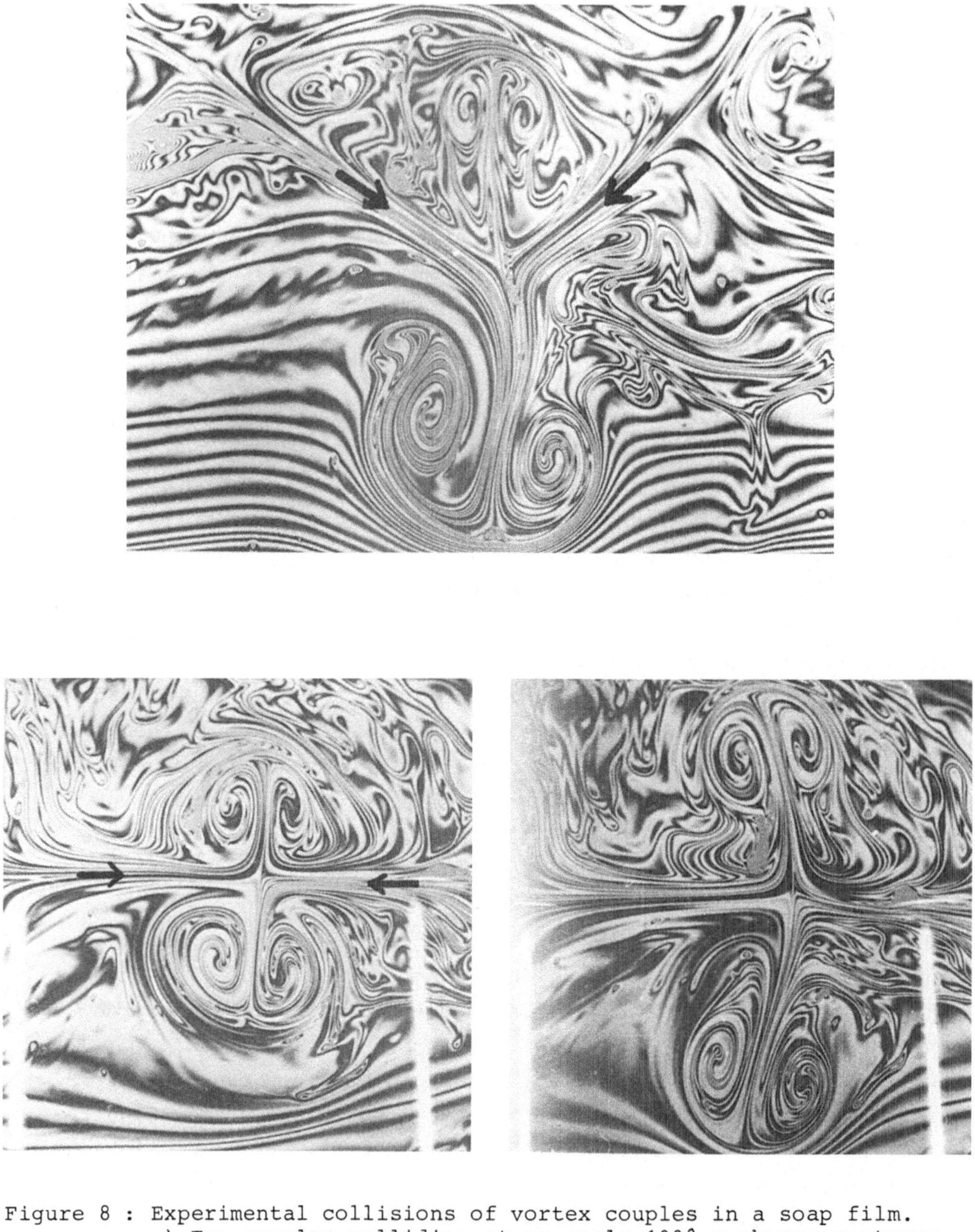

Figure 8 : Experimental collisions of vortex couples in a soap film.
a) Two couples colliding at an angle 100° exchange partners and form the couples shown on this photograph which can be compared to the final state of the simulated collision shown on Figure 7
b and c) Photographs after a head on collision

C. Discussion

The structure and properties of isolated couples of vortices have been widely investigated theoretically and numerically, particularly on the case of meteorological modons. Many characteristic features about their structure, their stability in the presence of neighbouring structures, their collisions, are known but the relevance to naturally occuring situations was still hypothetical due to the scarcity of their experimental observation.
Couples of vortices had been observed in the result of several previous numerical studies of 2-D developed turbulence (Basdevant et al 1981, Basdevant and Sadourny 1983, Mc Williams 1984 ...). In this case however interaction with neighbouring vortices prevented them from having the circular structure of isolated couples and also limited their lifetime (Mac Williams et al 1981). The only laboratory observations of structured couples is due to Flierl et al (1983) who generate isolated modons by means of impulsive jets of water in a rotating tank. In the present experiment we have shown that in the case of a two dimensional turbulent wake the instability leading to couple formation for Re > 300 is as common as the better known pairing instability.

The overall structure of a two dimensional turbulent wake consists in a central zone formed of isolated vortices and of couples that have previously travelled outside the central zone along an arc of a circle before coming back into it. This central zone is surrounded by an expanding cloud of couples. As a couple moves away it will not be destroyed unless it undergoes a collision with another couple. Even in such a case there is a strong probability that new couples will emerge out of the collision.

These couples will thus be the most active agents of the diffusion of turbulence into the quiescent zone. Measured by a local probe their passage will be observed as intermittent bursts.

Such bursts are observed near three dimensional turbulent wakes so that the question arises whether similar couples could exist in the presence of three dimensional distorsions. In early experiments by Grant (1958) Twonsend (1970) and LaRue et al (1974) the spatial structures giving rise to the bursts were analyzed with correlation techniques and temperature measurements. They showed that these bursts could be ascribed to travelling vortex couples. The unexpected direction of the axes of these couples that are perpendicular to the wake center-plane can be understood in the following way. If the two long vortices form a couple on only a fraction of their length only this part will move out of the wake and form a loop that will be stretched and distorted in the shape of a horseshoe. In this case however the couple cannot move far out of the wake. No information about its detailed structure is available.

II. Comparison of vorticity dynamics and passive scalar dynamics

The classical phenomenological theory of incompressible two-dimensional turbulence is based, as far as small scales are concerned, on the concept of enstrophy cascade (Kraichnan 1967, Leith 1968, Batchelor 1969, Basdevant, Lesieur et Sadourny, 1978). At asymptotically high Reynolds numbers, energy is conserved and cascades to larger and larger scales, while enstrophy cascades from the forcing scales to smaller and smaller scales until it reaches dissipation scales. The phenomenological spectrum of enstrophy Z(k) is then obtained, in the enstrophy inertial range, by equating the constant entrophy transfer rate η to the ratio of enstrophy available at the selected wavenumber k, by the caracteristic time of transfer $\tau(k)$:

$$\eta = kZ(k)/\tau(k). \tag{1}$$

The caracteristic time, $\tau(k)$, of the distorsion of structures of scale k^{-1}, is obtained from velocity gradients at larger scales :

$$\tau(k) \sim \left(\int_{k_o}^{k} Z(P)dp\right)^{-1/2} \tag{2}$$

Elimination of $\tau(k)$ between (1) and (2) leads to the classical k^{-1} enstrophy (or k^{-3}energy) spectrum with logarithmic correction (Kraichnan 1971) :

$$Z(k) = C\ \eta^{2/3}k^{-1}(\log k)^{-1/3} \tag{3}$$

The same argument applies for a passive scalar for which :

$$\mu = k\ X\ (k)/\tau(k) \tag{4}$$

where X(k) is the variance spectrum of the scalar and μ its transfer rate to smaller scales. Elimination of $\tau(k)$ between (2) and (4) leads to the same scale dependency (Lesieur, Sommeria and Holloway 1981) :

$$X(k) = C'\ \eta^{-1/3}\mu\ k^{-1}(\log k)^{-1/3}$$

Therefore, phenomenological theory does not distinguish the variance spectrum of a passive scalar from the enstrophy spectrum of the flow.

To test numerically these theoretical predictions we modelled a two-dimensional turbulent flow forced by a stationary energy source. The detailed description of the numerical model is given in Babiano et al 1984. Once a stationary regime is obtained, say at time t_o, a passive scalar is injected in the flow whose initial spectrum is obtained by adding a random phase to the vorticity spectrum ; at time t_o the enstrophy spectrum of the flow and the passive scalar variance spectrum are thus identical (Fig. 9), however, all organized

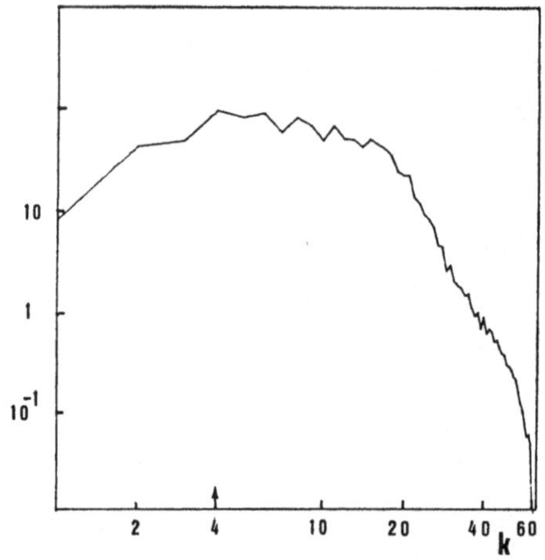

Figure 9 : Enstrophy spectrum Z(k) for the flow and variance spectrum X(k) for the passive scalar at $t = t_o$. The two spectra are identical. The arrow indicates the injection scale (log.log scale).

structures are destroyed in the pollutant field (fig. 10). Starting from t_o the advection equation of the scalar is integrated in parallel to the Navier-Stokes equation, with same dissipations and same injection. After an integration long enough to reach a stationary regime for the pollutant, the results are those displayed in Figures 11 and 12. We see that, contrary to the theoretical predictions, the spectra X(k) and Z(k) have quite different shapes. The pollutant spectrum is close to k^{-1}, while the enstrophy spectrum is much steeper. On figure 11 is indicated a phenomenological spectrum $\tilde{X}(k)$ or $\tilde{Z}(k)$ calculated from equation (4), where $\tau(k)$ has been estimated by (2) using the computed enstrophy spectrum. This theoretical spectrum is indeed close to the pollutant spectrum in a wavenumber range that can be identified to the inertial range ;

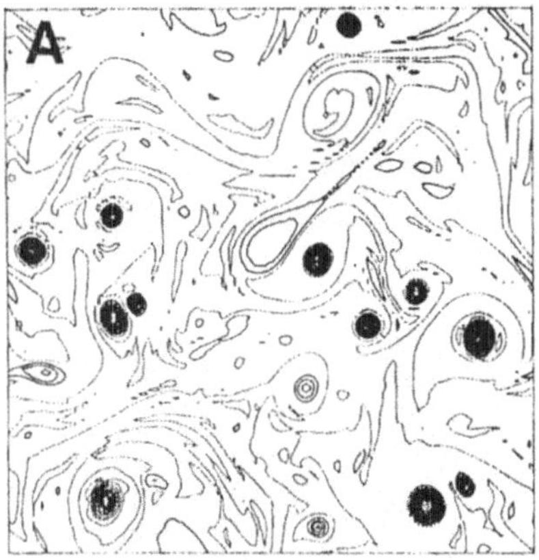

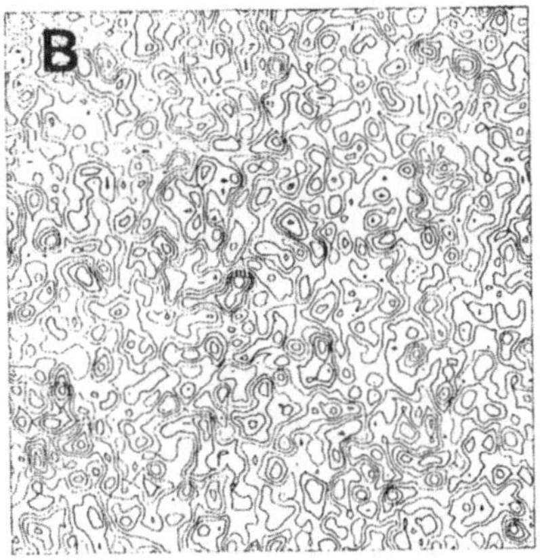

Figure 10 : A isolines of vorticity at $t = t_o$
B isolines of the passive scalar at $t = t_o$

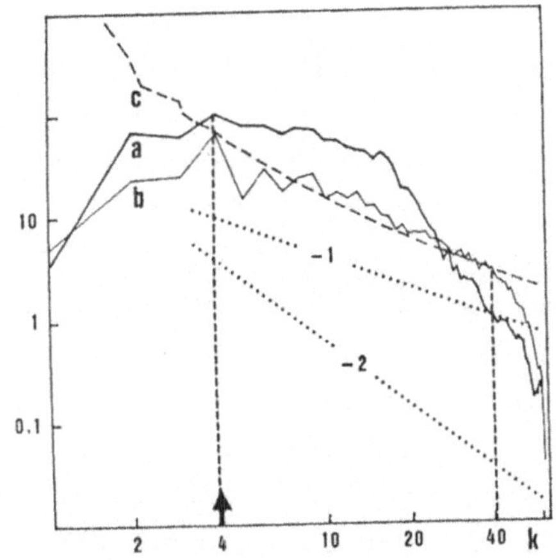

Figure 11 : After 4000 time steps : (a) enstrophy spectrum of the flow ; (b) variance spectrum of the passive scalar ; (c) spectrum obtained from phenomenological theory $\tilde{X}(k)$ or $\tilde{Z}(k)$. k^{-1} and k^{-2} slopes are indicated ; arrow designates injection scale ; the limits of the inertial range are marked by vertical discontinuous lines (log.log. scale).

but the enstrophy spectrum, obviously, does not follow the phenomenological law.

The comparison of vorticity and pollutant fields (Fig. 12 A,B) explains the failure of the theory. We see that the two fields

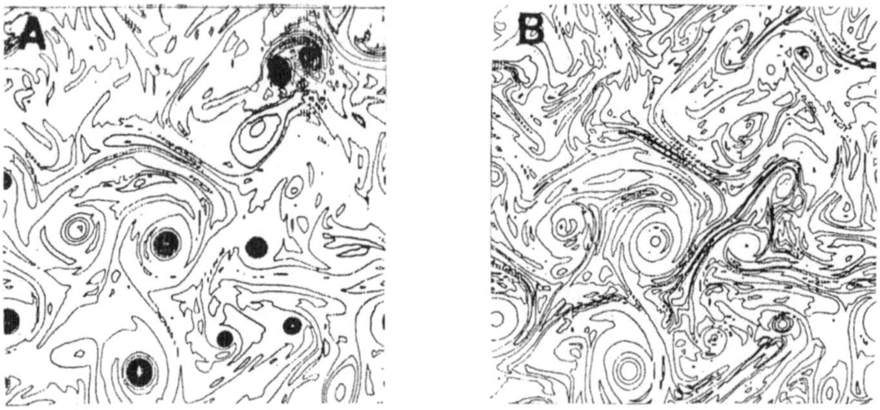

Figure 12 : At same time as in Figure 11
A isolines of vorticity
B isolines of the passive scalar

exhibit very similar structures, as these structures are induced by the same velocity shears ; the two fields are nevertheless completely decorrelated, because there is no systematic correspondence of sign between them. The fundamental difference between the vorticity field and the passive scalar field is the existence in the vorticity field of strong coherent circular vortices where an important proportion of vorticity is concentrated. Mc Williams (1984) has shown the stability of these coherent vortices which are robust enough to be unaffected by the cascade process. We see that the enstrophy of the flow can be schematically divided into two parts. One is the enstrophy concentrated within coherent structures : this enstrophy does not cascade, except when collisions occur, the remaining enstrophy is scattered in the vortex surroundings ; it cascades to smaller scales under its own velocity shears, but mostly under the stronger shears induced by the coherent structures.

The phenomenological law (1) is thus incorrect and should be replaced by

$$\eta = k\ Z^{*}(k)/\tau(k) \tag{6}$$

where $Z^{*}(k)$ is the enstrophy spectrum of the vortex background. Clearly (2) and (6) will lead to a universal spectrum only if there exist a universal relation between $Z(k)$ and $Z^{*}(k)$.

The failure of the classical phenomenological theory arises from the fact that it ignores the possibility for vorticity and stream-function to be locally correlated in coherent quasi-stationary structures (in other words, the existence of energy condensations within the enstrophy inertial range). The flow spectrum seems to be very sensitive to this existence of such coherent structures ; consequently, it may not even have a universal shape.

REFERENCES

1) AREF, H. and SIGGIA, E.D., 1981, J. Fluid Mech., 109, pp. 435-463

2) BABIANO, A., C. BASDEVANT, B. LEGRAS et R. SADOURNY, 1984 , C.R. Acad. Sci., Paris, 299, II, pp. 601-604

3) BATCHELOR, G.K., 1967, An introduction to fluid dynamics, Cambridge University Press, pp. 534-535

4) BATCHELOR, G.K., 1969, Phys. Fluids, Suppl. II, pp. 233-239

5) BASDEVANT, C., M. LESIEUR and R. SADOURNY, 1978, J. Atm. Sc., 35, pp. 1028-1042

6) BASDEVANT, C., B. LEGRAS, R. SADOURNY and M. BELAND, 1981, J. Atm. Sci., 38, pp. 2305-2326

7) BASDEVANT, C. et R. SADOURNY, 1983, J. Méc. Th. et Appl., n° spécial, pp. 243-269

8) CHRISTIANSEN, J.P. and N.J. ZABUSKY, 1973, J. Fluid Mech., 61, pp. 219-243

9) COUDER, Y., 1984, J. Phys. Lett., 45, pp. 353-360

10) COUDER, Y., C. BASDEVANT et H. THOME, 1984, C.R. Acad. Sc. Paris, 299, II, pp. 89-94

11) DEEM, G.S. and N.J. ZABUSKY, 1978 a, Phys. Rev. Lett., 40, pp. 859-862

12) DEEM, G.S. and N.J. ZABUSKY, 1978 b, in "Solitons in action", Edited by Lonngren and Scott, Academic Press, pp. 277-296

13) FABRE, H., B. LAVIRON, C. BASDEVANT and Y. COUDER, 1984, to be published

14) FLIERL, G., V. LARICHEV, J. Mc WILLIAMS and G. REZNIK, 1980, Dyn. Atm. and Oceans, 5, pp. 1-41

15) FLIERL, G., M. STERN and J. WHITEHEAD, 1983, Dyn. Atm. and oceans, 7, pp. 233-263

16) GRANT, H.L., 1958, J. Fluid Mech., 4, pp. 149-190

17) KRAICHNAN, R.H., 1967, Phys. Fluids, 10, pp. 1417-1427

18) KRAICHNAN, R.H., 1971, J. Fluid Mech., 47, pp. 525-535

19) LARICHEV, V. and G. REZNIK, 1976, Rep. USSR Acad. Sci., 231, (5), pp. 1077-1079

20) LA RUE J.C. and P.A. LIBBY, 1974, Phys. Fluids, 17, pp. 873-878

21) LEITH, C.E., 1968, Phys. Fluids, 10, pp. 1417-1423

22) LESIEUR, M., J. SOMMERIA and G. HOLLOWAY, 1981, C.R. Acad. Sci. Paris, 292, II, pp. 271-274

23) McWILLIAMS, J.C., G.R. FLIERL, V.D. LARICHEV, and G.M. REZNIK, 1981, Dyn. Atm. and Oceans, 5, p. 219-238

24) McWILLIAMS, J.C. and N.J. ZABUSKY, 1982, Geophys. Astrophys. Fluid Dyn., 19, pp. 207-227

25) McWILLIAMS, 1984, J. Fluid Dyn., 146, pp. 21-43

26) PIERREHUMBERT, R.T., 1980, J. Fluid Mech., 99, pp. 129-144

27) STERN, M.E., 1975, J. Mar. Res., 33, pp. 1-13

28) TANEDA, S., 1959, J. Phys. Soc. Japan, 14, pp. 843-848

29) THOMSON, W., 1867, Phil. Mag., 34, p. 20

30) TOWNSEND, A.A., J. Fluid Mech., 41, pp. 13-46

NUMERICAL SIMULATION OF DECAYING TWO-DIMENSIONAL TURBULENCE : COMPARISON BETWEEN GENERAL PERIODIC AND TAYLOR-GREEN LIKE FLOWS

M.E. Brachet[1] , M. Meneguzzi[2] and P.L. Sulem[3]

1 - CNRS, Observatoire de Nice, BP 139, 06007 Nice ,France

2 - Service d'Astrophysique,Centre d'Etudes Nucleaires de Saclay, 91191 Gif sur Yvette, and CNRS, France.

3 - School of Mathematical Sciences, Tel Aviv University,Israel and CNRS, Observatoire de Nice, France

Since the first calculations of Lilly[1] ,it has been recognized that high resolutions are required to properly simulate the small scale inertial range of incompressible two-dimensional Navier-Stokes turbulence[2] . Preliminary calculations at a 512^2 resolution presented by Orszag[3] showed that when the large scale Reynolds number is increased from 1100 to 25000, a distinct change is observed from a k^{-4} energy spectrum predicted by Saffman[4] to a spectrum roughly proportional to k^{-3} which is expected if an enstrophy cascade develops[5,6,7]. Spectral simulations whith 1024^2 collocation points were more recently reported[8,9]. In these calculations, aliasing was removed by spectral truncation at a wavenumber $k_M = 341$. To achieve this resolution on a 1 megaword Cray-1 computer, the "sparse mode technique" was implemented, i.e. the stream function has a Fourier representation

$$\psi(x,y,t) = \sum_{l,m=0}^{N/2} a_{lm}(t) \sin lx \sin my \qquad (1)$$

where the a_{lm} vanish unless l and m are both even or odd jointly. They are initialized as gaussian random variables with a covariance such that the energy spectrum is

$$E(k) = C_0\, k\, e^{-(k/k_0)^2} \qquad (2)$$

As in the Taylor-Green vortex, this representation implies symetries in the physical space, including reflectional invariance on the sides of an impermeable box defined by x=0 and π , y=0 and π . The inertial exponent n and the dissipative scale β were obtained by fitting the angle averaged energy spectrum with a function

$$E(k) = C(t)\, k^{n(t)}\, e^{-\beta(t)k} \qquad (3)$$

A transition from an early n = -4 energy spectrum to a persistent n = -3 spectrum was observed.

We first briefly report on new simulations of symmetric flows. We use $k_0 = 5$ (corresponding to an energy $\Sigma = .132$ and an enstrophy $\Omega = 3.264$), and two different viscosities $\nu = 7.\ 10^{-5}$ and $\nu = 2.35\ 10^{-5}$. With the former viscosity, the integral Reynolds number is the same as in the run reported in ref. 9, where $k_0 = 3.5$, $\nu = 3.33\ 10^{-5}$, $\Sigma = .132$ and $\Omega = 3.264$. Increasing k_0 at constant Reynolds number produces a sizable decrease of the overall time scale. The transition from the k^{-4} to the k^{-3} regime occurs around t = 4 instead of t = 6 in the run reported in ref. 9. The maximum enstrophy dissipation is around 4.5 for $\nu = 7.\ 10^{-5}$ and around 6 for $\nu = 2.35\ 10^{-5}$. However, although there is more space averaging in the present runs, the fluctuations on the inertial exponents are not appreciably decreased. Note that the fluctuations are smaller with smaller viscosity (figure 1). Figure 2 shows the contours of the vorticity and of the squared vorticity gradient for $\nu = 3.33\ 10^{-5}$ at t = 1, 2 and 4.

The above 1024^2 simulations were performed with symmetrical initial data. In order to investigate to what extent the observed small scale dynamics are generic, we turned to general periodic flows : we present here preliminary calculations made with a maximum wavenumber $k_M = 100$. Both spectral and pseudo-spectral computations (i.e. with and without dealiasing) have been performed at this resolution. The results are in complete agreement (five digits in the enstrophy dissipation). Figure 3 shows contours of vorticity and of squared vorticity gradient. The main observation is the strong similarity with the Taylor-Green like symmetric flows. Figure 4 shows the inertial exponent as a funtion of time. A transition to a value of the order of -3 is visible, although the fluctuations are relatively strong. Note that, because of our moderate Reynolds number, there is considerable overlap between the inertial and dissipative ranges. This may explain that n fluctuates slightly above -3 if, as indicated by dominant balance arguments, the algebraic dissipative range prefactor is higher than -3. Less noisy results are expected to be obtained at higher resolution. Although preliminary, this computation nevertheless suggests the general character of the small scale dynamics observed in Taylor-Green like flows. Note that in the simulations we are reporting here, we do not observe the emergence of isolated, coherent vortices[11,12,13]. The reason is probably that these structures develop for times much larger that the time of maximum enstrophy dissipation, while our integration time is only twice that time. Calculations with higher resolution, both with and without symmetries (2048^2 and 800^2 respectively) are under way.

We thank C. Temperton for his very efficient CRAY-1 Fast Fourier Transform routines, and the NCAR Scientific Computer Division for his Graphics Software.

R E F E R E N C E S

(1) - Lilly D.K., Phys. Fluids Suppl., 12, 240 (1969).

(2) - Herring J.R., Orszag S.A., Kraichnan H.R., Fox D.G., Jour. Fluid Mech. 66, 417 (1974)

(3) - Orszag S.A., Proc. 5th Inter. Conf. on Numerical Methods in Fluid Dynamics, Lect. Notes in Phys. 59, 32 (1977)

(4) - Saffman R.G., Stud. in Appl. Math. 50, 377 (1971)

(5) - Kraichnan H.R., Phys. Fluids 10, 1417 (1967)

(6) - Bachelor G.K., Phys. Fluids 12, 232 (1969)

(7) - Kraichnan H.R., J. Fluid Mech. 47, 525 (1971)

(8) - Brachet M.E., Sulem P.L., in Proc. 4th Beer Sheva Seminar on MHD flows and turbulence (1984), AIAA Progress in Astronautics and Aeronautics, in press.

(9) - Brachet M.E., Sulem P.L., in Proc. 9th Inter. Conf. on Numerical Methods in Fluid Dynamics, Saclay (1984), Springer Lect. Notes in Physics, in press.

(10) - Brachet M.E., Meiron D.I., Orszag S.A., Nickel B.G., Morf R.H. and Frisch U., J. Fluid Mech. 130, 411 (1983)

(11) - Basdevant C., Legras B., Sadourny R., Beland B., J. Atmos. Sci. 38, 2305 (1981)

(12) - McWilliams J.C., J. Fluid Mech., 146, 21 (1984)

(13) - Herring J.R., McWilliams J.C., "Comparison of direct numerical simulations of two-dimensional turbulence with two-point closure: the effect of intermittency." (preprint)

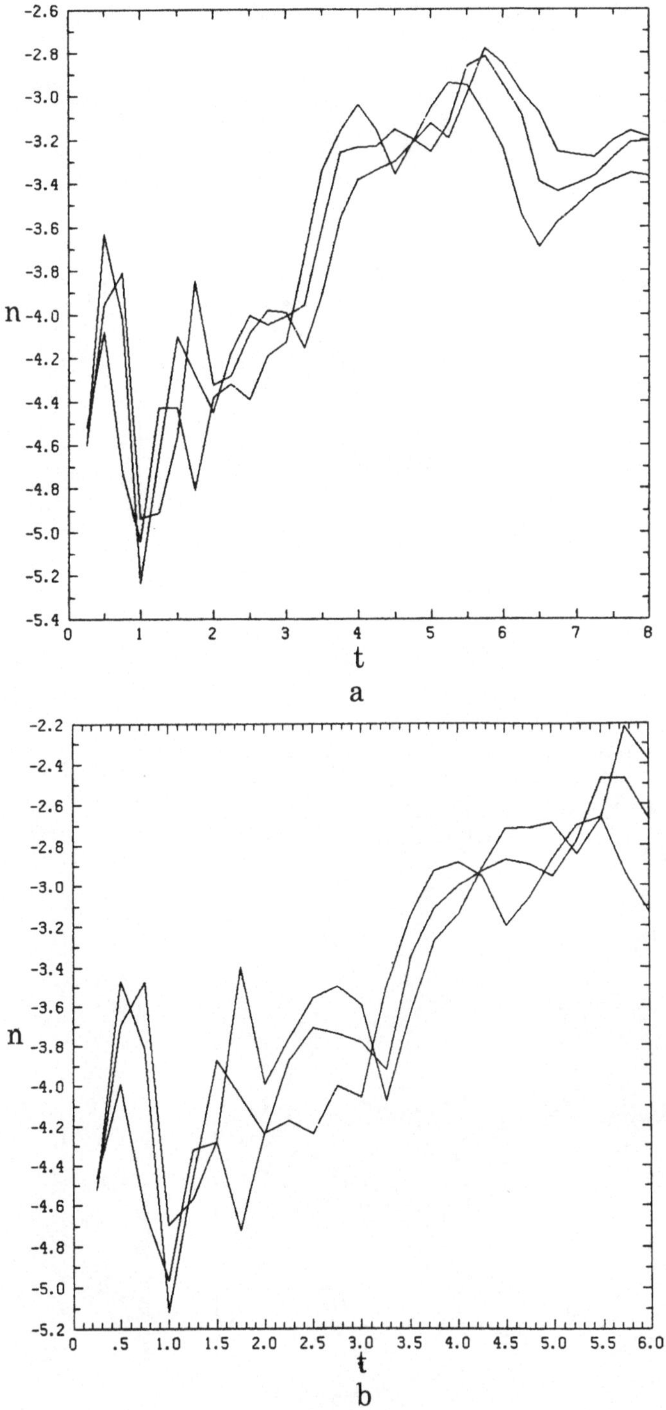

Figure 1

Inertial exponent n as a function of time for the symmetric flow with $k_0 = 5$ and (a) $\nu = 2.35\ 10^{-5}$, (b) $\nu = 7.\ 10^{-5}$. The k^{-4} regime occurs at t=1, when the inviscid exponential flattening of the vorticity gradient sheets is stopped by viscous effects.

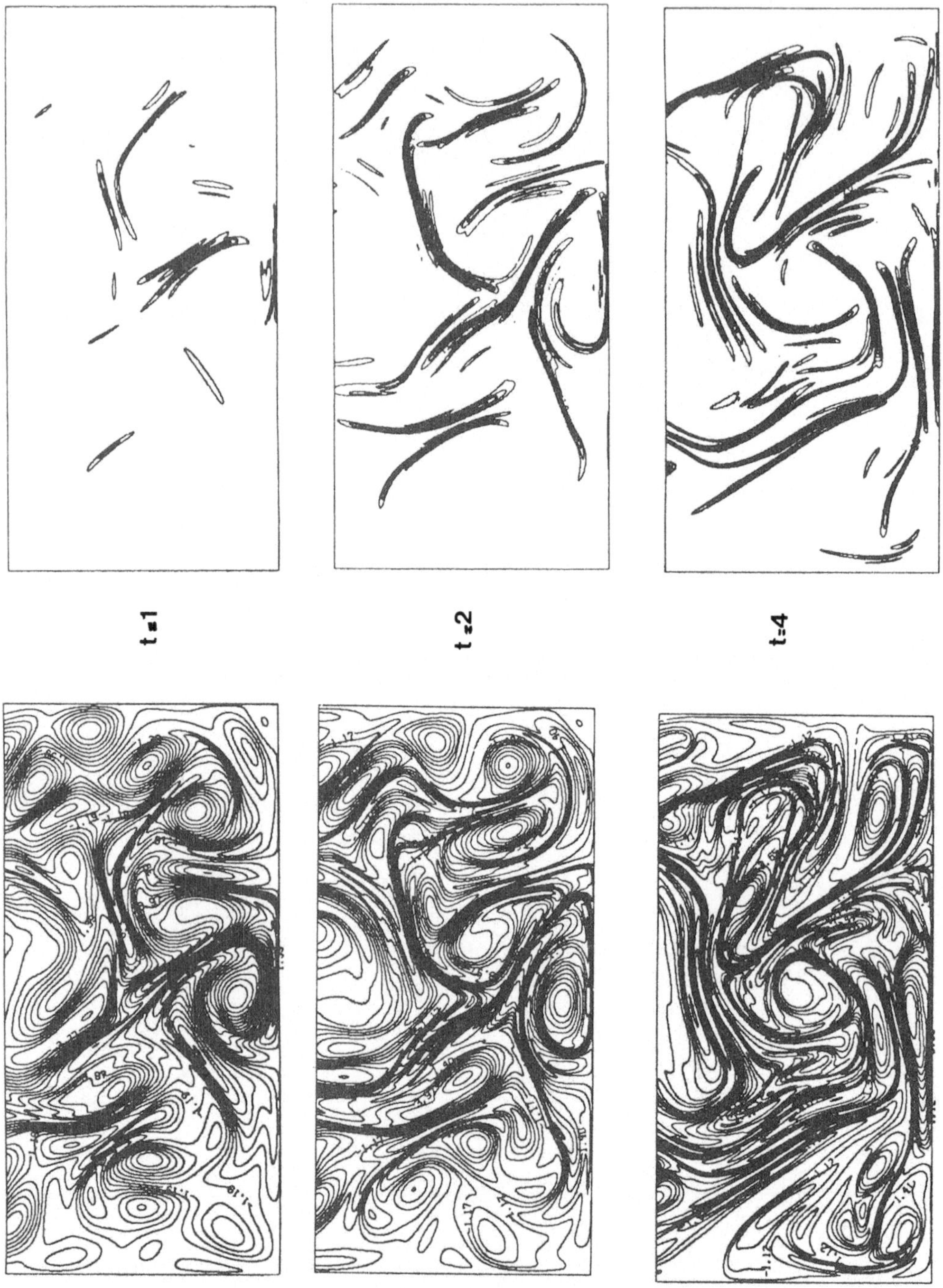

Figure 2

Contours of vorticity (left), and squared vorticity gradient (right) for the symmetric flow with $k_0 = 5$ and $\nu = 3.33\ 10^{-5}$ at t = 1, 2, 4 in the slab $0 < x < \pi$, $0 < y < \pi / 2$. The complete flow can be reconstructed by a rotation of 180 degrees around the point $(\pi / 2, \pi / 2)$ followed by mirror symmetries on the sides of the box $x = 0$, $x = \pi$, $y = 0$, $y = \pi$.

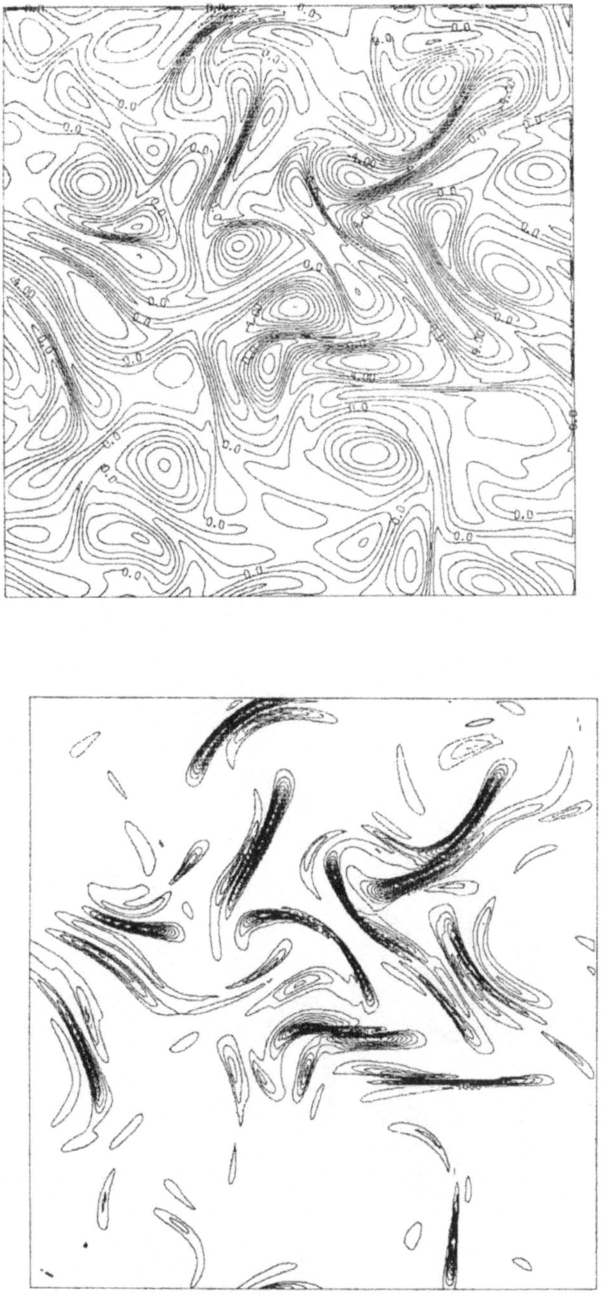

a

Figure 3

Contours of vorticity and squared vorticity gradient for a flow with $k_0 = 3.5$ and $\nu = 10^{-3}$ at (a) t = 1, (b) t = 5.

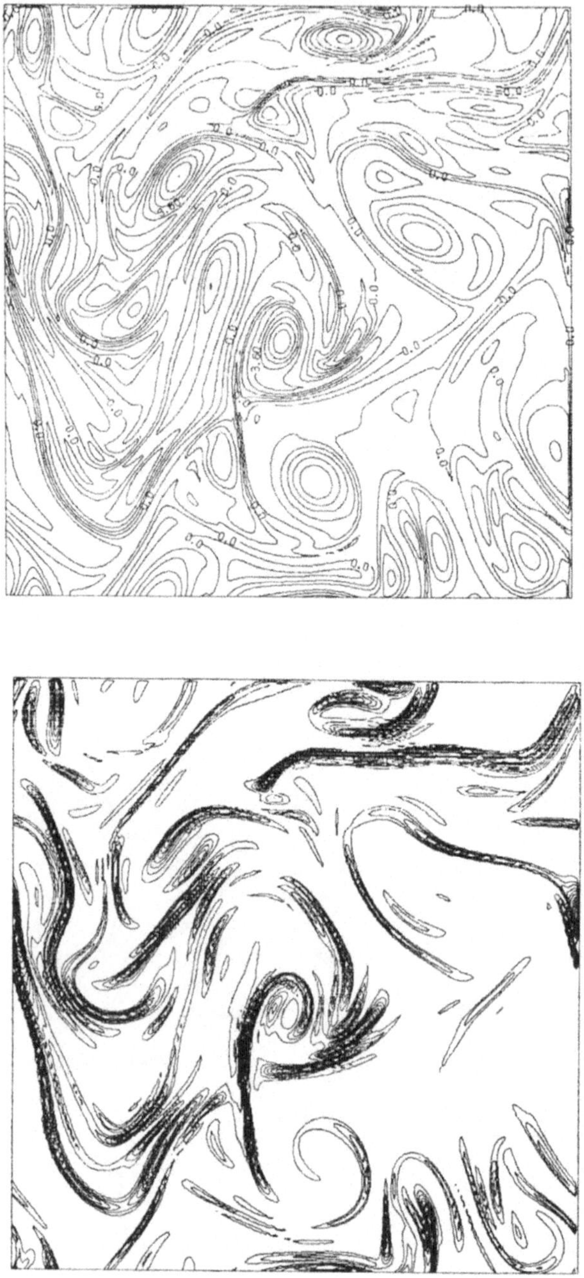

b

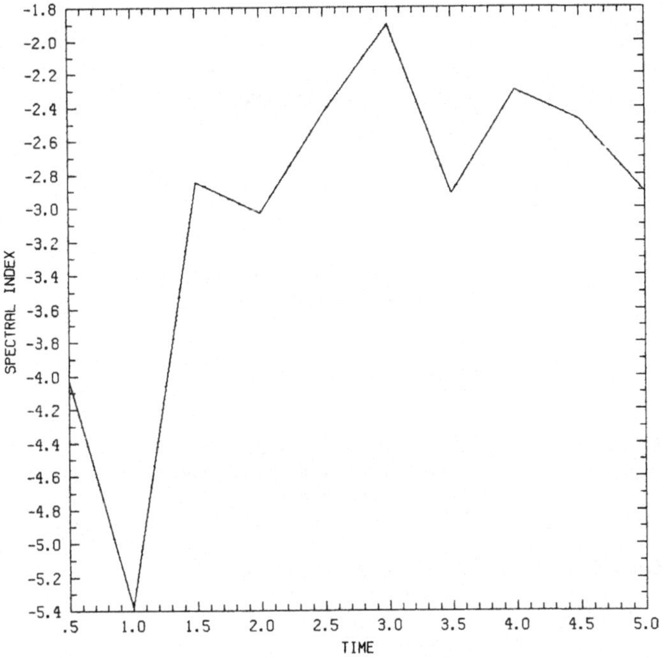

Figure 4

Inertial exponent n as a function of time for non symmetric flow with $k_0 = 3.5$ and $\nu = 10^{-3}$.

LIST OF PARTICIPANTS

Mr. **ANDRE** Jean-Claude	CENTRE NATIONAL DE RECHERCHES METEOROLOGIQUES Avenue Eisenhower prolongée 31057 TOULOUSE Cedex	FRANCE
Mr. **AUPOIX** G.	ONERA-CERT B.P. 4025 31055 TOULOUSE Cedex	FRANCE
Mr. **BASDEVANT** Claude	ECOLE NORMALE SUPERIEURE Lab. Météorologie Dynamique 24, rue Lhomond 75231 PARIS Cedex 05	FRANCE
Mlle **BEGUE** Catherine	AVIONS MARCEL DASSAULT Bréguet Aviation 78, Quai Carnot B.P. 300 92214 ST CLOUD Cedex	FRANCE
Mr. **BERTOGLIO** J.P.	LABORATOIRE DE MECANIQUE DES FLUIDES 36, avenue Guy de Collongue B.P. 163 69 ECULLY Cedex	FRANCE
Mr. **BRACHET** M.E.	OBSERVATOIRE DE NICE B.P. 139 06003 NICE	FRANCE
Mr. **CALFLISCH** Russel	NEW YORK UNIVERSITY Courant Institute 251 Mercer Street NEW-YORK, N.Y. 10012	USA
Mr. **CAMBON**	METRAFLU 64, Chemin des Mouilles 69130 ECULLY	FRANCE
Mr. **CHILDRESS** Steve	NEW YORK UNIVERSITY Courant Institure 251 Mercer Street NEW-YORK, N.Y. 10012	USA
Mr. **CHOLLET** J.P.	INSTITUT DE MECANIQUE DE GRENOBLE B.P. 68 38402 ST MARTIN D'HERES	FRANCE
Mr. **CLAVIN** Paul	UNIVERSITE DE PROVENCE Lab. Dynamique et Thermophysique Centre de St Jérôme 13997 MARSEILLE Cedex 13	FRANCE

Mr. **COULLET** P.	UNIVERSITE DE NICE Dept. de Mathématiques Parc Valrose 06034 NICE Cedex	FRANCE
Mr. **COURTY** Jean-Claude	AVIONS MARCEL DASSAULT BRÉGUET AVIATION 78, Quai Carnot B.P. 300 92214 ST CLOUD Cedex	FRANCE
Mr. **DANG TRAN** Khoa	ONERA 32, avenue de la Division Leclerc 92320 CHATILLON/BAGNEUX	FRANCE
Mme **FARGE** Marie	ECOLE NORMALE SUPERIEURE Lab. Météorologie Dynamique 24, rue Lhomond 75231 PARIS Cedex 05	FRANCE
Mr. **FERZIGER** Joel H.	STANFORD UNIVERSITY Mechanical Engineering Dept. Thermosciences Division STANFORD, CA 94305	USA
Mr. **FRISCH** Uriel	OBSERVATOIRE DE NICE B.P. 139 06003 NICE	FRANCE
Mme **FRISCH**	OBSERVATOIRE DE NICE B.P. 139 06003 NICE	FRANCE
Mr. **GENOUX**	DRET Groupe 6 26, Boulevard Victor 75996 PARIS ARMEES	FRANCE
Mr. **GRAPPIN** Roland	OBSERVATOIRE DE MEUDON DAF 92190 MEUDON	FRANCE
Mr. **GUILLARD**	INRIA Sophia-Antipolis Route des Lucioles 06560 VALBONNE Cedex	FRANCE
Mr. **HAUGUEL** Alain	E.D.F. Lab. National d'Hydraulique 6, Quai Watier 78400 CHATOU	FRANCE
Mr. **HERRING** Jackson R.	N.C.A.R. Mesoscale Research Section P.O. Box 3000 BOULDER, Col. 80307	USA
Mr. **HOPFINGER** Emile	IMG BP 53 Centre de Tri 38041 GRENOBLE Cedex	FRANCE

Mr. **KELLER** Joseph B.	STANFORD UNIVERSITY Mathematics Dept. STANFORD, CA 94305	USA
Mr. **KNIGHT** Doyle D.	RUTGERS UNIVERSITY College of Engineering Dept. Mech. and Aerospace Eng. P.O. Box 909 PISCATAWAY, N.J. 08854	USA
Mr. **LAURENCE** M.D.	E.D.F. Direction Etudes & Recherches 6, Quai Watier 78400 CHATOU	FRANCE
Mr. **LEITH** Cecil E.	UNIVERSITY OF CALIFORNIA Lawrence Livermore Lab. PO Box 808 LIVERMORE, Ca 94550	USA
Mr. **LEORAT** Jacques	OBSERVATOIRE DE MEUDON Place Janssen 92190 MEUDON	FRANCE
Mr. **MANLEY** Oscar	U.S. Dept. of Energy Washington, C.D.	USA
Mr. **MANNEVILLE** P.	CEN SACLAY Institut de Recherche Fondamentale 91191 GIF SUR YVETTE	FRANCE
Mr. **McLAUGHLIN** Dave W.	UNIVERSITY OF ARIZONA Department of Mathematics Building 89 TUCSON, Arizona 85721	USANCE
Mr. **McWILLIAMS** James	N.C.A.R. P.O. Box 3000 BOULDER, Col. 80307	USA
Mr. **MENEGUZZI** Maurice	CEN DPHEP Section Astrophysique BP 2 91191 GIF SUR YVETTE	FRANCE
Mr. **MORY** Mathieu	I.M.G. Domaine Universitaire B.P. 68 38402 ST MARTIN D'HERES	FRANCE
Mr. **ORLANDI** Paolo	UNIVERSITA DEGLI STUDI Dept. Maccanica a Aeronautica Via Eudossiana 18 00184 ROMA	ITALIE
Mr. **PAPANICOLAOU**	NEW YORK UNIVERSITY Courant Institute 251 Mercer Street NEW YORK, N.Y. 10012	USA

Mr. **PERRIER** Pierre	AVIONS MARCEL DASSAULT 78, Quai Carnot B.P. 300 92214 ST CLOUD Cedex	FRANCE
Mr. **PIRONNEAU** Olivier	INRIA Rocquencourt Domaine de Voluceau B.P. 105 78153 LE CHESNAY Cedex	FRANCE
Mr. **POMEAU**	CEN SACLAY Institut de Recherche Fondmentale 91191 GIF SUR YVETTE Cedex	FRANCE
Mlle **POUQUET** Annick	OBSERVATOIRE DE NICE B.P. 139 06003 NICE	FRANCE
Mr. **ROUX** Bernard	IMFM 1, rue Honorat 13003 MARSEILLE	FRANCE
Mr. **ROY** Philippe	ONERA OAT2 29, Avenue de la Div. Leclerc 92320 CHATILLON/BAGNEUX	FRANDE
Mr. **SHE**	OBSERVATOIRE DE NICE B.P. 139 06003 NICE	FRANCE
Mr. **SIGGIA** Eric	CORNELL UNIVERSITY Lab. of Atomic and Solid Stage Physics Clark Hall ITHACA, NY 14853	USA
Mr. **SOMMERIA** Joel	IMG BP 68 38402 ST MARTIN D'HERES	FRANCE
Mr. **SPIEGEL** E.A.	COLUMBIA UNIVERSITY Dept. of Astronomy NEW YORK, NY 10027	USA
Mr. **SULEM** P.L.	TEL AVIV UNIVERSITY School of Math. Sciences RAMAT AVIV 69978 TEL AVIV	ISRAEL
Mr. **TARTAR** Luc	CEA Service MA B.P. 27 94190 VILLENEUVE ST GEORGES	FRANCE
Mr. **THUAL**	OBSERVATOIRE DE NICE BP 139 06003 NICE	FRANCE
Mlle **TUCKERMAN** Laurette	CEN Saclay DPHG/PSRM Orme des Merisiers 91191 GIF SUR YVETTE	FRANCE

Mr. **VAN ATTA** C.W. | UNIVERSITY OF CALIFORNIA
Dept. of Appl. Mech. and Eng.
Sciences
Mail Code B-010
LA JOLLA, CA 92093 | USA

Mr. **ZALESKI** S. | ENS
Groupe Physique des Solides
24, rue Lhomond
75231 PARIS Cedex 05 | FRANCE

Vol. 195: Trends and Applications of Pure Mathematics to Mechanics. Proceedings, 1983. Edited by P. G. Ciarlet and M. Roseau. V, 422 pages. 1984.

Vol. 196: WOPPLOT 83. Parallel Processing: Logic, Organization and Technology. Proceedings, 1983. Edited by J. Becker and I. Eisele. V, 189 pages. 1984.

Vol. 197: Quarks and Nuclear Structure. Proceedings, 1983. Edited by K. Bleuler. VIII, 414 pages. 1984.

Vol. 198: Recent Progress in Many-Body Theories. Proceedings, 1983. Edited by H. Kümmel and M. L. Ristig. IX, 422 pages. 1984.

Vol. 199: Recent Developments in Nonequilibrium Thermodynamics. Proceedings, 1983. Edited by J. Casas-Vázquez, D. Jou and G. Lebon. XIII, 485 pages. 1984.

Vol. 200: H. D. Zeh, Die Physik der Zeitrichtung. V, 86 Seiten. 1984.

Vol. 201: Group Theoretical Methods in Physics. Proceedings, 1983. Edited by G. Denardo, G. Ghirardi and T. Weber. XXXVII, 518 pages. 1984.

Vol. 202: Asymptotic Behavior of Mass and Spacetime Geometry. Proceedings, 1983. Edited by F. J. Flaherty. VI, 213 pages. 1984.

Vol. 203: C. Marchioro, M. Pulvirenti, Vortex Methods in Two-Dimensional Fluid Dynamics. III, 137 pages. 1984.

Vol. 204: Y. Waseda, Novel Application of Anomalous (Resonance) X-Ray Scattering for Structural Characterization of Disordered Materials. VI, 183 pages. 1984.

Vol. 205: Solutions of Einstein's Equations: Techniques and Results. Proceedings, 1983. Edited by C. Hoenselaers and W. Dietz. VI, 439 pages. 1984.

Vol. 206: Static Critical Phenomena in Inhomogeneous Systems. Edited by A. Pękalski and J. Sznajd. Proceedings, 1984. VIII, 358 pages. 1984.

Vol. 207: S. W. Koch, Dynamics of First-Order Phase Transitions in Equilibrium and Nonequilibrium Systems. III, 148 pages. 1984.

Vol. 208: Supersymmetry and Supergravity/Nonperturbative QCD. Proceedings, 1984. Edited by P. Roy and V. Singh. V, 389 pages. 1984.

Vol. 209: Mathematical and Computational Methods in Nuclear Physics. Proceedings, 1983. Edited by J. S. Dehesa, J. M. G. Gomez and A. Polls. V, 276 pages. 1984.

Vol. 210: Cellular Structures in Instabilities. Proceedings, 1983. Edited by J. E. Wesfreid and S. Zaleski. VI, 389 pages. 1984.

Vol. 211: Resonances – Models and Phenomena. Proceedings, 1984. Edited by S. Albeverio, L. S. Ferreira and L. Streit. VI, 369 pages. 1984.

Vol. 212: Gravitation, Geometry and Relativistic Physics. Proceedings, 1984. Edited by Laboratoire "Gravitation et Cosmologie Relativistes", Université Pierre et Marie Curie et C.N.R.S., Institut Henri Poincaré, Paris. VI, 336 pages. 1984.

Vol. 213: Forward Electron Ejection in Ion Collisions. Proceedings, 1984. Edited by K. O. Groeneveld, W. Meckbach and I. A. Sellin. VII, 165 pages. 1984.

Vol. 214: H. Moraal, Classical, Discrete Spin Models. VII, 251 pages. 1984.

Vol. 215: Computing in Accelerator Design and Operation. Proceedings, 1983. Edited by W. Busse and R. Zelazny. XII, 574 pages. 1984.

Vol. 216: Applications of Field Theory to Statistical Mechanics. Proceedings, 1984. Edited by L. Garrido. VIII, 352 pages. 1985.

Vol. 217: Charge Density Waves in Solids. Proceedings, 1984. Edited by Gy. Hutiray and J. Sólyom. XIV, 541 pages. 1985.

Vol. 218: Ninth International Conference on Numerical Methods in Fluid Dynamics. Edited by Soubbaramayer and J. P. Boujot. X, 612 pages. 1985.

Vol. 219: Fusion Reactions Below the Coulomb Barrier. Proceedings, 1984. Edited by S. G. Steadman. VII, 351 pages. 1985.

Vol. 220: W. Dittrich, M. Reuter, Effective Lagrangians in Quantum Electrodynamics. V, 244 pages. 1985.

Vol. 221: Quark Matter '84. Proceedings, 1984. Edited by K. Kajantie. VI, 305 pages. 1985.

Vol. 222: A. García, P. Kielanowski, The Beta Decay of Hyperons. Edited by A. Bohm. VIII, 173 pages. 1985.

Vol. 223: H. Saller, Vereinheitlichte Feldtheorien der Elementarteilchen. IX, 157 Seiten. 1985.

Vol. 224: Supernovae as Distance Indicators. Proceedings, 1984. Edited by N. Bartel. VI, 226 pages. 1985.

Vol. 225: B. Müller, The Physics of the Quark-Gluon Plasma. VII, 142 pages. 1985.

Vol. 226: Non-Linear Equations in Classical and Quantum Field Theory. Proceedings, 1983/84. Edited by N. Sanchez. VII, 400 pages. 1985.

Vol. 227: J.-P. Eckmann, P. Wittwer, Computer Methods and Borel Summability Applied to Feigenbaum's Equation. XIV, 297 pages. 1985.

Vol. 228: Thermodynamics and Constitutive Equations. Proceedings, 1982. Edited by G. Grioli. V, 257 pages. 1985.

Vol. 229: Fundamentals of Laser Interactions. Proceedings, 1985. Edited by F. Ehlotzky. IX, 314 pages. 1985.

Vol. 230: Macroscopic Modelling of Turbulent FLows. Proceedings, 1984. Edited by U. Frisch, J. B. Keller, G. Papanicolaou and O. Pironneau. X, 360 pages. 1985.

Selected Issues from

Lecture Notes in Mathematics

Vol. 932: Analytic Theory of Continued Fractions. Proceedings, 1981. Edited by W.B. Jones, W.J. Thron, and H. Waadeland. VI, 240 pages. 1982.

Vol. 934: M. Sakai, Quadrature Domains. IV, 133 pages. 1982.

Vol. 935: R. Sot, Simple Morphisms in Algebraic Geometry. IV, 146 pages. 1982.

Vol. 936: S.M. Khaleelulla, Counterexamples in Topological Vector Spaces. XXI, 179 pages. 1982.

Vol. 937: E. Combet, Intégrales Exponentielles. VIII, 114 pages. 1982.

Vol. 938: Number Theory. Proceedings, 1981. Edited by K. Alladi. IX, 177 pages. 1982.

Vol. 942: Theory and Applications of Singular Perturbations. Proceedings, 1981. Edited by W. Eckhaus and E.M. de Jager. V, 363 pages. 1982.

Vol. 953: Iterative Solution of Nonlinear Systems of Equations. Proceedings, 1982. Edited by R. Ansorge, Th. Meis, and W. Törnig. VII, 202 pages. 1982.

Vol. 956: Group Actions and Vector Fields. Proceedings, 1981. Edited by J.B. Carrell. V, 144 pages. 1982.

Vol. 957: Differential Equations. Proceedings, 1981. Edited by D.G. de Figueiredo. VIII, 301 pages. 1982.

Vol. 963: R. Nottrot, Optimal Processes on Manifolds. VI, 124 pages. 1982.

Vol. 964: Ordinary and Partial Differential Equations. Proceedings, 1982. Edited by W.N. Everitt and B.D. Sleeman. XVIII, 726 pages. 1982.

Vol. 968: Numerical Integration of Differential Equations and Large Linear Systems. Proceedings, 1980. Edited by J. Hinze. VI, 412 pages. 1982.

Vol. 970: Twistor Geometry and Non-Linear Systems. Proceedings, 1980. Edited by H.-D. Doebner and T.D. Palev. V, 216 pages. 1982.

Vol. 972: Nonlinear Filtering and Stochastic Control. Proceedings, 1981. Edited by S.K. Mitter and A. Moro. VIII, 297 pages. 1983.

Vol. 978: J. Ławrynowicz, J. Krzyż, Quasiconformal Mappings in the Plane. VI, 177 pages. 1983.

Vol. 979: Mathematical Theories of Optimization. Proceedings, 1981. Edited by J.P. Cecconi and T. Zolezzi. V, 268 pages. 1983.

Vol. 982: Stability Problems for Stochastic Models. Proceedings, 1982. Edited by V.V. Kalashnikov and V.M. Zolotarev. XVII, 295 pages. 1983.

Vol. 989: A.B. Mingarelli, Volterra-Stieltjes Integral Equations and Generalized Ordinary Differential Expressions. XIV, 318 pages. 1983.

Vol. 994: J.-L. Journé, Calderón-Zygmund Operators, Pseudo-Differential Operators and the Cauchy Integral of Calderón. VI, 129 pages. 1983.

Vol. 999: C. Preston, Iterates of Maps on an Interval. VII, 205 pages. 1983.

Vol. 1000: H. Hopf, Differential Geometry in the Large, VII, 184 pages. 1983.

Vol. 1003: J. Schmets, Spaces of Vector-Valued Continuous Functions. VI, 117 pages. 1983.

Vol. 1005: Numerical Methods. Proceedings, 1982. Edited by V. Pereyra and A. Reinoza. V, 296 pages. 1983.

Vol. 1007: Geometric Dynamics. Proceedings, 1981. Edited by J. Palis Jr. IX, 827 pages. 1983.

Vol. 1015: Equations différentielles et systèmes de Pfaff dans le champ complexe – II. Seminar. Edited by R. Gérard et J.P. Ramis. V, 411 pages. 1983.

Vol. 1021: Probability Theory and Mathematical Statistics. Proceedings, 1982. Edited by K. Itô and J.V. Prokhorov. VIII, 747 pages. 1983.

Vol. 1031: Dynamics and Processes. Proceedings, 1981. Edited by Ph. Blanchard and L. Streit. IX, 213 pages. 1983.

Vol. 1032: Ordinary Differential Equations and Operators. Proceedings, 1982. Edited by W. N. Everitt and R. T. Lewis. XV, 521 pages. 1983.

Vol. 1035: The Mathematics and Physics of Disordered Media. Proceedings, 1983. Edited by B.D. Hughes and B.W. Ninham. VII, 432 pages. 1983.

Vol. 1037: Non-linear Partial Differential Operators and Quantization Procedures. Proceedings, 1981. Edited by S.I. Andersson and H.-D. Doebner. VII, 334 pages. 1983.

Vol. 1041: Lie Group Representations II. Proceedings 1982–1983. Edited by R. Herb, S. Kudla, R. Lipsman and J. Rosenberg. IX, 340 pages. 1984.

Vol. 1045: Differential Geometry. Proceedings, 1982. Edited by A.M. Naveira. VIII, 194 pages. 1984.

Vol. 1047: Fluid Dynamics. Seminar, 1982. Edited by H. Beirão da Veiga. VII, 193 pages. 1984.

Vol. 1048: Kinetic Theories and the Boltzmann Equation. Seminar, 1981. Edited by C. Cercignani. VII, 248 pages. 1984.

Vol. 1049: B. Iochum, Cônes autopolaires et algèbres de Jordan. VI, 247 pages. 1984.

Vol. 1054: V. Thomée, Galerkin Finite Element Methods for Parabolic Problems. VII, 237 pages. 1984.

Vol. 1055: Quantum Probability and Applications to the Quantum Theory of Irreversible Processes. Proceedings, 1982. Edited by L. Accardi, A. Frigerio and V. Gorini. VI, 411 pages. 1984.

Vol. 1057: Bifurcation Theory and Applications. Seminar, 1983. Edited by L. Salvadori. VII, 233 pages. 1984.

Vol. 1058: B. Aulbach, Continuous and Discrete Dynamics near Manifolds of Equilibria. IX, 142 pages. 1984.

Vol. 1059: Séminaire de Probabilités XVIII, 1982/83. Proceedings. Edité par J. Azéma et M. Yor. IV, 518 pages. 1984.

Vol. 1063: Orienting Polymers. Proceedings, 1983. Edited by J.L. Ericksen. VII, 166 pages. 1984.

Vol. 1065: A. Cuyt, Padé Approximants for Operators: Theory and Applications. IX, 138 pages. 1984.

Vol. 1066: Numerical Analysis. Proceedings, 1983. Edited by D.F. Griffiths. XI, 275 pages. 1984.

Vol. 1071: Padé Approximation and its Applications, Bad Honnef 1983. Prodeedings Edited by H. Werner and H.J. Bünger. VI, 264 pages. 1984.

Vol. 1072: F. Rothe, Global Solutions of Reaction-Diffusion Systems. V, 216 pages. 1984.

Vol. 1085: G.K. Immink, Asymptotics of Analytic Difference Equations. V, 134 pages. 1984.

Vol. 1086: Sensitivity of Functionals with Applications to Engineering Sciences. Proceedings, 1983. Edited by V. Komkov. V, 130 pages. 1984

Vol. 1100: V. Ivrii, The Precise Spectral Asymptotics for Elliptic Operators Acting in Fiberings over Manifolds with Boundary. V, 237 pages. 1984.